UCLA Symposia on Molecular and Cellular Biology, New Series

Series Editor, C. Fred Fox

RECENT TITLES

UCLA Symposia Board

Transcriptional Control Mechanisms

Transcriptional Control Mechanisms

Proceedings of a Cetus-UCLA Symposium
Held in Keystone, Colorado
April 6–13, 1986

Editors

Daryl Granner
Departments of Molecular Physiology and Biophysics
Vanderbilt University
Nashville, Tennessee

Michael G. Rosenfeld
Departments of Physiology and Pharmacology
University of California at San Diego
La Jolla, California

Shing Chang
Cetus Corporation
Emeryville, California

Alan R. Liss, Inc. • New York

Address all Inquiries to the Publisher
Alan R. Liss, Inc., 41 East 11th Street, New York, NY 10003

Copyright © 1987 Alan R. Liss, Inc.

Printed in the United States of America

Library of Congress Cataloging-in-Publication Data

Cetus— UCLA Symposium on Transcriptional Control
  Mechanisms (1986: Keystone, Colo.)
  Transcriptional control mechanisms.

Includes index.
    1. Genetic transcription—Congresses. 2. Cellular
control mechanisms—Congresses. 3. Genetic regulation—
Congresses. I. Granner, Daryl K. II. Rosenfeld, Michael G.
III. Chang, Shing, IV. Title.
QH450.2.C48   1985   574.87'322   86-27737
ISBN 0-8451-2651-2

Contents

Contributors

Renato J. Aguilera, Departments of Microbiology and Immunology, University of California, Berkeley, Berkeley, CA 94720 **[71]**

Lori A. Allison, Banting and Best Department of Medical Research, University of Toronto, Toronto M5G 1L6, Canada **[383]**

Jacques Archambault, Department of Medical Genetics, University of Toronto, Toronto M5S 1A8, Canada **[383]**

Lee E. Babiss, Laboratory of Molecular Cell Biology, Rockefeller University, New York, NY 10021 **[421]**

David Baltimore, Whitehead Institute for Biomedical Research, Cambridge, MA 02142, and Department of Biology, Massachusetts Institute of Technology, Cambridge, MA 02139 **[83]**

Derk J. Bergsma, Department of Cell Biology, Baylor College of Medicine, Houston, TX 77030 **[57]**

Albrecht Bindereif, Departments of Biochemistry and Molecular Biology, Harvard University, Cambridge, MA 02138 **[181]**

Chester A. Bisbee, Department of Cell Biology, Baylor College of Medicine, Houston, TX 77030 **[313]**

John E. Blume, Department of Biochemistry, University of Illinois, Urbana, IL 61801 **[259]**

Alan J. Cann, Department of Medicine, Division of Hematology-Oncology, School of Medicine, University of California, Los Angeles, Los Angeles, CA 90024 **[463]**

C. Thomas Caskey, Institute for Molecular Genetics, Baylor College of Medicine, Houston, TX 77030 **[45]**

Thomas R. Cech, Departments of Chemistry and Biochemistry, University of Colorado, Boulder, CO 80309-0215 **[171]**

Roger Chalkley, Departments of Molecular Physiology and Biophysics, Vanderbilt University, Nashville, TN 37232; present address: Department of Biochemistry, University of Iowa, Iowa City, IA 52240 **[221,247]**

Shing Chang, Cetus Corporation, Emeryville, CA 94617 **[xix]**

Irvin S.Y. Chen, Department of Medicine, Division of Hematology-Oncology, School of Medicine, University of California, Los Angeles, Los Angeles, CA 90024 **[463]**

A. Craig Chinault, Institute for Molecular Genetics, Baylor College of Medicine, Houston, TX 77030 **[45]**

Robert Chiu, Division of Pharmacology, School of Medicine, University of California, San Diego, La Jolla, CA 92093 **[295]**

The numbers in brackets are the opening page numbers of the contributors' articles.

David T.W. Chu, Departments of Molecular Physiology and Biophysics, Vanderbilt University, Nashville, TN 37232 **[275]**

David A. Clayton, Department of Pathology, Stanford University School of Medicine, Stanford, CA 94305 **[103,395]**

P. Colin, Department of Medicine, L.G.M.E. du C.N.R.S., 67085 Strasbourg Cédex, France **[209]**

Tracy Cooke, Division of Pharmacology, School of Medicine, University of California, San Diego, La Jolla, CA 92093 **[295]**

Michael G. Cordingley, Hormone Action and Oncogenesis Section, Laboratory of Experimental Carcinogenesis, National Cancer Institute, National Institutes of Health, Bethesda, MD 20892 **[333]**

Matt Cotten, Departments of Molecular Physiology and Biophysics, Vanderbilt University, Nashville, TN 37232; present address: Department of Biochemistry, University of Iowa, Iowa City, IA 52240 **[221,247]**

Nessly Craig, Department of Biological Science, University of Maryland Baltimore County, Catonsville, MD 21228 **[113]**

E. Bryan Crenshaw III, Department of Biology, School of Medicine, University of California, San Diego, La Jolla, CA 92093 **[369]**

Val Culotta, Department of Biological Chemistry, The Johns Hopkins University School of Medicine, Baltimore, MD 21205 **[113]**

Gokul C. Das, Laboratory of Molecular and Developmental Biology, National Eye Institute, National Institutes of Health, Bethesda, MD 20205; present address: Division of Molecular Biology and Biophysics, School of Basic Life Sciences, University of Missouri-Kansas City, Kansas City, MO 64110 **[405]**

Monique V. Davies, Genetics Institute, Cambridge, MA 02140 **[195]**

Gail Dinter-Gottlieb, Departments of Chemistry and Biochemistry, University of Colorado, Boulder, CO 80309-0215 **[171]**

Elaine A. Elion, Department of Cell Biology, Albert Einstein College of Medicine, Bronx, NY 10461 **[21]**

Ronald M. Evans, Molecular Biology and Virology Laboratory, The Salk Institute, La Jolla, CA 92138 **[369,379]**

Philip Feigelson, Department of Biochemistry, Columbia University College of Physicians and Surgeons, New York, NY 10032 **[343]**

Robert P. Fisher, Department of Pathology, Stanford University School of Medicine, Stanford, CA 94305 **[103]**

Rodrigo Franco, Department of Biology, School of Medicine, University of California, San Diego, La Jolla, CA 92093 **[379]**

Jeffrey M. Friedman, Howard Hughes Medical Institute, Rockefeller University, New York, NY 10021 **[5,421]**

James D. Friesen, Department of Medical Genetics, University of Toronto, Toronto M5S 1A8, Canada **[383]**

T. Venkat Gopal, Clinical Hematology Branch, National Heart, Lung, and Blood Institute, National Institutes of Health, Bethesda, MD 20892 **[135]**

Daryl K. Granner, Departments of Molecular Physiology and Biophysics, Vanderbilt University, Nashville, TN 37232 **[xix, 275]**

Michael W. Gray, Department of Pathology, Stanford University School of Medicine, Stanford, CA 94305; present address: Department of Biochemistry, Dalhousie University, Halifax, Nova Scotia B3H 4H7, Canada **[395]**

M.R. Green, Departments of Biochemistry and Molecular Biology, Harvard University, Cambridge, MA 02138 **[181]**

James M. Grichnik, Department of Cell Biology, Baylor College of Medicine, Houston, TX 77030 **[57]**

Gordon L. Hager, Hormone Action and Oncogenesis Section, Laboratory of Experimental Oncogenesis, National Cancer Institute, National Institutes of Health, Bethesda, MD 20892 **[333]**

Richard W. Hanson, Department of Biochemistry, Case Western Reserve University, Cleveland, OH 44106 **[1]**

Alois Haslinger, Division of Pharmacology, School of Medicine, University of California, San Diego, La Jolla, CA 92093; present address: Institute of Biochemistry, University of Vienna, Vienna, Austria **[295]**

Adriana Heguy, Division of Pharmacology, School of Medicine, University of California, San Diego, La Jolla, CA 92093 **[295]**

Sheryl Henderson, Department of Biological Chemistry, The Johns Hopkins University School of Medicine, Baltimore, MD 21205 **[113]**

Peter Herrlich, Division of Pharmacology, School of Medicine, University of California, San Diego, La Jolla, CA 92093; present address: Institute for Genetics and Toxicology, Karlsruhe University, Karlsruhe, Federal Republic of Germany **[295]**

James E. Hixson, Department of Pathology, Stanford University School of Medicine, Stanford, CA 94305; present address: Department of Genetics, Southwest Foundation for Biomedical Research, San Antonio, TX 78284 **[103]**

J.C. Homo, Department of Medicine, L.G.M.E. du C.N.R.S., 67085 Strasbourg Cédex, France **[209]**

Thomas J. Hope, Departments of Microbiology and Immunology, University of California, Berkeley, Berkeley, CA 94720 **[71]**

Masayoshi Imagawa, Division of Pharmacology, School of Medicine, University of California, San Diego, La Jolla, CA 92093 **[295]**

Richard J. Imbra, Division of Pharmacology, School of Medicine, University of California, San Diego, La Jolla, CA 92093 **[295]**

C. James Ingles, Banting and Best Department of Medical Research, University of Toronto, Toronto M5G 1L6, Canada **[383]**

Shunsuke Ishii, Laboratory of Molecular Biology, National Cancer Institute, National Institutes of Health, Bethesda, MD 20892 **[437]**

Alfred Johnson, Laboratory of Molecular Biology, National Cancer Institute, National Institutes of Health, Bethesda, MD 20892 **[437]**

Carsten Jonat, Division of Pharmacology, School of Medicine, University of California, San Diego, La Jolla, CA 92093; present address: Institute for Genetics and Toxicology, Karlsruhe University, Karlsruhe, Federal Republic of Germany **[295]**

James Kadonaga, Department of Biochemistry, University of California, Berkeley, Berkeley, CA 94720 **[437]**

Michael Karin, Division of Pharmacology, School of Medicine, University of California, San Diego, La Jolla, CA 92093 **[295]**

Susan Kass, Department of Biological Chemistry, The Johns Hopkins University School of Medicine, Baltimore, MD 21205 **[113]**

Randal J. Kaufman, Genetics Institute, Cambridge, MA 02140 **[11,195]**

Elliott S. Klein, Department of Pharmacology, Stanford University School of Medicine, Stanford, CA 94305 **[343]**

T.M. Kristie, Department of Molecular Genetics and Cell Biology, University of Chicago, Chicago, IL 60637 **[159]**

Stuart Leff, Department of Biology, School of Medicine, University of California, San Diego, La Jolla, CA 92093 **[369]**

Alex Lichtler, Hormone Action and Oncogenesis Section, Laboratory of Experimental Oncogenesis, National Cancer Institute, National Institutes of Health, Bethesda, MD 20892 **[333]**

Sergio Lira, Department of Biology, School of Medicine, University of California, San Diego, La Jolla, CA 92093 **[369, 379]**

John T. Lis, Section of Biochemistry, Molecular and Cell Biology, Cornell University, Ithaca, NY 14853 **[235]**

John Majors, Department of Biological Chemistry, Washington University School of Medicine, St. Louis, MO 63110 **[31]**

Glenn T. Merlino, Laboratory of Molecular Biology, National Cancer Institute, National Institutes of Health, Bethesda, MD 20892 **[437]**

Aaron P. Mitchell, Department of Biochemistry and Biophysics, University of California San Francisco, San Francisco, CA 94143 **[147]**

Matthew Moyle, Banting and Best Department of Medical Research, University of Toronto, Toronto M5G 1L6, Canada **[383]**

Patricia Murtha, Genetics Institute, Cambridge, MA 02140 **[195]**

Christian Nelson, Department of Biology, School of Medicine, University of California, San Diego, La Jolla, CA 92093 **[369,379]**

David A. Nielsen, Department of Biochemistry, University of Illinois, Urbana, IL 61801 **[259]**

Uri Nir, Hormone Research Institute, University of California, San Francisco, San Francisco, CA 94143-0534 **[123]**

Bert W. O'Malley, Department of Cell Biology, Baylor College of Medicine, Houston, TX 77030 **[357]**

P. Oudet, Department of Medicine, L.G.M.E. du C.N.R.S., 67085 Strasbourg Cédex, France **[209]**

Ira Pastan, Laboratory of Molecular Biology, National Cancer Institute, National Institutes of Health, Bethesda, MD 20892 **[437]**

Martine Pastorcic, Department of Cell Biology, Baylor College of Medicine, Houston, TX 77030 **[357]**

Pragna I. Patel, Institute for Molecular Genetics, Baylor College of Medicine, Houston, TX 77030 **[45]**

Joram Piatigorsky, Laboratory of Molecular and Developmental Biology, National Eye Institute, National Institutes of Health, Bethesda, MD 20205 **[405]**

Sharon L. Pitts, Department of Developmental Biology, Genentech, Inc., South San Francisco, CA 94080 **[451]**

E. Regnier, Department of Medicine, L.G.M.E. du C.N.R.S., 67085 Strasbourg Cédex, France **[209]**

Rosemary Reinke, Department of Biochemistry, Columbia University College of Physicians and Surgeons, New York, NY 10032; present address: Department of Biological Chemistry, University of California, Los Angeles, Los Angeles, CA 90024 **[343]**

Helene Richard-Foy, Hormone Action and Oncogenesis Section, Laboratory of Experimental Carcinogenesis, National Cancer Institute, National Institutes of Health, Bethesda, MD 20892 **[333]**

Gordon M. Ringold, Department of Pharmacology, Stanford University School of Medicine, Stanford, CA 94305 **[343]**

Bernard Roizman, Department of Molecular Genetics and Cell Biology, University of Chicago, Chicago, IL 60637 **[159]**

Jeffrey M. Rosen, Department of Cell Biology, Baylor College of Medicine, Houston, TX 77030 **[1,313]**

Joseph D. Rosenblatt, Department of Medicine, Division of Hematology-Oncology, School of Medicine, University of California, Los Angeles, Los Angeles, CA 90024 **[463]**

Michael G. Rosenfeld, Departments of Physiology and Pharmacology, School of Medicine, University of California, San Diego, La Jolla, CA 92093 **[xix,369,379]**

C. Ruhlmann, Department of Medicine, L.G.M.E. du C.N.R.S., 67085 Strasbourg Cédex, France **[209]**

Barbara Ruskin, Departments of Biochemistry and Molecular Biology, Harvard University, Cambridge, MA 02138 **[181]**

Andrew Russo, Department of Biology, School of Medicine, University of California, San Diego, La Jolla, CA 92093 **[369]**

William J. Rutter, Hormone Research Institute, University of California, San Francisco, San Francisco, CA 94143-0534 **[123]**

Hitoshi Sakano, Departments of Microbiology and Immunology, University of California, Berkeley, Berkeley, CA 94720 **[71]**

Kazuyuki Sasaki, Departments of Molecular Physiology and Biophysics, Vanderbilt University, Nashville, TN 37232; present address: Central Research Laboratory, Nisshin Flour Milling Company, Iruma-gun, 354 Saitama, Japan **[275]**

Sundramurthi Satyabhama, Division of Pharmacology, School of Medicine, University of California, San Diego, La Jolla, CA 92093 **[295]**

P. Schultz, Department of Medicine, L.G.M.E. du C.N.R.S., 67085 Strasbourg Cédex, France **[209]**

Robert J. Schwartz, Department of Cell Biology, Baylor College of Medicine, Houston, TX 77030 **[57]**

Linda Sealy, Departments of Molecular Physiology and Biophysics, Vanderbilt University, Nashville, TN 37232; present address: Department of Biochemistry, University of Iowa, Iowa City, IA 52240 **[221,247]**

Scott B. Selleck, Department of Biological Chemistry, Washington University School of Medicine, St. Louis, MO 63110 **[31]**

Ranjan Sen, Whitehead Institute for Biomedical Research, Cambridge, MA 02142, and Department of Biology, Massachusetts Institute of Technology, Cambridge, MA 02139 **[83]**

Neil P. Shah, Department of Medicine, Division of Hematology-Oncology, School of Medicine, University of California, Los Angeles, Los Angeles, CA 90024 **[463]**

David J. Shapiro, Department of Biochemistry, University of Illinois, Urbana, IL 61801 **[259]**

Barbara Sollner-Webb, Department of Biological Chemistry, The Johns Hopkins University School of Medicine, Baltimore, MD 21205 **[113]**

Timothy A. Stewart, Department of Developmental Biology, Genentech, Inc., South San Francisco, CA 94080 **[451]**

E. Brad Thompson, Department of Human Biological Chemistry and Genetics, University of Texas Medical Branch, Galveston, TX 77550-2779 **[5]**

Jeremy W. Thorner, Department of Biochemistry, University of California, Berkeley, Berkeley, CA 94720 **[325]**

Robert Tjian, Department of Biochemistry, University of California, Berkeley, Berkeley, CA 94720 **[437]**

John Tower, Department of Biological Chemistry, The Johns Hopkins University School of Medicine, Baltimore, MD 21205 **[113]**

Ming-Jer Tsai, Department of Cell Biology, Baylor College of Medicine, Houston, TX 77030 **[357]**

Sophia Y. Tsai, Department of Cell Biology, Baylor College of Medicine, Houston, TX 77030 **[357]**

Twee Y. Tsao, Institute for Molecular Genetics, Baylor College of Medicine, Houston, TX 77030 **[45]**

Scott W. Van Arsdell, Department of Biochemistry, University of California, Berkeley, Berkeley, CA 94720 **[325]**

William Wachsman, Department of Medicine, Division of Hematology-Oncology, School of Medicine, University of California, Los Angeles, Los Angeles, CA 90024 **[463]**

Michael D. Walker, Hormone Research Institute, University of California, San Francisco, San Francisco, CA 94143-0534 **[123]**

Heng Wang, Department of Cell Biology, Baylor College of Medicine, Houston, TX 77030 **[357]**

Jonathan R. Warner, Department of Cell Biology, Albert Einstein College of Medicine, Bronx, NY 10461 **[21]**

E. Weiss, Department of Medicine, L.G.M.E. du C.N.R.S., 67085 Strasbourg Cédex, France **[209]**

Janice L. Williams, Department of Medicine, Division of Hematology-Oncology, School of Medicine, University of California, Los Angeles, Los Angeles, CA 90024 **[463]**

Jolene Windle, Department of Biological Chemistry, The Johns Hopkins University School of Medicine, Baltimore, MD 21205 **[113]**

Jerry K.-C. Wong, Banting and Best Department of Medical Research, University of Toronto, Toronto M5G 1L6, Canada **[383]**

Preface

The 1986 Cetus-UCLA Symposium on "Transcriptional Control Mechanisms" was held at Keystone, Colorado, April 6–13. This conference, attended by some 500 scientists, covered a broad range of topics, with the main emphasis on DNA sequence-specific regulation of gene expression and an examination of proteins that bind to these regions. This was examined from the standpoint of chemical and physical analyses of DNA-protein interactions, a biochemical analysis of transcription factors, RNA processing, identification of enhancers and enhancer binding proteins and how complex chromatin structure affects transcription. A large portion of the conference dealt with a number of specific examples of transcription control. Examples discussed included those regulated by hormones, oncogenes, and heat shock. The systems explored ranged from cultured cells to transgenic mice, and to experimental model systems in which exploration of genetics was effectively utilized to identify functionally important gene products. This book reflects a sampling of the papers presented.

The analysis of eukaryotic transcription mechanisms is aimed at defining the elements necessary for reconstitution *in vitro* of a system that will accurately and efficiently transcribe a particular gene. The next step will be to control transcription of a specific gene as is done in the intact cell. In this way the essential features can be precisely defined.

Various components of this quest are falling into place. DNA regions, generally adjacent to the initiation site on the 5' side, that are necessary for transcription are being found in all genes. These include the TATA and CAAT boxes, but other regions are being defined. It appears that different genes use different combinations of these basal promoter elements, but as yet, no definite patterns for classes of genes have become apparent. Genes subject to regulation have additional regulatory elements. The best studied of these are enhancers, regions of DNA that act *cis* but are not strictly location and orientation dependent. Some hormones, glucocorticoids in particular, appear to interact with enhancer elements through their intracellular receptors. Most of these regions enhance transcription but suppressor regions also are recognized.

It is presumed that these DNA sequences bind proteins that influence transcription. In many cases these proteins have been identified, often by the band-retardation assay, and some (SP-1 and NF-1) have been partially purified. Rapid progress is being made in this area, due in no small part to the fact that the DNA sequence itself can be used to bind the protein and thereby remove it from other cellular proteins.

Once the specific DNA elements and associated binding proteins are available biophysical studies, similar to the elegant examples presented as studies of prokaryotic transcription, can be conducted and the physical basis of the regulation of transcription of eukaryotic genes can be approached. Already, concordance of the actions of prokaryotic and eukaryotic DNA binding protein actions are suggested.

Post-transcriptional control mechanisms are equally important as evidenced by a number of reports dealing with RNA processing. Assays for these processes are being developed and in combination with genetic approaches, isolation of a number of the critical regulatory proteins appears close at hand.

By mid-week it was obvious that the complex nature of eukaryotic transcription is going to require some time to sort out. This news, albeit frustrating at times, has a bright side for it ensures the need for more conferences of this sort.

Special thanks are due Cetus Corporation for the generous sponsorship of this meeting. We also gratefully acknowledge gifts from Genetics Institute. We sincerely appreciate and gratefully acknowledge the 1986 UCLA Symposia Director's Sponsors—Burroughs-Wellcome Company; Monsanta Corporation Research Laboratories; and The Upjohn Company. Additional support was provided by gifts from contributors to the UCLA Symposia Director's Fund—Biosciences Laboratory, Corporate Technology, Allied-Signal Corporation; AMOCO Corporation; Allelix, Inc., AMGen, Inc.; Boehringer Ingelheim Pharmaceuticals, Inc.; Hoechst-Roussel Pharmaceuticals, Inc.; New England Biolabs Foundation; Ayerst Laboratories Research, Inc.; Bristol-Myers Company; Celanese Research Company; Wyeth Laboratories; and Codon Corporation. Additionally, we wish to thank the UCLA Symposia staff, especially Betty Handy and Robin Yeaton, who ensured that things ran smoothly.

Daryl Granner
Geoff Rosenfeld
Shing Chang

Transcriptional Control Mechanisms, pages 1–4
© **1987 Alan R. Liss, Inc.**

WORKSHOP SUMMARY: HORMONES AND TRANSCRIPTION I

Jeffrey M. Rosen and Richard W. Hanson

Department of Cell Biology
Baylor College of Medicine
Houston, Texas 77030

Department of Biochemistry
Case Western Reserve University
Cleveland, Ohio 44106

The presentations at this workshop focused on the definition by DNA-mediated gene transfer of cis-acting DNA sequences responsible for hormonal regulation of several different peptide and steroid-responsive genes, and, in several cases, the identification of trans-acting proteins that interact with these sequences. A brief summary of the results presented at the workshop is given below. For more details on each presentation, consult the authors' abstracts published in the Journal of Cellular Biochemistry, Supplement 10D, 1986, and more detailed descriptions included in this volume.

C. Bancroft described his studies of the tissue-specific regulation of the rat prolactin (Prl) gene, and of the rat growth hormone (GH) gene by glucocorticoids and thyroid hormone. In both cases, flanking regions of these genes were linked to a chloramphenicol acetyltransferase (CAT) reporter gene. Protein blotting experiments were also performed to identify proteins interacting with specific DNA sequences. Tissue-specific expression of the rat Prl–CAT construction was observed in a pituitary tumor-derived cell line, GH_3, but not in a variety of other non-pituitary derived cell lines. Deletion analysis revealed a 70% decrease in CAT activity between −1957 and −614 bases upstream of the site of initiation of transcription (CAP site), a 20-fold drop between −614 and −374 (the presumptive location of a tissue-specific enhancer), an 8-fold increase between −374 and −228 (the presumptive location of a silencer), and regulatory sites for several hormones within the first few hundred bases of 5' flanking DNA. Protein blotting experiments identified two nuclear proteins of 44 and 48 kDa, which were both tissue-specific and displayed

selective binding to flanking sequences in the Prl gene.
Several isoelectric variants of these proteins were observed
by two-dimensional gel electrophoresis.   At least two
distinct binding regions were identified between −4.8 and
2.0 kb and −1270 and −620 bp in the 5' flanking region of
the Prl gene.   For the GH gene a tissue-specific enhancer
was observed much closer to the CAP site, and specificity
was only lost using a deletion between −200 and −100. Hormonal regulation of GH-CAT fusion genes was observed with
both dexamethasone and thyroid hormone, but the fold induction was less than that observed for GH mRNA.   Finally, the
dexamethasone-responsive element was partially localized.
Deletions to −580 bp still resulted in an 8-fold induction,
and to −236 bp in a 6-fold induction of CAT activity.

Paul Webb, working with Malcolm Parker, discussed the
hormonal regulation of the mouse mammary tumor virus (MMTV)
promoter in several breast cancer cell lines.   Based upon an
earlier observation that the MMTV provirus was induced by
both androgens and glucocorticoids in the Shionogi cell
line, S115, the response of an MMTV LTR-CAT fusion gene to
various steroid hormones was studied in S115, T-47D, clone
11, and ZR-75 cells.   Dihydrotestosterone (DHT)-induced CAT
expression in S115 and ZR-75, but not in T-47D cells,
despite the presence of functional DHT receptors demonstrated by the induction of a 43 kDa protein.   Dexamethasone
and progesterone were effective inducers of CAT in ZR-75
cells, but only progesterone was effective in T-47D cells.
Several deletion mutants were tested in both the ZR-75 and
T-47D cells.   Deletions up to −765 were fully induced, those
to −463 were still responsive, but deletion to −373 resulted
in the loss of steroid hormone inducibility.   One difference
between these breast tumor cell lines is the level of the
various steroid receptors, e.g., in T-47D cells the level of
the progesterone receptor is unusually high.   Thus, MMTV
gene expression can be regulated by steroid hormones other
than glucocorticoids, but the presence of the steroid
receptor while necessary is not sufficient to account for
steroid-responsiveness.

Richard Miksicek, working in the laboratory of Gunter
Schutz, described the definition of glucocorticoid response
elements (GRE) in the tyrosine aminotransferase (TAT) gene.
TAT gene expression is regulated by both cAMP and glucocorticoids in normal liver and a Reuber hepatoma cell line,
FTO-2B.   An analysis of DNase I hypersensitive sites in the
TAT gene had revealed numerous sites in a region spanning 40
kb from 20 kb upstream to 10 kb downstream of the gene.   Two
regions in particular displayed changes with hormonal treatment, one at −200 bp and the second at −2.5 kb.   These sites

were not effected by cAMP. A deletion series of mutant TAT-CAT fusion genes were assayed in mouse L-cells. Deletions to -2527 were fully active, but those at -2504 lost inducibility. Four potential GREs were identified in this region. Elements II and III were found to be critical for induction, while I and IV were less important. The DNase I footprints of the glucocorticoid receptor complex binding to regions II and III were related to the presence of 3 DNase I hypersensitive sites. Three regions of cleavage and 2 gaps were observed. Genomic sequencing was employed as well as detected sites of hormone-dependent protection and enhancement in the region of elements II and III. Elements II and III when linked to a heterologous promoter in a thymidine kinase (Tk)-CAT vector resulted in an 8.2-fold induction of CAT gene expression, while deletion of site II yielded only a 2.4-fold increase. Finally, a 22 bp synthetic oligonucleotide, corresponding to site II alone when linked to Tk-CAT, was able to elicit a 3.2-fold response to dexamethasone.

Y. Hod working with Richard Hanson described studies to define the sequences responsible for the hormonal regulation of the phosphoenolpyruvate carboxykinase (PEPCK) gene by cAMP and glucocorticoids. A PEPCK promoter fragment was fused to Tk and a series of deletion mutants generated and assayed for hormonal responsiveness after selection in Tk⁻ FTO-2B cells. This fusion gene responded to cAMP and glucocorticoids, but was not inhibited by insulin. A sequence positioned between -62 and -108 could confer induction of Tk in an orientation independent fashion to both cAMP and glucocorticoids. A sequence, -91 CTTACGTCAGAG -80, was identified as a potential cAMP regulatory element, and was homologous to sequences identified in several other cAMP-responsive genes, such as rat preprosomatostatin, human vasoactive intestinal polypeptide and porcine plasminogen activator. Sequence analysis of the PEPCK gene identified at least two GREs, two putative binding sites for the transcription factor SP 1 and TATA and CAAT box homologies. In addition to positive elements, a potential negative regulatory element was also identified in the PEPCK promoter at -150 to -175. An attempt was made to develop an in vitro transcription system from hepatoma cells that might permit further characterization of transcription factors. Using linear truncated templates, specific initiation was demonstrated. The reaction was linear for 2 1/2 hr and required both a nuclear and an S 100 fraction for maximal activity. No effect of cAMP has yet been demonstrated in the in vitro system.

J. Rosen described studies designed to elucidate the hormone- and tissue-specific elements regulating casein gene expression in mammary epithelial cells using casein-CAT fusion genes. Mammary epithelial cells were transfected on hydrated collagen gels and the hormonal regulation of casein gene expression studied on detached gels, because both cell-cell and cell-substratum interactions are required for the expression of differentiated function. No hormonal responsiveness of constructions containing 1 kb of 5' flanking DNA was observed in both transiently transfected and selected COMMA-1D cells under conditions where the endogenous β-casein gene was induced as much as 150-fold by prolactin and glucocorticoids. This was consistent with the observation that the primary effect of these hormones on casein gene expression appears to be at the post-transcriptional level involving changes in the rate of nuclear RNA turnover. However, when a series of casein-CAT fusion genes containing up to 1 kb of flanking DNA, exon I and a portion of intron A were assayed in both epithelial (COMMA-1D) cells and fibroblasts (NIH 3T3 cells), several interesting differences were observed. Constructions containing exon I and intron A displayed 10- to 20-fold more CAT activity than those containing only 5' flanking DNA sequences in COMMA-1D cells, but this difference was not as prominent in NIH-3T3 cells. Both a strong mammary-specific inhibition by proximal 5' flanking sequences and a weaker mammary-specific enhancement by internal sequences was observed.

The final presentation at this workshop dealt with the identification of tissue-specific enhancers in the albumin and α-fetoprotein (AFP) genes by L. Muglia. A series of AFP-CAT fusion genes were assayed in Hep G2 cells as well as CV 1 and Hepa 1 cells. A 2.2 kb 5' flanking AFP fragment was insufficient to direct efficient expression in Hep G2 cells, but sequences to −7 kb were sufficient. The enhancer element between −2.2 and −7 kb functioned in both orientations and activated the SV40 and albumin promoters from 5' and 3' positions. The SV40 enhancer was able to partially substitute for the AFP enhancer when linked to the 2.2 kb AFP promoter. A series of deletion mutants were assayed in Hep G2 cells and identified both negative and positive regulatory elements in the AFP flanking region. Thus, a tissue-specific promoter appears to exist in the sequences immediately proximal to the CAP site, a rather large positive enhancer region is present between −2.2 and −7 kb, and a negative element is present between −3.5 and −4 kb.

Transcriptional Control Mechanisms, pages 5–10
© **1987 Alan R. Liss, Inc.**

Workshop Summary: Hormones and Transcription II

Jeffrey M. Friedman[1] and E. Brad Thompson[2]

1 Howard Hughes Medical Institute
The Rockefeller University
New York, N. Y. 10021

2 Department of Human Biological Chemistry and Genetics
The University of Texas Medical Brance
Galveston, TX 77550-2779

Under the general rubric of hormones and transcription, several specific areas were discussed in this workshop. The first general topic concerns the structure of the recently cloned glucocorticoid receptor. The second topic dealt with assays of cis active transcriptional control elements in the vicinity of tissue specific and hormonally induced promoters. The specific subjects that were addressed included the structure of the rat glucocorticoid receptor (GR) and its gene; the use of the GR cDNA to examine GR receptor mutants; an adenoviral vector that can score the transcriptional activity of recombinant DNAs; the negative regulation of the pro-opiomelanocortin (POMC) gene by glucocorticoids; negative regulatory elements in the long terminal repeat of the mouse mammary tumor viral gene; and the special requirements for obtaining expression of the whey acidic protein gene.

Rusconi (University of California, San Franscisco) discussed the rat glucocorticoid receptor gene and compared it with the human glucocorticoid receptor structure that has been published and that had been discussed previously at this meeting in plenary session. The two clones of human glucocorticoid receptor cDNA have been isolated and sequenced. They differ from one another in several ways, but they are identical throughout their coding sequence up to a point near the 3' end. From nucleotide 2313, they diverged. One had continued coding sequence for 150 nucleotides followed by a long non-translated 3' sequence of approximately 2.4 kb. The other, from the point of divergence, had a unique coding sequence of 45 nucleotides followed by a completely different 3' sequence which ran on an additional 1.5 kb. Rusconi reported that the coding sequence of the rat gene was extremely similar to that of the human. The 3' untranslated region contained sequences highly homologous to each of the two human untranslated regions. But in the rat gene, they occurred consecutively on the same DNA. Rusconi reported that the rat receptor gene had been transcribed from a vector and the transcripts translated in vitro to produce a protein of the correct molecular weight for wild type receptor, much as had been done for the human gene. He further said that the rat gene was being used to map mouse cell mutant receptor mRNAs from cultured cell lines, and described deletions that defined the 3' end of the rat gene as the coding sequence for the steroid binding site. More central regions in the coding sequence appeared to be important for the DNA binding. He stated that transfection of the gene into r^- rat cells had successfully produced rat^+ cells both in transient and long-term assays. However, in the stable transfectants which became r^+, only one of three showed receptor function as shown by glucocorticoid induction of an endogenous cell gene.

E. B. Thompson (Department of Human Biological Chemistry and Genetics, University of Texas Medical Branch, Galveston) reported on the use of human glucocorticoid receptor cDNAs to explore r^- mutants obtained from the human leukemic cell line CEM. His laboratory has produced monoclonal antibodies to the human glucocorticoid receptor. With these they had compared the expression of receptor peptide with the binding-site analysis of receptors in the r^- cells. By competitive binding

assay of binding sites, these cells contained
about 10% wild type receptors. These residual
receptors had been labeled with ^{3}H dexamethasone
mesylate and shown to be of normal size. Immunob-
lotting with antireceptor monoclonal antibody,
however, revealed normal size receptor protein in
quantity similar to that of wild type cells.
Northern blot analysis showed that the r$^-$ cells
produced the same 7s, 4.5s and 2.5s mRNAs as do
wild type cells, both similar in quantity to wild
type. In addition, the mutant cells displayed a
smaller ~1.8s RNA that hybridized with the receptor cDNA. They concluded that this r$^-$ clone is
predominantly making a non-glucocorticoid binding
form of receptor since transcription and message
stability appear to be normal. Analysis of this
clone should be valuable in defining the part of
the protein responsible for the steroid binding
activity.

     Jeff Friedman described work done with Lee
Babiss in which an adenoviral vector was developed
as a method of introducing chimaeric DNA into a
wide range of continuous and primary cultured
cells. In this vector, recombinant adenovirus
have been made, using standard techniques, in
which the E1A enhancer and promoter, coding
sequence and the E1B promoter have been deleted
and replaced with either the rat albumin, mouse
globin or mouse immunoglobulin heavy chain promoters. Thus in each virus a tissue specific promoter is directing the synthesis of E1B RNAs.
These viruses contain deletions such that no synthesis of E1a proteins occurs. As a consequence
the viral recombinants will not replicate efficiently and exhibit extended eclipse periods.
These viruses have been used to introduce the
chimaeric DNA into all the cells in populations of
Hep G2 (hepatoma) cells, mpc 11 (myeloma) cells
and 1° liver cells. In infected hepatoma cells,
E1B mRNA and transcription can be scored only from
the virus containing the albumin promoter whereas
in myeloma cells E1B RNA can only be detected in
cells infected with the virus containing the immunoglobulin promoter. Moreover, in cultured primary hepatocytes, only the virus containing the
albumin promoter directs the synthesis of E1B

mRNAs, and the activity of the viral promoter
declines with time in culture in parallel with the
decline seen for the endogenous gene. Thus the
adenovirus system allows the introduction of
chimaeric DNA into a wide range of mammalian
cells, including primary cells, and reveals the
presence of tissue specific transcription factors
in hepatic cells for the albumin promoter. This
vector may ultimately be useful for defining the
activity of hormone responsive cis elements in
cultured cells.

Lee Babiss described additional experiments
with these viruses in which the effect of an
enhancer element was probed by inserting the E1A
enhancer element upstream of the globin and albumin promoters in the viral recombinant. It was
initially found that the enhancer, by itself,
increases the transcription rate of the albumin
promoter three fold in hepatoma cells but has no
effect on the globin promoter. The viruses as
constructed are deficient for the E1A protein so
Babiss next considered the effect of E1A protein
on the rate and specificity of transcription of
the virally encoded promoters. This was accomplished by coinfecting Hep G2 cells with the test
viruses and dl 313, an adenovirus that synthesizes
E1A and provides the E1A protein in trans. This
virus is deficient for E1B, so E1B synthesis can
still be scored as an index of promoter activity
on the test virus. Babiss found that the E1A protein, in trans, increased the transcription rate
from the albumin promoter containing the E1A
enhancer 100 fold in hepatoma cells but had no
effect on the globin promoter in these cells even
with the E1A enhancer in cis. Moreover the E1A
protein has no effect on these promoters if the
E1A enhancer was absent. Thus the E1A promoter in
combination with the enhancer will boost the rate
of transcription as long as an albumin specific
factor is present. These data raise the possibility that enhancer effects may be mediated either
by direct or indirect interactions with other factors necessary for transcription complex formation. Babiss considered the possibility that hormone inducible enhancers function in a similar

way, and speculates that enhancer-promoter interactions may explain how the same hormone can activate different genes in different tissues.

J. Drouin (Molecular Biology Institute of Montreal, Montreal, Canada) reported on the pro-opiomelanocortin gene. He stressed two themes, the tissue specificity of this gene and its down-regulation by glucocorticoids. In these studies, as much as 4.5 kb of upstream sequence for the rat POMC gene was fused to the neomycin resistance gene and transfected into primary pituatry cells as well as AT-20 cells. Neomycin resistant transformants were studied further and it was found that in primary anterior pituitary cells and AT-20 cells the gene was expressed and down regu-lated by glucocorticoids (as is the endogenous POMC gene). However, in L-cells there was very low expression. Deletion analysis showed that the steroid regulation was still present when as lit-tle as 700 base pairs of the flanking region was used to drive the neomycin resistance gene. Further mapping is underway. A glucocorticoid response element was found in several places in the POMC gene: at two sites in the first intron in the middle of the first exon, at $^-64$ prior to the first exon and at $^-580$ prior to the first exon. These sites were being footprinted with purified rat liver glucocorticoid receptor. A site at -74 to -51 base pairs footprinted strongly with glucocorticoid receptor. This region con-tained three CCAAT boxes, but no TGTTCT sequences. The -580 site and one of the intron sites have slight homology to the consensus hexonucleotide for genes positively regulated by glucocorticoid, but the other had little resemblance.

M. G. Cordingly (Hormone/Oncogenesis Labora-tory, National Cancer Institute, Bethesda, Mary-land) discussed negative regulatory elements in the MMTV LTR. Sequences within 225 base pairs of the cap region in the LTR are required for hormone responsiveness. In addition, sequences between $^-$107 and $^-$371 interfered with the activation of the MTMV promoter in the absence of hormone. Mapping experiments with oligonucleotide mutations through this region were carried out. They found that

with the wild type LTR, basal expression was low,
but glucocorticoid-induced expression was high.
On the other hand, when the negative element was
mutated, basal expression was high and there was
less induction. They concluded that an overlapping area has both positive and negative control
properties. They are using their bovine papilloma
virus mini chromosome system to study this process, cloning into it the MMTV LTR driving the 30s
v ras gene. Cycloheximide and dexamethasone
experiments were being used to show that cycloheximide itself is a partial inducer of this construct. Dexamethasone was a better inducer, and
in combination, the two were synergistic inducers.
This could be shown both by measurements of both
stedy state mRNA and transcription. Cordingly
invoked nucleosome structure as possibly having
relevance for regulation, calling attention to the
positioning of regulatory elements prior to the
start site of transcription. He concluded that
their results implied that action of additional
protein(s) besides the glucocorticoid receptor was
involved in regulation of expression of the MMTV
LTR.

    J. M. Rosen (Department of Cell Biology, Bay-
lor College of Medicine, Houston) discussed the
expression of whey acidic protein (WAP). This
mammary specific protein is inducible by a combination of insulin, cortisol, and prolactin in
organ cultures of intact mammary glands. However,
primary cultures of cells in monolayer would not
express WAP. A WAP expression vector, containing
1.5 kb of 5' flanking sequences as well as considerable 3' flanking DNA was transfected into
several cell lines. Eventually lines containing
1-50 copies per line in stable transfectants were
isolated. However, there was no expression of WAP
until the transfected cells were placed in the
cleared mammary fat pad, whereupon re-expression
of both the endogenous mouse gene and the
transfected WAP gene were observed. Rosen concluded that the cellular environment surrounding
the transfected cells affected the expression of
transfected and endogenous genes. By extension,
one might anticipate that this also signifies
local extracellular control over expression of
endogenous genes in vivo.

Transcriptional Control Mechanisms, pages 11–19
© **1987 Alan R. Liss, Inc.**

WORKSHOP SUMMARY: MAMMALIAN
EXPRESSION SYSTEMS

Randal J. Kaufman

Genetics Institute
Cambridge, MA 02140

The focus of this workshop was on the
utility of different expression systems
available for the high level production of
proteins from heterologous genes in mammalian
cell hosts. Particular emphasis was placed on
the production of proteins as pharmaceutical
therapeutics. There are several reasons to
use a mammalian cell host to produce a protein
pharmaceutical. First, the proteins are
secreted in biologically active conformations
and, if applicable, with the appropriate
disulphide bond formations. O-linked
glycosylation and either simple, hybrid, or
complex asparagine N-linked glysocylation
occurs at normal positions. Other
post-translational modifications such as
gamma-carboxylation of glutamic acid residues
occurs at correct positions. Finally,
mammalian DNA sequences are readily expressed
in a variety of mammalian cell hosts. The
primary disadvantage results from the high
cost of mammalian cell culture. However,
advances in the development of serum-free
media and in the large scale growth of
mammalian cells in cell fermenters are
dramatically reducing this cost making it more
competitive with yeast and bacterial
production systems.

Although a variety of systems are
potentially available for the expression of
heterologous genes in mammalian cells, this
workshop focused on two systems most useful
for the high level expression of heterologous
proteins: First, cotransfection of Chinese

hamster ovary cells with a heterologous gene and a dihydrofolate reductase (DHFR) gene and subsequent selection for methotrexate resistance via gene amplification and second, bovine papilloma virus vectors for the introduction of multiple DNA copies into mouse fibrolblasts. The development of other mammalian expression systems used for vaccine development was also discussed. Two examples use recombinant adenoviruses and vaccinia viruses to produce hepatitis B surface antigen and feline leukemia virus antigen, respectively.

R. Kaufman described the cloning and expression of human granulocyte-macrophage colony stimulating factor (GM-CSF) (1). The cloning was accomplished by directly screening cDNA libraries constructed in a monkey COS cell expression vector which yielded GM-CSF in the conditioned media of transfected cells. The vector utilized [p91023(B)] was optimized for high level expression of heterologous genes in COS cells. One unique feature of the vector was the presence of the adenovirus virus-associated (VA) genes which potentiate translation of mRNAs derived from the expression plasmid. DNA was prepared from approximately 200 pools of 400-500 recombinants each and was transfected into COS monkey cells. Conditioned media was assayed after transfection, and pools yielding biologically active GM-CSF were subdivided and retransfected into COS cells. Six positive clones were obtained, all of which encoded the same amino acid sequence. The expression cloning approach is useful for cloning genes for which little protein sequence data is available and for which there exist sensitive assays for the presence of the desired protein. The GM-CSF expression plasmid was introduced with a DHFR expression plasmid into DHFR deficient CHO cells (DUKX-BII) and selection applied for the DHFR positive phenotype (growth in the absence of added

nucleosides) as described (2). Selection for
growth in increasing concentrations of
methotrexate resulted in cells containing 500
copies of the GM-CSF gene which expressed 10 -
20 ug/ml of GM-CSF. GM-CSF represented over
90% of the total radiolabeled secreted protein
present in the conditioned medium. The GM-CSF
had an identical molecular weight, apparently
identical glycosylation pattern, and an
identical specific activity to that derived
from activated human T cells. These results
demonstrate the capacity of CHO cells to
efficiently synthesize and secrete large
quantities of a glycosylated protein expressed
from a heterologous gene.
     R. Kingston described an approach to
express potentially toxic proteins in CHO
cells utilizing a drosophila heat shock
protein 70 (hsp70) inducible promoter and
coamplification with DHFR (3). The coding
region of the mouse c-myc gene (the second and
third exons) was fused to the drosophila hsp
70 promoter and the chimaeric gene was
introduced into DHFR deficient CHO cells by
cotransfection with a selectable DHFR gene.
The resultant cell lines were grown in
increasing concentrations of methotrexate
which resulted in amplification of the
transfected DNA. Cells lines obtained contain
up to 2000 copies of the introduced c-myc
gene. Since the drosophila heat shock
promoter has an extremely low basal level of
expression, integration and amplification of
the recombinant c-myc gene occurred while the
gene was in an "off" state. Incubation of
these cells at $43^{o}C$ resulted in at least a
100-fold induction of the c-myc mRNA.
Translation of this mRNA occured when the
cells were returned to $37^{o}C$, and after 3-4
hours the c-myc protein levels reached
approximately 1 mg per 10^9 cells. These
results demonstrate that 2000 copies of the
heat shock promoter are highly inducible
suggesting heat shock induction may not
involve a negative factor or that such a

factor is not limiting in CHO cells. More
importantly, whatever positive factor is
involved in the induction is not limiting even
at such high gene amplification. The c-myc
protein produced from these recombinant cell
lines has the characteristics described
previously for the endogenous mouse c-myc
protein. The products are phosphoproteins
with relative mobilities of 64,000, 66,000,
and 75,000 and are associated with the nuclear
matrix. The induced cells die, suggesting
that high levels of c-myc are cytotoxic. The
amplification of genes placed under control of
the drosophila hsp 70 promoter may provide a
general method for high levels of inducible
expression of proteins in mammalian cells.
        B. Felber described success with bovine
papilloma virus (BPV) derived vectors for
expression of heterologous genes. BPV is a
small circular DNA virus that morphologically
transforms a variety of cells. Vectors
containing the entire BPV genome or a 69%
subgenomic transforming fragment, in many
cases, are stable as multicopy (20 - 100
copies/cell) extrachromosoml elements in
transformed cells. This is especially true
for vectors which contain the 100% BPV
genome. In other cases the vector sequences
are maintained as oligomers integrated into
the host chromosome in a head to tail tandem
array. Derivatives of BPV vectors contain
selectable markers which obviate relying on
morphoogical transformation to obtain cells
harboring exogenous DNA. One particularly
useful vector was described which contained
the 100% BPV genome with a human
metallothionine gene as a selectable marker
and a mouse metallothionine gene promoter to
drive transcription of human growth hormone
(HGH) (4). Morphological transformants
derived after transfection of mouse C127
fibroblasts expressed HGH and its expression
increased 5 - 7 fold upon addition of cadmium
chloride. Selection of transformants in 20 uM
cadmium chloride produced transformants which

contain 50 - 100 episomal copies. Growth in
increasing concentrations of cadmium chloride
did not result in further increases in the
copy number. Cells obtained in this way
express HGH at 100 ug/ml/day in roller
bottles. When a mouse metallothionine gene
was used as a selectable marker, the gene
frequently deleted due to homologous
recombination within duplicated sequences
present on the plasmid. Expression of other
heterologous genes in this system include HTLV
I and III trans-acting transcriptional (TAT)
activators, human beta and gamma interferons,
human chorionic gonadotropin, tissue
plasminogen activator, and calcitonin.

N. Sarver discussed the role of the BPV
enhancer and of 3' noncoding sequences in the
transcription unit for the expression of
heterologous genes in BPV derived vectors in
murine C127 cells. Vectors containing the
100% BPV genome contain an enhancer which is
lacking in vectors containing the 69%
subgenomic fragment since it is just 3' of the
69% fragment endpoint. The enhancer is
required for stable episomal maintenance and
for efficient expression of heterologous
genes. The insertion of the murine sarcoma
virus enhancer element can substitute for the
BPV enhancer. Experiments were also described
which demonstrate that expression of human
gamma-interferon can be modulated by sequences
in the 3' untranslated region of the
transcription unit. The vector utilized in
these experiments was constructed in a pML2
derivative (which has deleted pBR322 sequences
which are detrimental to DNA replication in
mammalian cells) containing, in a 5' to 3'
direction, the murine sarcoma virus enhancer,
the mouse metallothionine promoter, the human
gamma-interferon coding region, a 3' noncoding
region from the bacterial neomycin
phosphotransferase gene, and the SV40 small t
antigen intron and the SV40 early
polyadenylation signal. Progressive removal
of the of the neomycin phosphotransferae gene

resulted in increased interferon expression
that correlated with the size of the
deletion. Aditional removal of the SV40 small
t antigen intron resulted in a 10 - fold
decrease in the level of interferon
expression. The optimal vector contained the
SV40 intron and SV40 polyadenylation signal
and yielded approximately 10 ug/ml of
gamma-interferon.

     V. B. Reddy described the expression of
human tissue plasminogen activator (t-PA) in
BPV derived vectors in mouse C127 cells. The
t-PA transcription unit contained the mouse
metallothionine 1 promoter and the t-PA coding
region with diffferent 3' ends consisting of
either two introns and polyadenylation signal
from MT-1 DNA, the SV40 early intron and
polyadenylation signal, or the SV40 early
polyadenylation signal alone. After
transfection and selection for foci in C127
cells, it was found that t-PA protein and mRNA
was expressed 100 fold higher in the
expression vector containing the SV40 early
polyadenylation signal alone. Since nuclear
run-off experiments indicated similar rates of
transcription, the increased t-PA expression
resulted from increased t-PA mRNA stability.
The best vector expressed 0.3 ug/ml/day of
human t-PA. These results are in contrast to
those presented by N. Sarver, where higher
levels of expression require the SV40 small t
intron, and point to the principle that
similar noncoding regions can behave quite
differently in different expression systems.
Although there has been considerable success
with BPV expression systems, the biology of
BPV expression, and the requirements for its
plasmid maintenance and regulation of copy
number are poorly understood (See Botchan et.
al., this volume). As more features of the
biology of BPV become understood, BPV will
become a more useful system for expression of
a wide variety of heterologous genes.

J. Morin described the expression of
hepatitis B surface antigen (HBsAg) in a
recombinant adenovirus (5). The early region
1 of the adenovirus type 5 genome was replaced
with the coding region for the HBsAg flanked
by the adenovirus 2 major late promoter and an
intron and polyadenylation signal from the
early region of SV40. Two different viruses
were constructed and contained either 33 bp of
the first adenovirus late leader or almost the
entire tripartite leader (170 bp) preceeding
the coding region of HBsAg. Equivalent
amounts of RNA were produced late in
infection. However, HBsAg protein was 70-fold
higher late in infection with the virus
containing the full tripartite leader
sequence. One likely interpretation is that
the tripartite leader is critical for
translation of mRNAs in the late stage of
adenovirus infection. The HBsAg produced was
glycosylated and secreted into the medium as
particles that were essentially
indistinguishable from the 22-nm particles
found in human serum. The level of expression
of HBsAg from the tripartite leader virus was
low (14 ng/5×10^6cells). Thus, this is low
compared to levels obtained with other
systems. However, the use of adenovirus as a
vector to transmit viral antigens provides a
novel and interesting approach to live vaccine
development.

J. Nunberg presented the development of
a vaccinia recombinant virus encoding the
expression of the feline leukemia virus
envelope protein. The advantages of a live
recombinant vaccinia virus as a vaccine are
that a strong immune response can be elicited,
virus can spread and protect unvaccinated
individuals by herd immunity, and the
protection is long lived. Using the
techniques of B. Moss and colleagues a variety
of vaccinia recombinants have been constructed
encoding the influenza virus hemagglutinin,
the herpes simplex virus glycoprotein G, the
VSV G glycoprotein, the HBsAg, the sindbis

virus polyprotein, the EBV gp 340 and 220, the antigens of plasmodium knowles, and the human respiratory syncytial virus (RSV) G glycoprotein (for example see 6). The majority of these viruses elicit an immune response in infected animals. The envelope (env) gene of the Gardner-Arnstein strain FeLV (subgroup B) under the transcriptional control of the vaccinia early '7.5 K gene' promoter was inserted into the vaccinia virus genome. Studies on the expression of the FeLV env gene in vaccinia recombinant infected cells have revealed no qualitative differences in the processing and intracellular transport of the FeLV env gene product compared to that of FeLV. Proteolytic processing of Pr85, the env precursor protein, occurs to yield mature gp70 and p15 which accumulate on the cell surface. However, upon infection of cats and mice with the recombinant virus no antibody could be detected. A possible explanation for this may be that vaccinia replication is reduced 100-fold in cats and in mice. Attempts are now being directed to alter the host range of the vaccinia recombinants.

REFERENCES

1. Wong, G.G., J.S. Witek, P.A. Temple, K.M. Wilkens, A.C. Leary, D.P. Luxenberg, S.S. Jones, E.L. Brown, R.M. Kay, E.C. Orr, C.S. Shoemaker, D.W. Golde, R.J. Kaufman, R.M. Hewick, E.A. Wang, S.C. Clark (1985) Human GM-CSF: Molecular cloning of the complementary DNA and purification of the natural and recombinant proteins. Science 228: 8100-815.

2. Kaufman, R.J., L. Wasley, L. Spiliotes, S. Gossels, S. Latt, G. Larsen, R. Kay (1985) Coamplification and coexpression of human tissue-type plasminagen activator and murine dihydrofolate reductase sequences in Chinese hamster ovary cells. Mol Cell. Biol. 5: 1750-1759.

3. Wurm, F.M., K.A. Gwinn, and R.E. Kingston (1986) Inducible overexpression of the mouse c-myc protein in mammalian cells. Proc. Natl. Acad. Sci. In press.

4. Pavlakis, G.N. and D.H. Hamer (1983) Regulation of a metallothionine-growth hormone hybrid gene in bovine papilloma virus. Proc. Natl. Acad, Sci. 80: 397-401.

5. Davis, A.R., B. Kostek B.B. Mason, C.L. Hsiao, J. Morin, S.K. Dheer, and P.P. Hung (1985) Expression of hepatitis B surface antigen with a recombinant adenovirus. Proc. Natl. Acad. Sci. 82: 7560-7564.
6. Elango, N., G.A. Prince, B.R. Murphy, S. Venkatesan, R.M. Chanock, and B. Moss (1986)Resistance to human respiratory syncytial virus (RSV) infection induced by immunization of cotton rats with a recombinant vaccinia virus expressing the RSV G glycoprotein. Proc. Natl. Acad. Sci. 83: 1906-1910.

Transcriptional Control Mechanisms, pages 21–29
© 1987 Alan R. Liss, Inc.

CHARACTERIZATION OF A YEAST RNA POLYMERASE I ENHANCER[1]

Elaine A. Elion and Jonathan R. Warner

Department of Cell Biology
Albert Einstein College of Medicine
Bronx, New York 10461

ABSTRACT We have shown that the rDNA repeat of the
yeast Saccharomyces cerevisiae contains a sequence that
is a transcriptional enhancer for RNA polymerase I with
many of the same properties as the RNA polymerase II
enhancers. We now show that this enhancer is active
when placed either upstream or downstream of the
transcription unit in the genome. Its activity does not
depend on the circularity of a plasmid. The enhancer is
extensive; its activity is dependent upon sequences
spread over a range of 150 or more nucleotides.
Finally, recent data of Klootwijk and Planta (pers.
comm.) suggest that the termination site of rRNA
transcription lies within the enhancer sequences. These
observations suggest a model for the role of an enhancer
in a tandemly repeated set of genes.

INTRODUCTION

The ribosomal RNA of the yeast Saccharomyces cerevisiae
is transcribed from a tandem array of 100 nearly identical
genes. We have studied the sequences involved in regulating
this transcription by constructing an artificial rRNA gene
containing unique sequences from the E. coli bacteriophage
T7. On introduction of this gene into yeast we found that
it was transcribed from the correct initiation site by an α-

---

[1]This work has been supported by GM25532 from the NIH
and Grant NP-474-P from the American Cancer Society.
General support is provided by NIH CA13330.

amanitin resistant polymerase, presumably RNA polymerase I
(1,2). This transcription can be increased 10 to 15 fold if
a 190 bp EcoRI-HindIII fragment is present on the same
plasmid in either orientation, upstream or downstream of the
transcription unit (2).

We now demonstrate that the enhancer element is
functional when placed downstream of the transcription unit
within the chromosome, and present a hypothesis describing
how an enhancer might work in the unique environment of a
tandemly repeated transcription unit.

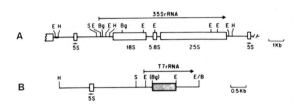

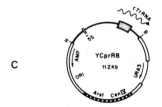

FIGURE 1. **Relevant Genes.** a) The rDNA repeat of S.
cerevisiae. b) The T7rDNA construct rR8 (1). c) rR8 in
YCp50 (1).

RESULTS AND DISCUSSION

The Experimental Design

The rDNA genes of S. cerevisiae reside on chromosome
XII (3) as a set of approximately 100 head-to-tail tandem
repeats each of which encompasses the divergently

transcribed 5S RNA and 35S pre-rRNA genes (Fig. 1a). Our
studies of rRNA transcription in yeast have been possible
through the use of an artificial rRNA gene, rR8 (Fig. 1b)
(1) that consists of rDNA sequences allowing proper
initiation and termination of 35S rRNA transcription and a
fragment of unique DNA from E. coli bacteriophage T7 that
allows measurement of transcription from the gene in vivo.

The rDNA Enhancer Functions from a Downstream Position Even Within the Genome

Although our previous results showed that the enhancer
functions both upstream and downstream of the transcription
unit in a plasmid, the interpretation of this result is made
ambiguous by the circular nature of the plasmid. To
determine whether the enhancer truly functions both 5' and
3' to a 35S rRNA transcription unit, we integrated several
artificial genes into a nonribosomal locus, the ura3 gene on
chromosome V (4). A nonribosomal locus was chosen to avoid
the effects of neighboring rDNA sequences.

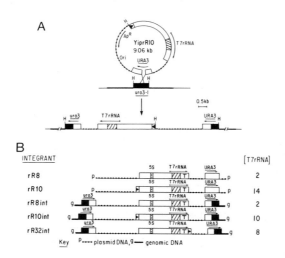

FIGURE 2. **Integration of rDNA constructs into the
ura3-1 gene of strain W303-a** (R. Rothstein). a) Homologous
recombination between plasmid and genomic URA3 sequences.
b) Diagram of rDNA constructs within the URA3 locus and the
relative amount of T7rRNA produced from each. Sequences
flanking constructs are the same whether on plasmids or
integrated.

The artificial genes were cloned into a plasmid, YIp5, which contains a wild type URA3 gene but lacks sequences that allow its replication in yeast. Plasmid integration was directed to the ura3-1 locus on chromosome V by linearizing the plasmids within its URA3 gene to produce recombinogenic double stranded ends (Fig. 2a) (5). Southern analysis of mitotically stable Ura$^+$ transformants was done to identify clones in which integration had occurred at the ura3 rather than the rDNA loci. The resulting integrant constructs, shown in Fig. 2b, are flanked by the same vector sequences previously tested on a circular centromere plasmid (Fig. 1c).

Transcription of the integrated constructs was determined by densitometric analysis of quantitative slot blots of total RNA using a single stranded probe specific for the T7 DNA within the artificial gene. Samples were normalized for input by reprobing slot blots with ribosomal protein gene, TCM1 or CYH2. The same relative amounts of T7rRNA are produced from rR8, rR10 and rR32 when they reside within the URA3 locus as when they are on a centromere plasmid (Fig. 2b). This result demonstrates that the enhancer is effective in a linear arrangement within the genome both upstream and downstream of the site of transcription initiation. Interestingly, it also suggests that the enhancer may function bidirectionally within the rDNA repeat in the genome.

Sequences at Both 5' and 3' Boundaries of the Enhancer are Necessary for Its Function

The sequence of the E-H fragment has been determined for 21 independently isolated copies of the rDNA repeat (Fig. 3; 1,6,7). The sequence is characterized by an AT-rich center that shows substantial sequence polymorphism and by 5' and 3' borders that are highly conserved.

A computer search of the E-H fragment revealed two sets of overlapping inverted repeats. One set (thick arrows) overlaps the HindIII site. Since all subcloning of the 190 bp fragment had been done after filling in the HindIII site, this repeat is intact in all constructs. To determine whether this repeat is important for enhancer activity, we made a 5 bp deletion at the HindIII site of rR10 using mung bean nuclease. The resulting construct, rR51, yields much lower levels of T7rRNA (Fig. 4a) suggesting that the inverted repeat in the intact gene serves a role in enhancer function.

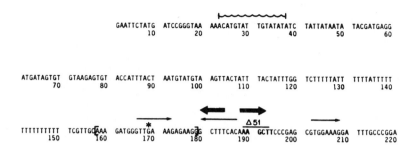

FIGURE 3.   **Sequence of the EcoRI-HindIII fragment and adjacent nucleotides.** The wavy line indicates an alternating Pu/Py tract, the thin arrows repeats showing homology to the SV40 core consensus sequence 5' GTGG$^{AAAG}_{TTTT}$G (8) and the thick arows an inverted repeat. The HindIII site is shown in bold and above it is the 5 bp deletion in rR51.

A second set of three 7 bp repeats (small arrows) includes two to the left and one to the right of the HindIII site. Surprisingly, this GTGAAAG sequence is similar to the SV40 enhancer core consensus sequences GTGG$^{AAAG}_{TTTT}$G (8). These repeats may not be crucial for enhancer activity since the removal of one of them in a deletion mutant that keeps the HindIII site intact does not abolish enhancer activity (rR36, Fig. 4b). However, the level of transcription is decreased 2-fold in this construct. Perhaps these repeats function in an additive manner such that the loss of one repeat leads to partial loss of enhancer activity. Such additive interactions of short repeated sequences have been shown to be important for other enhancer elements (9,10).

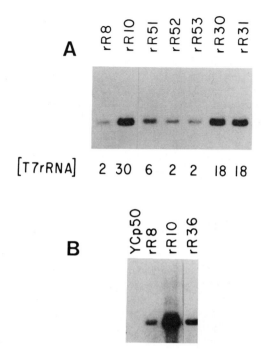

FIGURE 4. **Northern Analysis of Deletion Mutants.** a)
rR51 contains a 5 bp deletion at the HindIII site of rR10,
rR52 and rR53 contain a 140 bp FokI/ScrFI fragment
(nucleotides 77-217 in Fig. 3) cloned in either orientation
at the PvuII site of rR8. b) rR36 contains a 0.73 kb
deletion from the HindIII site to the PvuII site of rR10
that keeps the HindIII site intact.

Finally, to determine whether the conserved sequence
near the EcoRI site, which encompasses a purine-pyrimidine
tract indicated by the wavy line, is important for enhancer
function, a 140 bp FokI-ScrFI fragment encompassing
nucleotides 77-190 of the E-H fragment plus 27 nucleotides
downstream of the HindIII site was subcloned in both
orientations into the PvuII site of rR8 (rR52 and rR53, Fig.
4a). The transcription of rR52 and rR53 was compared to

that of rR30 and rR31 in which the E-H fragment is also
inserted in both orientations at the same PvuII.
Surprisingly, the FokI - ScrFI fragment did not function in
either orientation at the PvuII site. This result
demonstrates that sequences from both ends of the EcoRI-
HindIII fragment are necessary for the function of the
enhancer. The size of the enhancer recalls the RNA
polymerase II enhancer of SV40, but is very different from
the enhancer-like UASs (upstream activating sequences)
identified upstream of several yeast genes, which are
usually 10-15 nucleotides long.

Role of the rDNA Enhancer

It is premature to propose a detailed model for the
role of the rDNA enhancer. Nevertheless two observations
suggest a hypothesis for the functioning of an enhancer
within a repetitive transcription unit. (1) The enhancer
can work from either upstream or downstream of the rRNA
transcription initiation site. (2) The sequences necessary
for enhancer function are extensive and flank, or perhaps
include, the sequences now thought to be involved in the
termination of transcription of 35S rRNA (Klootwijk, J. and
R.J. Planta, pers. comm.). We propose that the enhancer
serves a dual function that coordinates transcription
termination with reinitiation as follows (see Fig. 5):

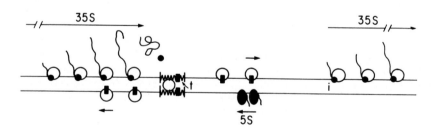

Key

○ Nontranscribing RNA PolI   ■ Shuttle factors   i Initiation site
◑ Transcribing RNA PolI   ◕ Shuttle complex   t Termination site
• Transcription factors   ● Transcribing RNA PolIII   ⋈ rDNA enhancer

FIGURE 5. **The rDNA Enhancer is Tailored for Repeated
Genes.** A model that couples transcription termination of
rRNA with reinitiation by RNA polymerase I. See text for an
explanation.

A transcribing polymerase that terminates within the EcoRI-HindIII fragment binds to a shuttle factor that is associated with this sequence. Once bound, the shuttle complex moves in a 3' to 5' direction either downstream on the transcribed strand to the next 35S initiation site or upstream on the nontranscribed strand, returning to the previous initiation site. In the former case, the polymerase would pass a transcribing 5S gene while in the latter instance it would pass a transcribing 35S gene. However, work by Wolffe et al. (11) has demonstrated that a polymerase can cross a transcription complex. This mechanism would serve to orient the movement of RNA polymerase I and thus increase its local concentration at the 35S initiation site, augmenting the efficiency of transcription initiation. The fact that the enhancer can function in either orientation suggests that it is functionally symmetrical with respect to activating transcription and may bind a dimer, either monomer of which could be recognized by RNA polymerase I. Finally, the coordination between termination and reinitiation would be a means to concentrate RNA polymerase I in the nucleolus, as certainly occurs in mammalian cells (12).

REFERENCES

1. **Elion EA, Warner JR** (1984). The major promoter element of rRNA transcription in yeast lies 2 kb upstream. Cell 39:663.
2. **Elion EA, Warner JR** (1986). An RNA polymerase I enhancer in yeast. Molec. Cell. Biol., in press.
3. **Petes T** (1979). Yeast ribosomal DNA genes are located on chromosome XII. Proc. Natl. Acad. Sci. USA 76:410.
4. **Mortimer RK, Schild D** (1981) Genetic Map of Saccharomyces cerevisiae in The Molecular Biology of the Yeast Saccharomyces. Cold Spring Harbor Laboratory, Cold Spring Harbor, NY, p. 641.
5. **Orr-Weaver TL, Szostak JW, Rothstein R** (1983). Genetic applications of yeast transformation with linear and gapped plasmids. Meth. Enz. 101:228.
6. **Skryabin KG, Eldarov MA, Larionov VL, Bayev AA, Klootwikj J, de Regt HF, Veldman GM, Planta RJ, Georgiev OI, Hadjiolov AA** (1984) Structure and function of the nontranscribed spacer regions of yeast rDNA. Nucl. Acids Res. 12:2955.

7. Swanson ME, Yip M, Holland MJ (1985). Characterization of an RNA polymerase I-dependent promoter within the spacer region of yeast ribosomal cistrons. J. Biol. Chem. 260:9905.
8. **Weiher H, Konig M, Gruss P** (1983). Multiple point mutations affecting the Simian virus 40 enhancer. Science 219:626.
9. **Herr W, Gluzman Y** (1985). Duplications of a mutated Simian virus 40 enhancer restore its activity. Nature 313:711.
10. **Gluzman Y** (ed) (1985). Current Communications in Biology Eukaryotic Transcription. The role of Cis- and trans- acting elements in initiation. Cold Spring Harbor Laboratory, Cold Spring Harbor, NY.
11. **Wolffe AP, Jordan E, Brown DD** (1986). A bacteriophage RNA polymerase transcribes through a Xenopus 5S gene transcription complex without disrupting it. Cell 44:381.
12. **Scheer U, Rose KM** (1984). Localization of RNA polymerase I in interphase cells and mitotic chromosomes by light and electron microscopic immunocytochemistry. Proc. Natl. Acad. Sci. USA 81:1431.

Transcriptional Control Mechanisms, pages 31–44
© 1987 Alan R. Liss, Inc.

PHOTOFOOTPRINTING IN VIVO OF THE YEAST GAL 1-10 PROMOTER[1]

Scott B. Selleck and John Majors

Department of Biological Chemistry
Washington University School of Medicine
St. Louis, MO 63110

ABSTRACT To elucidate critical steps in the
transcription initiation process we have devised a
protocol for obtaining information about DNA structure
and DNA-protein interactions at nucleotide level
resolution from intact yeast cells. Our procedure
combines the UV light "footprinting" method developed
by Becker and Wang(1) with the "genomic sequencing"
technique described by Church and Gilbert(2). We have
identified transcription dependent changes in
sensitivity of DNA to UV induced covalent modification
at several positions between the upstream activating
sequence (UAS$_G$) and the transcription initiation sites
of the GAL 1 and GAL 10 genes. The most prominent of
these changes occur at a common site within the
putative "TATA" boxes of the two genes. UV
modification at this site is enhanced only in
transcriptionally active promoters.

INTRODUCTION

The yeast Saccharomyces cerevisiae possesses several
attributes that make it a useful organism in which to study
eukaryotic transcription regulation. Genetic selection
schemes permit the identification and characterization of
genes that encode trans acting regulatory factors. Gene

[1]This work was supported by PHS grant CA38994 awarded
by the National Cancer Institute, a grant from the
Mallinkrodt Foundation and by NIH training grants GM 07200
and GM 07067.

"transplacement" technology allow promoter mutants generated in vitro to be studied within their native chromosomal context(3). These approaches have provided critical information about both cis and trans acting elements of several yeast regulatory systems(4,5,6).

Despite these advances, a detailed understanding of the ways in which specific effector molecules cause dramatic changes in transcription initiation rates is hampered by the lack of a system capable of accurate transcription in vitro from yeast promoters. To circumvent this problem we have devised a protocol for obtaining information about DNA structure and DNA-protein interactions at nucleotide level resolution from intact yeast cells. We are using this technique to investigate transcriptional regulation of the galactose inducible genes, GAL 1 and GAL 10.

METHODS

Photofootprinting

For DNA to be modified by light an incoming photon must be absorbed and the resulting excited state relaxed by chemical reaction. Events that influence either one of these processes will result in changes in the rate of photoproduct formation. Becker and Wang showed that protein binding to specific sequences can influence the level of photoproduct formation at discrete positions within the recognition site(1). The differences probably result from changes in structure or flexibility of the DNA and not from quenching of incoming photons(1). Photoproducts along a DNA sequence are detected by selective chemical cleavage of the phosphodiester bond at photomodified residues(1). The distance of the cut from a known restriction site as determined by electrophoresis on polyacrylamide gels, provides the exact location of the modified residue. This method has been used to detect the binding of E. coli RNA polymerase holoenzyme and lac repressor to the lacUV5 promoter and operator in vitro. Irradiation of intact E. coli demonstrated similar repressor dependent changes in sensitivity to photomodification within the binding site(1).

To extend this footprinting method to eukaryotic cells it is necessary to map breaks in the phosphodiester chain along a unique DNA sequence, from a sample of total genomic DNA. The "genomic sequencing" technique provides this capability(2). This method employs electrophoretic transfer

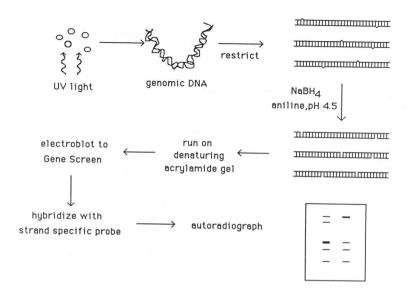

FIGURE 1. Photofootprinting Protocol. NaBH₄ and
aniline, pH 4.5 treatment placed in the figure following
restriction endonuclease digestion refer to the chemical
cleavage steps that selectively cut the phosphodiester bond
at photomodified positions.

of DNA from sequencing gels onto nylon membranes. Detection
of unique sequences is achieved by hybridization of the
immobilized DNA with a strand specific probe complementary
to one end of the restriction fragment that encompasses the
target.
　　　An outline of our experimental protocol(7) is presented
in figure 1. This procedure employs the following steps:
Cells are grown to mid log phase and irradiated with UV
light. Genomic DNA is isolated and cut with an appropriate
restriction enzyme prior to conducting the chemical
reactions that selectively cleave the phosphodiester bond at
photomodified positions. The DNA is then run on a
sequencing gel and transferred to GENE SCREEN membrane by
electroblotting as described by Church and Gilbert(2). The

"genomic blot" is then hybridized with a radiolabelled, strand specific probe.

Synthesis of RNA probes

Initially we chose to use the SP6 in vitro transcription system to generate high specific activity RNA probes. However, we were unable to reliably obtain incorporation of labelled NTPs into RNA molecules of sufficiently high specific activity for our purposes. We have since determined conditions for transcription in vitro by phage T7 RNA polymerase(8) that consistently provide 70-80% incorporation of [α-^{32}P] UTP (3000 Ci/mMole) into 140 nucleotide long transcripts. An autoradiograph of the in vitro transcription products separated on an 8% acrylamide/7M urea gel is presented in figure 2a. One of these RNA samples was used as a hybridization probe for a genomic blot. The autoradiograph obtained from this blot after a 16 hour exposure with intensifying screens is shown in figure 2b. Each lane was loaded with approximately 5 μg of yeast genomic DNA subjected to Maxam and Gilbert sequencing reactions(9). The sensitivity of the polymerase to the concentration of the limiting (radiolabelled) ribonucleotide is demonstrated by the results presented in figure 2c.

RESULTS

Photofootprinting of the GAL 1-10 Promoter

S. cerevisae grown on galactose demonstrate at least a 1000 fold transcriptional activation of those genes required for its transport and catabolism(10,11). GAL 1,7,10 constitute three coordinately regulated but independent transcription units(12). The cap sites of the divergently transcribed GAL 1 and 10 genes are separated by approximately 600 bp(13). The sequences within this region required for normal regulation have been extensively characterized through the efforts of several laboratories(13,14,15).
        Two independent loci that are involved in regulation of transcription from GAL 1,7,10 have been identified. GAL 4 encoded protein is required for induction and is therefore believed to be a transcriptional activator(10). Two groups

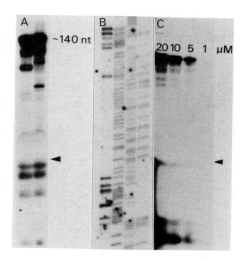

Figure 2.   Synthesis of RNA Probes with T7 RNA Polymerase.
A. Autoradiograph of radiolabelled products of in vitro
transcription from two different DNA templates, separated
on 8% acrylamide/7M urea gel. Material below the arrow
represents unincorporated α-^{32}P UTP. These reactions were
carried out at 37°C for 60 min in 10μl volume and included:
0.5 μg cut DNA template, 3 μM [α-^{32}P] UTP (3000 Ci/mMole),
7U RNAsin (Promega), 40 mM Tris, pH 8.0, 8 mM MgCl$_2$, 1 mM
spermidine, 25 mM NaCl, 5 mM dTT, 500 μM ATP,CTP,GTP, and
approximately 100 U T7 RNA polymerase (USBC units).
B. Autoradiograph of genomic blot probed with one of the RNA
samples shown in A. Each lane of a 8% acrylamide/7M urea
gel was loaded with approximately 5 μg yeast genomic DNA
subjected to Maxam and Gilbert sequencing reactions(9).
Hybridization and washing was performed according to Church
and Gilbert(2). Approximately 10^8 cpm of 4x10^9 cpm/μg RNA
probe was used for the hybridization.
C. Sensitivity of in vitro transcription reaction to UTP
concentration. Autoradiograph of in vitro transcription
products from reactions with different UTP concentrations,
separated on an acrylamide gel. Reactions were carried out
as in (A) except low specific activity UTP was used. UTP of
identical specific activity was included in all samples and
an equal volume of the reaction mix loaded on the gel.
Therefore, the band intensities accurately represent the
amount of RNA synthesized. Unincorporated material ran
below the arrow. Lanes 1-4: 20,10,5,and 1 μM UTP.

have shown by different methods that GAL 4 protein binds a
specific DNA sequence within $UAS_G(4,16)$. Recessive
mutations in GAL 80 result in constitutive expression of GAL
1,7,10(17). GAL 80 encoded protein is therefore believed to
interact with GAL 4 protein to prevent activation of
transcription in the absence of galactose. GAL 1,10 are
also subject to catabolite repression(15,18). Cells grown
on galactose and glucose show approximately a 200 fold
decrease in expression compared to cells grown on galactose
alone(15).

Using the protocol for photofootprinting in vivo we
have examined selected sequences between the UAS_G and the
GAL 1,10 transcription intiation sites. This region
includes the TATA sequence, which has been demonstrated to
be critical for normal transcription of a variety of yeast
genes(14,19-22). Figure 3 shows the locations of
restriction sites and hybridization probes used for this
study. Figure 4 presents the results from a footprinting
experiment. This autoradiograph provides information from
sequences -270 to -70 bp, upstream of the major GAL 1
initiation site. Lanes 1 and 2 show the extent of
background cleavage that results from the chemical reactions
used to detect the photomodifications. The location and
extent of UV dependent modifications obtained by irradiation
of protein free genomic DNA is indicated by the pattern of

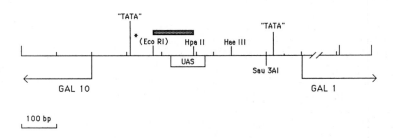

FIGURE 3: Map of GAL 1-10 Control Region. The
positions of the upstream activating sequence (UAS), TATA
boxes, and GAL 1,GAL 10 transcription initiation sites are
indicated. The rectangle above the sequence represents the
DNA fragment cloned into a T7 promoter bearing vector used
for RNA probe synthesis. The EcoRI site marked with an
asterisk is found only in those strains used for the
footprinting experiment in figure 4.

A   DNA
    -hv +hv G C + -
    1 2 3 4 5 6 7 8

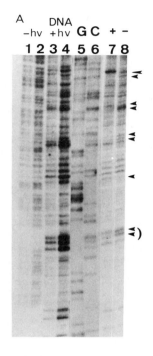

Figure 4: In vivo photofootprint from UAS$_G$ to TATA box of the GAL 1 promoter. A. The autoradiograph shown here was obtained by hybridization of a genomic blot with an RNA probe complementary to the lower strand of the GAL 1 promoter(see figure 3). The yeast strains used for this experiment bear a gal 10 promoter deletion constructed by M. Johnston(DEL 122) and contain an EcoRI linker at position -442 5' to the major GAL 1 initiation site(13). Expression of GAL 1 is unaffected by this mutation. All of the cells for this experiment were grown on 1% yeast extract,2% peptone (YP)5% glycerol. The genomic DNA was cut with EcoRI. For the purpose of this presentation non-neighboring gel lanes (4,5 and 6,7) have been juxtaposed. All yeast strains were obtained from M. Johnston. Lanes 1,3,7: DNA from strain bearing gal 80 deletion. Lanes

2,4,8: DNA from GAL 4/ GAL 80 strain. Lanes 1,2: DNA from unirradiated cells carried through chemical cleavage steps specific for photomodified residues. Lanes 3,4: Protein free genomic DNA irradiated for 25s prior to chemical cleavage reactions. Lanes 5,6: DNA subjected to G and C specific Maxam and Gilbert cleavage reactions. Lanes 7,8: DNA from irradiated cells with induced (GAL 4/gal 80 grown on glycerol) or uninduced (GAL 4/ GAL 80 grown on glycerol) GAL 1 gene. The arrows mark the transcription dependent changes indicated in figure 4b.
B. Sequence of the GAL 1 promoter and location of transcription dependent changes in sensitivity to photomodification. The numbers above the sequence are assigned according to Johnston and Davis(13). The horizontal arrow marks the direction of transcription for GAL 1 and the number below the sequence indicates the number of bp to the major initiation site. The box enclosing both strands denotes the hexanucleotide conserved for GAL 1 and GAL 10 TATA regions. Circled and boxed residues represent the locations of enhancements and repressions in sensitivity to photomodification respectively, that occur

B

```
520          530          540          550          560
GTTATGAAGA   GGAAAAATTG   GCAGTAACCT   GGCCCCACAA   ACCTTCAAAT
CAATACTTCT   CCTTTTTAAC   CGTCATTGGA   CCGGGGTGTT   TGGAAGTTTA
         9        8                                       7
570          580          590          600          610
GAACGAATCA   AATTAACAAC   CATAGGATGA   TAATGCGATT   AGTTTTTTAG
CTTGCTTAGT   TTAATTGTTG   GTATCCTACT   ATTACGCTAA   TCAAAAAATC
                                    6       5
620          630          640          650          660
CCTTATTTCT   GGGGTAATTA   ATCAGCGAAG   CGATGATTTT   TGATCTATTA
GGAATAAAGA   CCCCATTAAT   TAGTCGCTTC   GCTACTAAAA   ACTAGATAAT
                 4    3
670          680
ACAGATATAT   AAATGCAAAA
TGTCTATATA   TTTACGTTTT
    2          1
                 -80
          ⟶
```

upon stimulation of transcription. The number of residues
enclosed indicates the accuracy to which these modifications
have been mapped. The number below each modification refers
to the sites marked in figure 4a, where the uppermost arrow
is position 1.

bands in lanes 3,and 4. DNA samples from irradiated cells
that were transcriptionally active(GAL 4/gal 80 grown on
glycerol) or quiescent (GAL 4/GAL 80 grown on glycerol) are
shown in lanes 7 and 8 respectively. DNA subjected to
Maxam and Gilbert G and C specific reactions (lanes 5,6)
serve as size markers(9).
    Transcription dependent changes in sensitivity to
photomodification are marked in figure 4a and their
positions within the GAL 1 regulatory sequence(13) are shown
in figure 4b. The large arrow in figure 4a indicates the
position of the most prominent difference between the two
promoter states. The enhancement seen at this site for the
induced gene, as mapped more precisely in other experiments,
occurs at a T residue within the putative GAL 1 TATA box. A
nearby modification site is repressed for the active
promoter(small arrow). Duplicate experiments have shown
that each of the transcription dependent changes seen in
fig. 4a is reproducible, including such subtle differences
as those that result in a change in the relative intensities
of the three bands marked by the bracket in figure 4a. Note
that the pattern changes observed for irradiated cells are
not found in protein free irradiated DNA taken from these
same cells(compare lanes 3 and 4 to lanes 7 and 8),

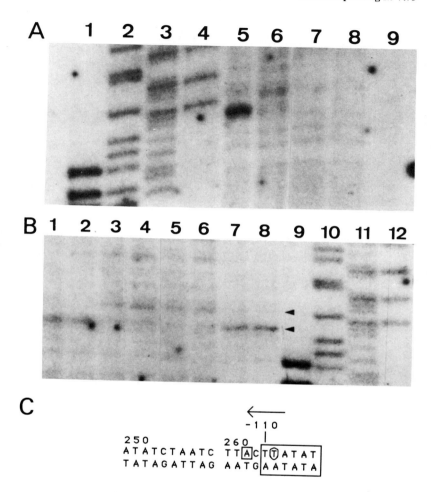

FIGURE 5: Photofootprint across the GAL 10 TATA box.
Autoradiographs of "genomic blots" probed with an RNA
complementary to the top strand of sequence listed below.
All of the strains in this experiment have wild type GAL 1-
10 genes.　Cells were grown on either YP2% galactose or YP5%
glycerol.　The genomic DNA was cut with Hpa II (Msp I).
A.　Lanes 1-4: Genomic DNA subjected to G, A+G, C+T, and C
specific Maxam and Gilbert sequencing reactions,
respectively.　Lanes 5,6: DNA from irradiated cells with
induced (lane 5, GAL 4/ GAL 80 grown on galactose) or

uninduced (lane 6: GAL 4 /GAL 80 grown on glycerol) GAL 10
gene. Lanes 7,8: DNA from unirradiated cells (GAL 4/GAL
80)grown on galactose (lane 7) or glycerol (lane 8) that was
subjected to photomodification specific cleavage reactions.
Lane 9: DNA from irradiated cells that received no chemical
cleavage treatment.
B.   Lanes 1-8: Genomic DNA from irradiated cells carried
through photomodification specific cleavage reactions.
Lanes 9-12: Maxam and Gilbert sequencing markers, G,
A+G,C+T,C. Cells received 80s(lanes 1,3,5,7) or 160s (lanes
2,4,6,8) light per aliquot. Lanes 1,2: GAL 4/gal 80 cells
grown on glycerol. Lanes 3,4: GAL 4/GAL 80 grown on
glycerol. Lanes 5,6: gal 4/gal 80 grown on glycerol. Lanes
7,8: GAL 4/GAL 80 grown on galactose. Photomodifications
that display transcription dependent changes are indicated
by arrows.
C.   Sequence of the GAL 10 TATA region. The boxed segment
represents the hexanucleotide also found for the GAL 1 TATA.
The location of transciption dependent changes in
sensitivity to photomodification are marked as in figure 4.

indicating that these changes are not due to covalent DNA
modifications in vivo that influence the sensitivity of
these residues to UV light.
     The transcription dependent enhancement described for
the GAL 1 TATA occurs at the underlined T residue within the
hexanucleotide 3' TATATT 5', a sequence that is also found
110 bp upstream of the major GAL 10 initiation site. Figure
5a shows that a transcription dependent enhancement of
photomodification at the same T residue is observed for the
GAL 10 sequence (lane 5; GAL 4/GAL 80 grown on galactose,
lane 6; GAL 4/GAL 80 grown on glycerol). A modification
site 3 bp downstream from the enhancement is repressed for
the transcriptionally active promoter.   These features are
light dependent since they are not found in DNA samples from
unirradiated cells carried through the chemical cleavage
steps (lanes 7,8). Figure 5b shows that this pattern,
enhancement at a T residue within TATA and repression at a
neighboring modification, is found in cells activated either
by growth on galactose(lanes 7,8 versus 3,4), or deletion of
the GAL 80 gene(lanes 1,2). These changes also require a
functional GAL 4 gene (lanes 5,6 versus 1 and 2).

DISCUSSION

Several studies indicate that the transcription dependent alteration in light sensitivity at the hexanucleotide 3' TATATT 5' marks the functional TATA sequence. 1) Successive deletions extending from UAS$_G$ toward the GAL 1 cap site have no effect on the induced level of expression unless they encompass the hexanucleotide at -80(15). 2) The sequence 3' TATATT 5' is found within the only TATA homologous sequence required for normal HIS 4 expression(21). 3) An identical sequence found in the CYC 1 promoter directs initiation at four of the six mRNA start sites and is the only TATA homology in which mutations severely depress transcription levels(22).

The promoter regions of many yeast genes contain several sequences homologous to the canonical TATA(23). Three(15) or four(22) potential TATA boxes have been proposed for the GAL 10 promoter. The photofootprinting technique revealed transcription dependent changes at the only one of these whose sequence is identical to those which are essential for CYC 1 and HIS 4 expression(21,22). We believe that the altered sensitivity to photomodification within the TATA control element results from a DNA structure change mediated by binding of a protein specific for this sequence. Proteins that are necessary for transcription in vitro and bind to the TATA region have been isolated from human and Drosophila cells(24,25). Presumably, the changes we have detected in vivo within TATA are mediated by the yeast counterpart of these transcription factors. From our data we cannot say whether the effect we observe is the result of de novo association of the protein with its recognition site or merely a change in its conformation.

Although we see multiple changes between UAS$_G$ and the GAL 1 TATA (see figure 4) these sequences can be removed without affecting the induced level of GAL 1 expression. However, deletions that include a segment of this region permit abnormally high levels of expression for uninduced and catabolite repressed cells(15). The transcription dependent repression of photomodification at position 630 occurs within this sequence. It is possible that specific DNA-protein contacts at this site are involved in determining the basal level of expression. An identical pentanucleotide 5' ACCCC 3' is found upstream of the GAL 10 TATA box and preliminary data show a transcription dependent repression at this sequence as well.

Recently a method for footprinting in vivo with dimethylsulfate in eukaryotic cells has been developed. This technique has been used successfully to detect protein binding within an immunoglobulin gene enhancer(26) and GAL 4 protein binding to regions within UAS_G(4). These experiments measure the relative accessibility of the reporter molecule to the major groove at G residues and the minor groove at A residues. The photofootprinting method monitors protein binding in a fundamentally different way; DNA-protein contacts are evidenced by changes in DNA structure that alter its sensitivity to photomodification. These two in vivo footprinting methods therefore provide distinct, yet complementary means of studying promoter organization. Used together, the variety of technologies now available will lead to a more sophisticated understanding of those processes that control transcription initiation rates.

ACKNOWLEDGEMENTS

We would like to thank P. Burgers and M. Johnston for sharing their yeast expertise as well as for numerous discussions and suggestions. We are indebted to M. Johnston for providing the yeast strains and DNA clones. We would also like to express our appreciation to M. Becker for many helpful conversations and for sharing data prior to publication. These experiments would not have been possible without the generous loan of equipment by E. Elson.

REFERENCES

1. Becker MM, Wang JC (1984). Use of light for footprinting DNA in vivo. Nature 309:682.
2. Church GM, Gilbert W (1984). Genomic sequencing. Proc Natl Acad Sci USA 81:1991.
3. Scherer S, Davis RW (1979). Replacement of chromosome segments with altered DNA sequences constructed in vitro. Proc Natl Acad Sci USA 76:4951.
4. Giniger E, Varnum SM, Ptashne M (1985). Specific DNA binding of GAL 4, a positive regulatory protein of yeast. Cell 40:767.
5. Johnson AD, Herskowitz I (1985). A repressor (MATα2 product) and its operator control expression of a set of cell type specific genes in yeast. Cell 42:237.

6. Hope IA, Struhl K (1985). GCN 4 protein, synthesized
    in vitro, binds HIS3 regulatory sequences: implications
    for general control of amino acid biosynthetic genes
    in yeast. Cell 43:177.
7. Selleck SB, Majors J (1986). Submitted for publication.
8. Tabor S, Richardson CC (1985). A bacteriophage T7 RNA
    polymerase/promoter system for controlled exclusive
    expression of specific genes. Proc Natl Acad Sci USA 82:
    1074.
9. Maxam AM, Gilbert W (1977). A new method for sequencing
    DNA. Proc Natl Acad Sci USA 74:560.
10. Douglas HC, Hawthorne DC (1964). Enzymatic expression
    and genetic linkage of genes controlling galactose
    utilization in Saccharomyces. Genetics 49:837.
11. St. John TP, Davis RW (1979). Isolation of galactose-
    inducible DNA sequences from Saccharomyces cerevisiae
    by differential plaque filter hybridization. Cell
    16:443.
12. St. John TP, Davis RW (1981). The organization and
    transcription of the galactose gene cluster of
    Saccharomyces. J Mol Biol 152:285.
13. Johnston M, Davis RW (1984). Sequences that regulate
    the divergent GAL1-10 promoter in Saccharomyces
    cerevisiae. Mol Cell Biol 4:1440.
14. Guarente L, Yocum RR, Gifford P (1982). A GAL10-CYC1
    hybrid yeast promoter identifies the GAL 4 regulatory
    region as an upstream site. Proc Natl Acad Sci USA
    79:7410.
15. West RW, Yocum RR, Ptashne M (1984). Saccharomyces
    cerevisiae GAL1-10 divergent promoter region: location
    and function of the upstream activating sequence UAS_G.
    Mol Cell Biol 4:2467.
16. Bram RJ, Kornberg RD (1985). Specifc protein binding to
    far upstream activating sequences in polymerase II
    promoters. Proc Natl Acad Sci USA 82:43.
17. Douglas HC, Pelroy G (1963). A gene controlling
    inducibility of the galactose pathway enzymes in
    Saccharomyces. Biochim biophys Acta 68:155.
18. Adams BG (1972). Induction of galactokinase in
    Saccharomyces cerevisiae: kinetics of induction and
    glucose effects. J Bacteriol 111:308.
19. Struhl K (1982). The yeast HIS3 promoter contains at
    least two distinct elements. Proc Natl Acad Sci USA
    79:7385.

20. Guarente L, Mason T (1983). Heme regulates transcription of the CYC1 gene of S. cerevisiae via an upstream activation site. Cell 32:1279.
21. Nagawa F, Fink GR (1985). The relationship between the "TATA" sequence and transcription initiation sites at the HIS4 gene of Saccharomyces cerevisiae. Proc Natl Acad Sci USA 82:8557.
22. Hahn S, Hoar ET, Guarente L (1985). Each of three "TATA elements" specifies a subset of the transcription initiation sites at the CYC-1 promoter of Saccharomyces cerevisiae. Proc Natl Acad Sci USA 82:8562.
23. Breathnach R, Chambon P (1981). Organization and expression of eucaryotic split genes coding for proteins. A Rev Biochem 50:349.
24. Sawadogo M, Roeder RG (1985). Interaction of a gene-specific transcription factor with the adenovirus major late promoter upstream of the TATA box region. Cell 43:165.
25. Parker CS, Topol J (1984). A Drosophila RNA polymerase II transcription factor containing a promoter-region-specific DNA-binding activity. Cell 36:357.
26. Ephrussi A, Church GM, Tonegawa S, Gilbert W (1985). B lineage-specific interactions of an immunoglobulin enhancer with cellular factors in vivo. Science 227:134.

Transcriptional Control Mechanisms, pages 45–55
© 1987 Alan R. Liss, Inc.

5'–REGULATORY ELEMENTS OF THE HUMAN HPRT GENE[1]

Pragna I. Patel, Twee Y. Tsao,
C. Thomas Caskey, and A. Craig Chinault

Institute for Molecular Genetics, and Howard Hughes
Medical Institute, Baylor College of Medicine
Houston, Texas 77030

ABSTRACT The 5'-flanking sequences influencing expression of the human HPRT gene have been examined by introducing various HPRT minigenes into HPRT-deficient hamster fibroblasts by $Ca(PO_4)_2$-mediated gene transfer followed by quantitation of stable $HPRT^+$ transformants. The promoter element has been localized within a region 210 bp upstream from the ATG codon. In addition, a negative regulatory element (NRE) has been identified upstream from the promoter. Deletion of this NRE causes a ten-fold increase in the frequency of $HPRT^+$ transformants. A transient expression assay involving microinjection of the minigenes into the $HPRT^-$ cells and monitoring of tritium incorporation into nucleic acid following uptake of labelled hypoxanthine suggested that this effect may be the result of increased transcriptional activity.

INTRODUCTION

Recent studies on the mechanism of transcription of eukaryotic protein-coding genes by RNA polymerase II have elucidated various classes of DNA sequences that play

[1]PIP was supported by fellowships from the Robert A. Welch Foundation and the Arthritis Foundation. This work was supported by the Howard Hughes Medical Institute, Public Health Service grant AM31428 to CTC from the National Institutes of Health, and American Cancer Society grant CD-171 to ACC.

specialized roles in determining the accuracy and the efficiency of transcription. The primary approach to identifying these sequences has been to mutagenize or delete DNA
sequences, typically located near the transcription initiation site, and to test the altered template for production
of RNA or protein either by reintroduction into the cell or
by in vitro transcription. Such studies have shown that
many eukaryotic promoters have an element called a TATA box,
with a consensus sequence TATAAA, located 25-30 bp upstream
from the transcription initiation site, which appears to
control the accuracy of site-specific transcription initiation (1,2). An additional conserved element (frequently
referred to as the CAAT box), which has the consensus
sequence GGTCAATCT and is found 70-100 bp upstream from the
cap site in many Pol II promoters, governs the efficiency of
transcription initiation (3). In addition, the immediate
upstream regions of genes regulated by hormones, metals or
heat contain relatively short sequence elements that play an
important role in the environmental response (4,5). Enhancers are a class of cis-acting DNA sequences that are strong
activators of general or tissue-specific transcription and
which function in an orientation and position-independent
manner (6). DNase footprinting has shown the interaction of
specific protein factors, some of which are required for
transcription by Pol II in vitro, with a few of the afore-
mentioned regulatory DNA sequence elements (7). Repressors
that interact with specific DNA sequences close to the promoter and play an important role in the control of gene
expression have been extensively studied in bacterial systems (8). However, to date, few examples of such negative
regulatory sequences in eukaryotic systems have been found
(see Discussion) (9-11).

Most of the information on regulatory DNA sequences has
come from genes with abundant mRNAs, most of which are
active in specialized cells, or from viruses that infect
eukaryotic cells and utilize their transcriptional machinery. Recently, a number of constitutively expressed genes
that are expressed at relatively low levels in most tissues,
so-called "housekeeping genes", have been cloned and characterized. These genes include hypoxanthine phosphoribosyltransferase (HPRT) (12,13), adenosine deaminase (14), dihydrofolate reductase (15), phosphoglycerate kinase (16) and
hydroxymethyl glutaryl CoA reductase (17). The 5'-flanking
regions of these genes are extremely GC-rich, lack a consensus TATA or CAAT sequence and contain several copies of
the sequences 5'-CCGCCC-3' or its inverted complement. The

latter sequence is also present in several viral promoters including that of the herpes thymidine kinase gene (18). In the latter they have been shown to function as flexible distal transcriptional signals in which the distance from the cap site, orientation and number of copies can vary, although changes in these parameters do influence the magnitude of the distal signal effect. Functional analysis of the promoters of the various housekeeping genes will determine if a distinctive transcriptional mechanism is used for this class of genes.

Our laboratory has cloned and characterized the human gene encoding HPRT, an enzyme that catalyzes the conversion of hypoxanthine and guanine to their respective 5'-nucleotides (19). The gene is expressed at low levels in all tissues except in the brain, where it is expressed at an ~7-fold higher level (19). The mechanism of the increased expression in brain is unknown. This report describes ongoing studies of the human HPRT 5'-flanking region by deletion analysis that have allowed dissection of the promoter from an apparent negative regulatory element (NRE).

RESULTS

Ca$(PO_4)_2$ Mediated Gene Transfer of HPRT Minigenes.

A human HPRT minigene, pHPT36, with 1600 bp of 5'-flanking sequences ligated to human HPRT cDNA via a Xma III site in the first exon and a polyadenylation signal from hamster HPRT cDNA, was used for functional analysis of the human HPRT promoter (13; Figure 1). This minigene was introduced by Ca$(PO_4)_2$ mediated gene transfer (20) into RJK88, a hamster cell line with a total deletion of the HPRT gene (21). At 48 hours after transfection, the cells were split 1:10 into HAT medium (1 x 10^{-3}M hypoxanthine, 1 x 10^{-5}M aminopterin and 1 x 10^{-4}M thymidine) to select for cells expressing HPRT. After 10 days in selection, surviving colonies were quantitated by staining with 0.1% methylene blue. HPRT$^+$ cells were produced at a frequency of ~1.7 x 10^{-6} per µg of pHPT36 DNA. This frequency, although relatively low, was at least ten-fold higher than that obtained with cDNA constructs that lacked 5'-flanking sequences. This suggested that pHPT36 had sufficient sequence to allow production of functional HPRT protein.

To further delineate sequences responsible for the promoter activity, deleted derivatives of pHPT36 containing

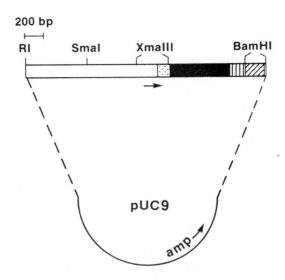

FIGURE 1.   Structure of a human HPRT minigene, pHPT36.
☐ 5'-untranscribed sequence;   ▦ 5'-untranslated sequence;
■ coding sequence;   ⦀ 3'-untranslated sequence;   ▨ polyadenylation signal from hamster cDNA.   The arrows indicate
direction of transcription.

different lengths of 5'-sequence were constructed by Bal 31
digestion from the unique Sma I site located ~930 bp upstream from the ATG codon in pHPT36, as previously described (13).   Selected derivatives of pHPT36 were introduced
into RJK88 cells by $Ca(PO_4)_2$-mediated gene transfer (20)
and transformants of the HPRT[+] phenotype quantitated as
described above.   Table 1 shows the extent of sequence upstream from the ATG codon in the derivatives of pHPT36
studied and the frequency of HPRT[+] transformants obtained
upon their transfection.   It is apparent that deletion of
the sequences lying between 356 bp and 245 bp upstream from
the ATG codon caused an ~ten-fold increase in the frequency of HPRT[+] transformants.   This suggests the presence
of a negative regulatory element (NRE) in this region.   In
addition, these studies indicate that the promoter element
required for basal levels of transcription lies within
210 bp upstream from the ATG codon.
       The morphology of HAT-resistant (HAT[R]) colonies

TABLE 1

FREQUENCY OF HPRT[+] TRANSFORMANTS UPON TRANSFECTION
WITH HPRT MINIGENES[a]

Plasmid	bp from ATG codon	# of HAT[R] colonies	HAT[R] colonies/µg
pHPT30 (cDNA)	---	3	2.0×10^{-7}
pHPT36	~1600	25	1.7×10^{-6}
pHPT36-A	~960	22	1.5×10^{-6}
pHPT36-W3	~560	50	3.3×10^{-6}
pHPT36-D1	356	24	1.6×10^{-6}
pHPT36-A25	246	960	6.5×10^{-5}
pHPT36-D2	234	1000	6.7×10^{-5}
pHPT36-A23	210	750	4.8×10^{-5}
pHPT36-D10	~200	825	5.5×10^{-5}
pHPT36-D3	157	9	6.0×10^{-7}
pHPT36-D4	125	17	1.1×10^{-6}
pHPT36-D6	49	3	2.0×10^{-7}

[a]Deleted derivatives of pHPT36 were made by
Bal 31 digestion from the Sma I site in this minigene
(Figure 1). Shown above is the extent of 5'-flanking
sequence in these derivatives. Five µg of DNA from
each minigene was transfected into 3×10^6 RJK88
(HPRT[-]) cells. Forty-eight hours after transfection,
the cells were split into five 100 mm dishes in HAT
medium. After 10 days in selection, the number of
HAT-resistant (HAT[R]) colonies were quantitated by
staining with 0.1% methylene blue.

obtained upon transfection of representative minigenes is
shown in Figure 2 and further serves to illustrate these
points. Variations in colony morphology as were seen, for
example, between pHPT36-A23 and pHPT36-D10 make definitive
quantitation difficult. Although the reason for these differences is not clear, they may be related to purity of the
plasmid DNA used for transfection or intrinsic variation in
DNA uptake. Such effects may be internally controlled for
by cotransfection with another selectable marker such as the
neomycin-resistance gene, and such experiments are in progress.

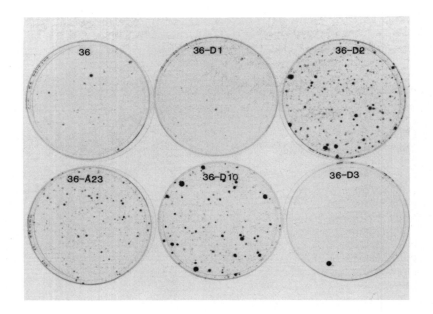

FIGURE 2. Morphology of HAT-resistant colonies obtained after transfection with the HPRT minigene, pHPT36 and its
deleted derivatives. The minigenes 36, 36-D1, 36-D2, 36-A23,
36-D10 and 36-D3 have ~1600, 356, 234, 210, ~200 and
157 bp, respectively, of sequence upstream from the ATG
codon.

Transient Expression of HPRT Minigenes.

The DNA transfection assay used above involves stable
integration of the minigenes into the genome of the recipient cells. To rule out the possibility that the difference
in the frequencies of HAT^R colonies obtained was due to a
difference in the ability of the minigenes to integrate into
genomic DNA, a microinjection assay that measures transient
expression of HPRT was used (13). Three minigenes that were
selected for this assay, (pHPT36, pHPT36-D2 and pHPT36-D6)
have 1600 bp, 234 bp and 49 bp, respectively, of sequence
upstream from the ATG codon. Approximately 250 copies of
either supercoiled or linearized minigene plasmid DNA were
microinjected into the nuclei of 100 HPRT⁻ RJK88 cells

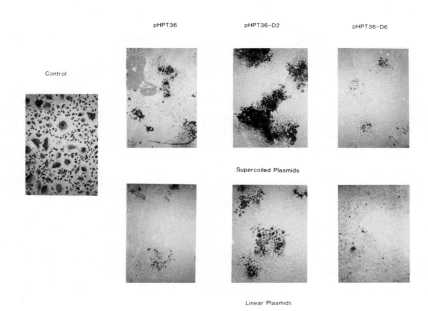

FIGURE 3. Autoradiographs of untreated control HPRT[+] mouse cells and of RJK88 (HPRT[−]) cells that were micro-injected with ~250 copies of either supercoiled or linearized pHPT36, pHPT36-D2 or pHPT36-D6 plasmids and monitored for incorporation of label after incubation in HAT medium containing [3H] hypoxanthine. The minigenes pHPT36, pHPT36-D2 and pHPT36-D6 contain ~1600 bp, 234 bp and 49 bp, respectively, of 5'-flanking sequence. Plasmids were linearized at the EcoR I site which lies at the extreme 5'-end of the minigenes (see Figure 1).

grown on coverslips. After recovery for 24 hours following injection, the cells were incubated for 16 hours in HAT medium containing [3H] hypoxanthine, fixed with 2% glutaraldehyde and subjected to autoradiography. Cells expressing HPRT incorporated label derived from [3H] hypoxanthine into nucleic acid as evidenced by the appearance of silver grains on the cells. Figure 3 shows an autoradiograph of control wild-type HPRT[+] cells and of RJK88 cells micro-injected with the various minigenes. This assay illustrates

that pHPT36-D2 allows maximal incorporation of [^{3}H] hypoxanthine, presumably because it contains the promoter element which is lacking in pHPT36-D6, but lacks the NRE that
is present in pHPT36. In addition, this effect is seen with
both supercoiled and linear plasmid DNA, although overall
higher incorporation is seen with supercoiled DNA. The
persistence of the negative regulatory effect with linearized plasmid DNA rules out the possibility that the closer
proximity of vector DNA sequences to the promoter element in
pHPT36-D2 is in some manner responsible for the increased
expression seen with this minigene.

DISCUSSION

Functional analysis of the GC-rich immediate 5'-flank-
ing region of the human HPRT gene has delineated promoter
activity to within 210 bp of the ATG codon. This region
includes three of the four GGGCGG sequence motifs seen in
the 5'-flanking regions of several "housekeeping
genes" (12-17). Deletion analysis of the mouse HPRT promoter has shown that 166 bp of sequence upstream from the ATG
codon is required for transcription and this includes one of
two GGGCGG motifs in the 5'-flanking region (22). Further
analysis of the human HPRT promoter will determine if there
are shared sequences of importance for transcriptional initiation between these promoters. Surprisingly, a negative
regulatory element has been identified immediately upstream
from the promoter. The dramatic effect of the NRE was first
seen in gene transfer experiments measuring stable HPRT$^+$
transformants (Figure 2) and has been localized to a region
between 356 bp and 245 bp upstream from the ATG codon using
this assay. It appears to function at the transcriptional
level as evidenced by its effect in transient expression
assays.
    Among the few "repressors" of gene function known in
eukaryotic systems is a "silencer" sequence described at the
mating type locus of the yeast Saccharomyces cerevisiae (9).
This cis-acting repressor sequence can, like an enhancer,
function in either orientation, relatively independently of
its position with respect to the regulated promoter and can
act on promoters 2600 bp away. Transcription of the CYC7
gene of Saccharomyces cerevisiae, encoding the iso-2-cyto-
chrome c protein, is controlled by a positive and a negative
element located ~240 bp and 300 bp respectively upstream
from the ATG codon (10). Deletion of the positive and nega-

tive element leads to decreased and increased CYC7 expression respectively, while deletion of both results in low wild-type-like expression of the gene. Thus, these sites appear to act antagonistically to give the low wild-type levels of CYC7 expression. Another example of a cis-acting repressor locus is the upstream mouse sequence (UMS) that prevents activation of the c-mos oncogene by a 3'-long terminal repeat (LTR) (11). Insertion of UMS 5' to the v-mos coding region also prevents 3'-LTR enhancement of its transforming activity but this inhibition is position-dependent and functions only when inserted between v-mos and its putative promoter.

Experiments in progress will determine if the NRE upstream from the HPRT promoter will function with heterologous promoters and whether its action is dependent on its position and orientation with respect to the promoter. DNA footprinting studies will determine if cellular proteins interact with the NRE and whether there is any correlation between the presence of such proteins and the level of expression. The in vivo significance of the NRE is unknown. It is possible that the antagonistic effect of the NRE on the promoter maintains low-level expression in all tissues except in brain tissue, where the NRE is ineffective due to lack of expression of the appropriate trans-acting factor. One approach to ascertaining if the NRE is in any manner related to increased HPRT expression in the brain would be to create transgenic mice bearing a reporter gene under the control of the HPRT promoter with and without the NRE and to examine expression of this gene in the brain and other tissues in these animals.

REFERENCES

1. Corden J, Wasylyk B, Buchwalder A, Sassone-Corsi P, Kedinger C, Chambon P (1980). Promoter sequences of eukaryotic protein-coding genes. Science 209:1406.
2. Breathnach R, Chambon P (1981). Organization and expression of eukaryotic split genes coding for proteins. Ann Rev Biochem 50:349.
3. Benoist C, O'Hare K, Breathnach R, Chambon P (1980). The ovalbumin gene-sequence of putative control regions. Nucl Acids Res 8:127.
4. Karin M, Haslinger A, Holtgreve H, Richards RI, Krauter P, Westphal HM, Beato M (1984). Characterization of DNA sequences through which cadmium and glucocorticoid

hormones induce human metallothionein-IIA gene. Nature 308:513.

5. Pelham HRB (1982). A regulatory upstream promoter element in the Drosophila hsp 70 heat-shock gene. Cell 30:517.

6. Gluzman Y, Shenk T, eds (1983). "Enhancers and Eukaryotic Gene Expression." Cold Spring Harbor, NY: Cold Spring Harbor Laboratory Press.

7. Dynan WS, Tjian R (1985). Control of eukaryotic messenger RNA synthesis by sequence-specific DNA binding proteins. Nature 316:774.

8. Miller JH, Reznikoff WS, eds (1978). "The Operon." Cold Spring Harbor, NY: Cold Spring Harbor Laboratory Press.

9. Brand AH, Breeden L, Abraham J, Sternglanz R, Nasmyth K (1985). Characterization of a "silencer" in yeast: A DNA sequence with properties opposite to those of a transcriptional enhancer. Cell 41:41.

10. Wright CG, Zitomer RS (1984). A positive regulatory site and a negative regulatory site control the expression of the Saccharomyces cerevisiae CYC7 gene. Mol Cell Biol 4:2023.

11. Wood TG, McGeady ML, Baroudy BM, Blair DG, Vande Woude GF (1984). Mouse c-mos oncogene activation is prevented by upstream sequences. Proc Natl Acad Sci USA 81:7817.

12. Melton DW, Konecki DS, Brennand J, Caskey CT 1984. Structure, expression and mutation of the hypoxanthine phosphoribosyltransferase gene. Proc Natl Acad Sci USA 81:2147.

13. Patel PI, Framson PE, Caskey CT, Chinault AC (1986). Fine structure of the human hypoxanthine phosphoribosyltransferase gene. Mol Cell Biol 6:393.

14. Valerio D, Duyvesteyn MGC, Dekker BMM, Weeda G, Berkens TM, van der Voorn L, van Ormondt H, van der Eb AJ (1985). Adenosine deaminase: Characterization and expression of a gene with a remarkable promoter. EMBO J 4:437.

15. Yang JK, Masters JN, Attardi G (1984). Human dihydrofolate reductase gene organization. Extensive conservation of the G + C-rich 5' non-coding sequence and strong intron size divergences from homologous mammalian genes. J Mol Biol 176:169.

16. Singer-Sam J, Keith DH, Tani K, Simmer RL, Shively L, Lindsay S, Yoshida A, Riggs AD (1984). Sequence of the promoter region of the gene for human X-linked 3-phosphoglycerate kinase. Gene 32:409.

17. Reynolds GA, Baser SK, Osborne TK, Chin DJ, Gil G,

Brown MS, Goldstein JL, Luskey KL (1984). HMGCoA reductase: A negatively regulated gene with unusual promoter and 5' untranslated regions. Cell 38:275.

18. McKnight SL, Kingsbury RC, Spence A, Smith M (1984). The distal transcription signals of the herpesvirus HR gene share a common hexanucleotide control sequence. Cell 37:253.

19. Kelley WN, Wyngaarden JB (1983). Clinical syndromes associated with hypoxanthine-guanine phosphoribosyltransferase deficiency. In Stanbury JB, Wyngaarden JB, Fredrickson DS, Goldstein JL, Brown MS (eds): "The Metabolic Basis of Inherited Disease," 5th edition, New York: McGraw-Hill, p 1115.

20. Wigler MA, Pellicer A, Silverstein S, Axel R (1978). Biochemical transfer of single-copy eukaryotic genes using total cellular DNA as donor. Cell 14:725.

21. Fuscoe JC, Fenwick RG, Ledbetter DH, Caskey CT (1983). Deletion and amplification of the HGPRT locus in Chinese hamster cells. Mol Cell Biol 3:1086.

22. Melton DW, McEwan C, McKie AB, Reid AM (1986). Expression of the mouse HPRT gene: Deletional analysis of the promoter region of an X-chromosome linked housekeeping gene. Cell 44:319.

Transcriptional Control Mechanisms, pages 57–69
© 1987 Alan R. Liss, Inc.

Delineation of Structural Features of the Chicken
Sarcomeric α-Actin Gene Promoters[1]

James M. Grichnik, Derk J. Bergsma,
and Robert J. Schwartz

Department of Cell Biology, Baylor College of Medicine,
Houston, Texas 77030

ABSTRACT   A 190 base pair fragment of the α-skeletal
actin gene promoter directs the transcription of the
chloramphenicol acetyl-transferase gene in a tissue
restricted and stage specific manner (1,2). Analysis
of this region revealed a partial dyad symmetry about
a central axis (-108) just upstream from the "CCAAT"
sequence (-91). A 411 nucleotide fragment which
contains the symmetrical promoter region functions in
both 5'-3' and 3'-5' orientations in transient
transfection expression experiments. However,
hybridization assays could not detect divergent
transcripts from the endogenous α-skeletal actin
gene. Sequence analysis of the promoter region of the
α-cardiac actin gene revealed similarities to its
skeletal counterpart. Although the induction of
cardiac actin mRNA preceeds the appearance of α-
skeletal mRNA during myogenesis in culture, transfection assays indicated co-expression of CAT from
both α-striated promoters. Thus, additional intragenic sequences may be involved with the temporal
expression of the α-cardiac actin gene in primary myoblast cultures. We showed that these two sarcomeric
actin promoters, in native 5'-3' direction, induce CAT
activity during myogenesis and are restricted in
expression in non-myogenic cells.

This work was supported by NIH grant NS15050,
Muscular Dystrophy Association of America and American
Heart Association, Texas affiliate.

INTRODUCTION

Our laboratory (3,4), as well as others (5,6), has described a switch from the non-muscle actins (β and γ) to the α-cardiac and α-skeletal actin mRNAs which are sequentially induced during myogenesis (4). In order to test for gene elements required for muscle cell specific expression, DNA sequences containing the 5'-flanking regions of the chicken α-skeletal actin gene were linked to the coding sequences of the chloramphenicol acetyltransferase (CAT) gene and transfected into myogenic and non-myogenic cells. The α-skeletal actin CAT constructions displayed stage specific activation during myoblast differentiation in vitro (9-15 fold increase in activity), and restricted expression in non-myogenic cells (1). Through a series of 5' deletions of the α-skeletal promoter, we have shown that tissue restricted and stage specific activation is maintained within a 190 base pair fragment (2).

We present here structural features identified within the 190 base pair promoter region of the α-skeletal actin gene which appear to be correlated with transcriptional activity. In addition, similar sequence and structural features were identified in the promoter region of the closely related α-cardiac actin gene.

RESULTS

Dyad symmetry in the α-skeletal actin promoter.

Through a series of 5' deletions of the α-skeletal actin gene's 5' flanking region, we have been able to localize the 5' border of the sequences responsible for activation of CAT activity (2). The smallest fragment which retains full tissue restricted and stage specific transcriptional promoting activity is a fragment extending from -200 to -11 upstream from the cap site. Previous reports, have noted the high degree of homology of the 5' flanking region of the chicken α-skeletal gene with rat and mouse α-skeletal actin genes (7-9). The 5' end of our alignment differs from that of Nudel et al. (8) and Chien-Tsung Hu et al. (9) in that sequences upstream of -194 in the rat sequence can be aligned directly with chicken sequences at -158 rather than introducing a spacer region and aligning at -212. In addition to a tighter region of homology, this alignment also extends the significant homologies shared between the sequences further upstream. Most of the identified homologous nucleotides are retained within this 190 base pairs of chicken α-skeletal sequence, 73% of which can be aligned with rat α-skeletal promoter region (figure 1b). Within this 190 nucleotide fragment we have identified a partial dyad symmetry about a central axis at -108 nucleotides upstream from the native transcription initiation site (figure 1a). This symmetry can also be identified in rat (figure 1b) and mouse (not shown) sequences. Although not all the regions are conserved, two pairs of sequence elements have been identified which are in chicken, rat, and mouse sequences (figure 2). The most striking of these symmetrical and homologous sequences has been termed the "internal element" for the purpose of discussion. This "internal element" [consensus (G/A)(G/C)CCAAA(T/G)A(A/T)GG(C/A)(G)] contains sequences resembling the "CCAAT" sequence which, in addition to the "TATA" sequence, has been identified in many promoters (10,11). The other symmetrical sequence (CGGGC(C/G)GT) termed the "external element" falls between the "internal elements" and "TATA" sequences (figure 1).

In the chicken promoter, the upstream symmetrical elements are located within sequences found to be important in the expression of the α-skeletal actin gene in transfection experiments. The removal of a 66 base pair fragment from -200 to -144 which contains one "external element" decreases activity by 50%, removing an additional 11 nucleotides to -133, which cuts directly into the "internal element", results in a 93% decrease in activity (2). If deletions are continued to -76 which removes the second "internal element", which contains a "CCAAT" sequence, activity drops to less than 1%. In an additional experiment, the region between the CCAAAT and ATAAAA sequences, at -76 and -32, was replaced with sequences from pBR322. This region, in addition to containing one of the "external element" sequences, contains a CATTCCT sequence which has been identified in many muscle regulated genes (12), and tentative SP1 (13) binding sequences. The activity of this sequence replacement construct dropped to less than 5% of that of the native sequence in myoblast transfection assays. The removal of the 5' sequences upstream of the CCAAAT sequence at -107 and the replacement of the sequences between the CCAAAT and ATAAA boxes did not decrease transcriptional activity in a <u>Xenopus</u> oocyte transcription system (2). It appears that the loss of activity at -133

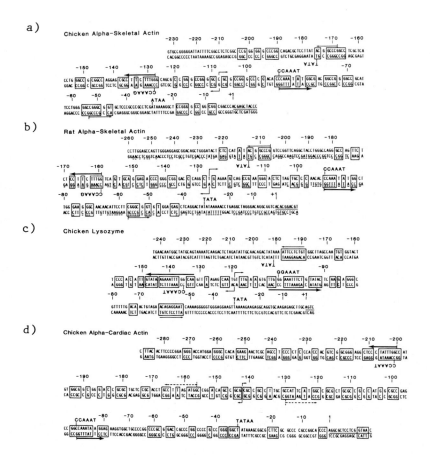

FIGURE 1. Symmetrical models of the chicken α-skeletal actin (a), rat α-skeletal actin (b), chicken lysozyme (c), and chicken α-cardiac actin (d) gene promoter regions. The native transcription initiation site of each gene is labeled as +1. "CCAAT" and "TATA" sequences are indicated on the appropriate strand. The thick arrows indicate the direction and location of the sequences which have been termed "internal elements", the dotted arrows indicate the weakly homologous "internal elements", and the fine arrows indicate the "external elements". The underlined bases of the rat α-skeletal actin represent the nucleotides with which the chicken sequence can be aligned homologously.

and between the CCAAAT and ATAAAA sequences may be due to specific interactions in myoblasts that are not required in a <u>Xenopus</u> transcription system. Certainly further investigation of these sequences will be necesary before we will be able identify specific effects.

Upon identifying the 190 base pair symmetrical structure of the α-skeletal actin promoter, we examined several other cellular genes to identify similar structures spanning the promoter regions. We have identified such structures in the sequences of chick lysozyme (14), silkworm fibroin (not shown;15), and chicken α-cardiac actin (16; figure 1). In the α-cardiac actin promoter region, we have also identified "internal element" sequences, of which the most 5' and most 3' are highly homologous to those identified in the α-skeletal sequence (figure 2). In addition the α-cardiac 5' flanking region appears to have a second pair of weakly homologous "internal elements" (figure 1d,2). These four regions have also been identified in the human α-cardiac gene (17; figure 2).

ALPHA-SKELETAL ACTIN

	INTERNAL	EXTERNAL
CHICKEN		
	5'(- 93)ACCCAAATATGGCG(- 80)3'	5'(- 54)CCGGGCGGTG(- 45)3'
	3'(-125)GCCCAAAGAAGGCG(-138)5'	3'(-167)CCGGGCCGTA(-176)5'
RAT		
	5'(- 94)ACCCAAATATGGCT(- 81)3'	5'(- 58)TCGGGCGGTG(- 49)3'
	3'(-157)GACCAAAGAAGGAG(-170)5'	3'(-205)CCGGGCCGTA(-214)5'
MOUSE		
	5'(- 98)ACCCAAATATGGCT(- 85)3'	5'(- 62)CGGGGCGGTG(- 53)3'
	3'(-165)GACCAAAGAAGGAG(-178)5'	3'(-212)CCAGGCCGTA(-221)5'

ALPHA-CARDIAC ACTIN

	INTERNAL	WEAK-INTERNAL
CHICKEN		
	5'(- 92)GGCCAAATAAGGAG(- 79)3'	5'(-132)CGCCATTCATGGCC(-119)3'
	3'(-199)GGCCAAATAGGGAG(-212)5'	3'(-154)GGCCATCTAAGGCA(-167)5'
HUMAN		
	5'(-101)GACCAAATAAGGCA(- 88)3'	5'(-151)CTCCATGAATGGCC(-148)3'
	3'(-226)GGCCAAATAGGGAG(-239)5'	3'(-189)GACCATGTAAGGAA(-202)5'

FIGURE 2. Sequence structure and location of the symmetrical and homologous "internal" and "external elements" of the chicken (2), rat (8), and mouse α-skeletal actin (9), chicken (16) and human α-cardiac actin (17).

Bidirectional promoter activity.

The identification of the symmetrical region in the location of the promoter led us to investigate whether

this region may function in a bidirectional manner.   In
addition to the noted symmetry, the 3'-5' oriented α-
skeletal promoter also has CCAAAG and TATAA sequences
spaced similarly to the CCAAAT and ATAAA sequences of the
5'-3' oriented promoter fragment (figure 1a).   The sequences are also present in similar positions in the rat
α-skeletal actin and chicken lysozyme promoter regions
(figure 1b,1c).   The "CCAAT" and "TATA" sequences have
been identified in several gene promoters and are believed
to play an essential role in transcription (10,11).   Test

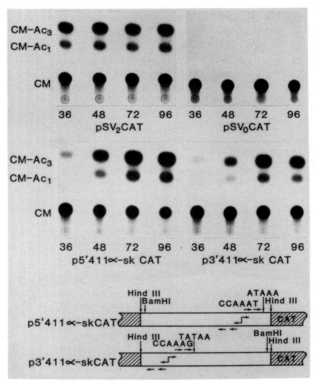

FIGURE 3. Bidirectional CAT activity.   Primary myoblasts from the same pooled population of dissociated
cells were transfected with p5'411α-skCAT and p3'411α-
skCAT at 24 hours after plating.   Cells were harvested at
36, 48, 72, and 96 hours postplating and assayed for CAT
activity.   The [^{14}C] chloramphenicol (CM) and acetylated
products (CM-AC1 and CM-AC3) were separated by thin layer
chromatography.

constructs were made by isolating a 411 nucleotide fragment, which includes the symmetrical region, inserted in both orientations upstream from the CAT gene (figure 3b). These CAT gene fusions, p5'411α-sk CAT and p3'411α-sk CAT, were transfected into myoblasts and chicken primary brain cells. Both constructs were restricted in CAT activity in the non-myogenic cell type (data not shown). At the 36 hour time point in culture, CAT activity isolated from cells transfected with either construct resulted in less than 2% acetylation of chloramphenicol, while cells transfected with pSV2-CAT resulted in 24% conversion. By 72 hours in culture, the myoblasts had differentiated into myotubes. At this time, the CAT activity from the cells transfected with pSV2-CAT had changed only slightly to 19% conversion, while p5'411α-sk CAT and p3'411α-sk CAT were now at levels of 66% and 37% conversion respectively (figure 3a). Direct comparison of the RNA transcripts will be required to determine the absolute activity of transcription in both orientations. However, it is evident from CAT activity that divergent transcription is quite strong. We do not know exactly which promoter sequences are directly responsible for the divergent transcripts. Preliminary data results suggest that divergent transcripts can still be transcribed from constructs containing only the 190 base pair symmetrical region (-200 to -11) in the 3'-5' orientation.

RNA dot blot analysis.

Since bidirectional activity could be identified from α-sk promoter CAT constructs, we examined the possibility that the endogenous chicken α-skeletal actin gene promoter might transcribe in a divergent direction in vivo. Uniformly labeled, radiolabeled stands in both the 3'-5' and 5'-3' direction which included sequences between -422 and -190 up stream from the transcription initiation start site of the α-skeletal actin gene were used as hybridization probes in RNA blots. We anticipated that the 5'-3' radiolabeled strand (5'3'PS5'α-sk) would hybridize with the divergent transcript and the 3'-5' radiolabeled strand (3'5'PS5'α-sk) would function as a negative control or identify any transcripts directed toward the α-skeletal actin gene. As controls, 3' specific UTR probes to β-cytoplasmic, α-skeletal, and α-cardiac actin were included to verify that the switching of actin genes in these transfected cultures proceeded as previously identified in non-transfected cells (4). The results of the RNA analysis confirmed our earlier findings on the actin mRNA levels (4) and showed, that no endogenous divergent transcript could be identified by either Northern (data not shown) or dot blot analysis (figure 4). We can not eliminate the possibility of a short lived, or highly transient transcript which would not be detected by our methods.

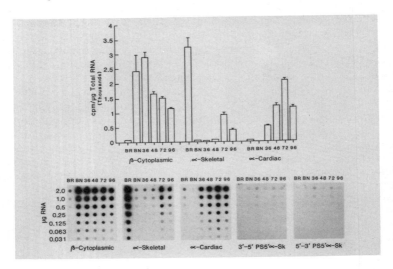

FIGURE 4. RNA dot blot analysis of actin mRNAs in adult breast muscle (BR), cultured chick embryonic brain (BN), and staged myoblasts at 36, 48, 72, and 96 hours in culture. RNA was dotted at concentrations ranging from 31 pg to 2 ug. The β-cytoplasmic, α-skeletal, and α-cardiac blots were hybridized with 3' specific probes (4). The divergent transcript probe and control were constructed as described above. The values were determined by linear regression analysis of each dot's radioactivity as determined by liquid scintillation counting.

Stage specific control of the α-cardiac promoter.

The RNA blot analysis demonstrated the sequential apearance of the α-cardiac mRNA and α-skeletal mRNA in cultured primary myoblasts (figure 4). This prompted us to investigate whether the sarcomeric actin promoters may

function in a sequential manner as well. We cloned a 2 kb
fragment of the α-cardiac 5' flanking region, including
the native transcription initiation start site and 17 base
pairs of the transcript leader sequence, upstream from the
CAT gene. This construct, p2.0α-HTCAT, and other promo-
ter-CAT constructs, were transfected into developing primary myoblasts (figure 5a). To control for the general
background increase in activity in all the constructs
transfected in this particular culture (possibly due to
continued division of the cultured cells initially), we
have set pSV2-CAT as a standard for baseline transcriptional activity levels (figure 5b). Once standardized, the
minimum 20 fold induction of the α-sarcomeric promoter
constructs is clearly marked in contrast to the β-cytoplasmic and histone promoter constructs. Since the expression of cardiac precedes skeletal α-actin mRNA during
myoblast differentiation and promoter region induced CAT
activity occurs coincidently, additional intragenic
sequences may be involved in their temporal expression.

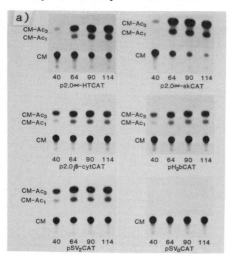

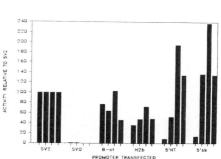

FIGURE 5. Cardiac CAT activity   Primary myoblasts
were transfected with p2.0α-HTCAT, p2.0α-SkCAT, p2.0β-
cytCAT, pH2bCAT, pSV0-CAT, and pSV2-CAT at 24 hours after
plating. Cells were harvested at 40, 64, 90, and 114
hours postplating and assayed for CAT activity (a). In
the lower panel, the data is expressed as bar graph of the
values standardized to the level of CAT activity of pSV2-
CAT at that time point (b).

Restricted cardiac promoter activity.

Previously, we have shown that the α–skeletal promoter is restricted in activity when transfected into chick primary brain cells (1). To identify if the α–cardiac promoter was also restricted, we transfected constructs which included promoters from β–cytoplasmic actin, histone H2b, α–cardiac actin, α–skeletal actin, pSV2–CAT, and pSV0–CAT into chick primary brain cells (figure 7). The α–cardiac and α–skeletal promoters were restricted in activity with CAT conversion levels below 1%, whereas the β–cytoplasmic, H2b histone, and SV40 promoter were found to be quite active with conversion levels of 17.6%, 32.1%, and 70.5% respectively (figure 6).

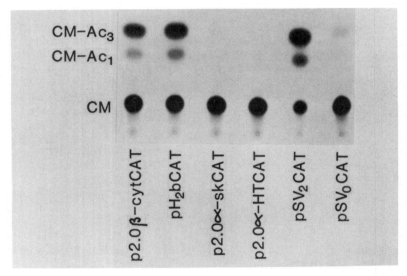

Figure 6. Tissue restricted cardiac CAT activity. Cultures of 11 day chick embryonic brain cells were transfected with 4 ug of p2.0β–cytCAT, pH2bCAT, p2.0α–skCAT, p2.0α–HTCAT, pSV2–CAT and pSV0–CAT. These cells were harvested 24 hours post–transfection and assayed for CAT activity.

DISCUSSION

Our preliminary studies with the α–cardiac actin gene promoter suggests similar structural and transcriptional characteristics to those of the α-skeletal actin gene promoter. The induction of CAT activity during myogenesis occurs coordinately. In addition, we have shown that both promoters are restricted in activity when transfected into non-myogenic chick primary brain cells.

One interesting feature of the α-skeletal actin promoters, is the appropriately spaced "CCAAT" and "TATA"-like sequences which can be identified in the divergent orientation. We have shown that a fragment containing this symmetrical region isolated from the chicken α-skeletal actin gene works perfectly well in either orientation, upstream of CAT, to support transcription in a tissue restricted and stage specific manner. Divergent transcripts from viral promoter (18), and cellular genes (19,20), have been identified. However, we have not been able to identify an endogenous divergent transcript originating from the promoter region of the endogenous α-skeletal actin gene. This could be due to a highly transient RNA product or possibly, due to sequences which prevent divergent transcription that have been removed during cloning.

Recently, other promoter sequences have been shown to drive tissue specific transcription in a bidirectional manner: a 520 base pair fragment upstream of the rat insulin II gene is sufficient to mediate cell-type specific expression of hybrid insulin/SV40 large-T antigen in both possible orientations (21). In addition, the rotation of a fragment of the fibroin promoter −234 to −66, which is required for transcriptional control and contains symmetrical sequences between −203 and −102, still modulates transcription (14). Symmetrical sequences can also be identified in the 208 nucleotides required for transcriptional activity of the chicken lysozyme gene (figure 1c). The "external elements" of this structure are located within regions protected by the binding of the glucocorticoid receptor (14).

The validity of this symmetrical sequence structure for the α-skeletal actin promoter has not yet been proven. However, it does appear that the promoters in which we have identified this structure can be activated to extremely high levels. Increased transcriptional activity could be accomplished by increased local concentrations of regulatory factors binding to the additional sites, or by cooperative binding of a transcription complex which would require interaction with symmetrical elements.

REFERENCES

1.   Grichnik JM, Bergsma DJ, Schwartz RJ (1986). Tissue restricted and stage specific transcription is maintained within 411 nucleotides flanking the 5' end of the chicken α–skeletal actin gene. Nucl Acids Res 14:1683–1701.

2.   Bergsma DJ, Grichnik JM, Gossett LMA, Schwartz RJ Delimitation and Characterization of cis–acting DNA sequences required for the regulated and transcriptional control of the chicken skeletal α–actin gene. Submitted

3.   Schwartz RJ, Rothblum KN (1981). Gene switching in myogenesis: Differential expression of the chicken actin multigene family. Biochemistry 20:4122–4129.

4.   Hayward LJ, Schwartz RJ (1985). Sequential expression of chicken actin genes during myogenesis. J Cell Biol In press.

5.   Bains W, Ponte H, Blau H, Kedes L (1984). Cardiac actin is the major actin gene product in skeletal muscle cell differentiation in vitro. Molec Cell Biol 1449–1453.

6.   Minty AJ, Alonso S, Caravatti M, Buckingham ME (1982). A fetal skeletal muscle actin mRNA in the mouse and its identity with cardiac actin mRNA. Cell 30, 185–192.

7.   Bergsma D, Hayward L, Grichnik J, Schwartz RJ (1985). Regulation of actin gene expression during chick myogenesis. Roche UCLA Symposium on Molecular Biology of Muscle Development, Eds Emerson C, Fischman DA, Nadal–Ginard B, Siddiqui MAQ, In press.

8.   Nudel U, Greenberg D, Ordahl CP, Saxel O, Neuman S, Yaffe D (1985). Developmentally regulated expression of a chicken muscle–specific gene in stably transfected rat myogenic cells. Proc Natl Acad Sci USA 82:3106–3109.

9.   Hu MC–T, Sharp, SB, Davidson N (1986). The complete sequence of the mouse skeletal α–actin gene reveals several conserved and inverted repeat sequences outside of the protein–coding region. Molec Cell Bio 6:15–25.

10. Breathnach R, Chambon P (1981). Organization and expression of eucaryotic split genes coding for proteins. Annu Rev Biochem 50:349–383.

11. Shenk T (1981). Transcription control regions: nucleotide sequence requirements for initiation by RNA polymerase II and III. Curr Topics Microbiol Immunol 93:25-46.

12. Ordal CP, Cooper TA (1983). Strong homology in promoter and 3'-untranslated regions of chick and rat α-actin genes. Nature 303:348-349.

13. Jones KA, Yamamoto KR, Tjian R (1985). Two distinct transcription factors bind to the HSV thymidine kinase promoter in vitro. Cell 42:559-572.

14. Renkawitz R, Schutz G, von der Ahe D, Beato M (1984). Sequences in the promoter region of the chicken lysozyme gene required for steroid regulation and receptor binding. Cell 37:503-510.

15. Tsuda M, Suzuki Y (1983). Transcription modulation in vitro of the fibroin gene exerted by a 200-base pair region upstream from the "TATA" box. Proc Nat Acad Sci USA 80:7442-7446

16. Chang KS, Rothblum KN, Schwartz RJ (1985). The complete sequence of the chicken α-cardiac actin gene: a highly conserved vertebrate gene. Nucleic Acids Res 13:1223-1237.

17. Minty A, Kedes L (1986). Upstream regions of the human cardiac actin gene that modulate its transcription in muscle cells: Presence of an evolutionarily-conserved repeated motif. Molec Cell Bio, In Press.

18. Saffer JD, Singer MF (1984). Transcription from SV40-like monkey DNA sequences. Nucleic Acids Res 12:4769-4788.

19. Crouse GF, Leys EJ, McEwen RN, Frayne EG, Kellems RE (1985). Analysis of the mouse dhfr promoter region: existence of a divergently transcribed gene. Molec Cell Bio 5:1847-1858.

20. Farnham PJ, Abrams JM, Schimke RT (1985). Opposite-strand RNAs from the 5' flanking region of the mouse dihydrofolate reductase gene. Proc Natl Acad Sci USA 82:3978-3982.

21. Hanahan D (1985). Heritable formation of pancreatic β-cell tumours in transgenic mice expressing recombinant insulin/simian virus 40 oncogenes. Nature 315:115-122.

Transcriptional Control Mechanisms, pages 71–81
© **1987 Alan R. Liss, Inc.**

IMMUNOGLOBULIN ENHANCER DELETIONS IN MURINE PLASMACYTOMAS[*]

Renato J. Aguilera, Thomas J. Hope, and Hitoshi Sakano

Department of Microbiology and Immunology, University
    of California, Berkeley, California 94720

ABSTRACT   We have analyzed immunoglobulin enhancer
deletions in murine plasmacytomas by DNA cloning.
This analysis revealed that the deletions occurred
between the J_H region and the switch region, removing
the Ig heavy-chain enhancer. We found that the loss of
the enhancer did not significantly affect the level of
heavy-chain mRNA expression. Nucleotide sequence
analysis of the area involved in one of the deletions
revealed that an inverted heptamer (GTGACAC) was
adjacent to the 5' recombination site. This along with
the presence of nucleotide insertion at the deletion
recombination site may suggest that the deletion was
mediated at least in part by the same recombination
machinery used for V-D-J joining. Both the J_H and
switch germline sequences involved in the deletion were
further analyzed by an <u>in vitro</u> DNA cleavage system
with an endonucleolytic activity purified in our laboratory from mouse fetal liver nuclear extracts. It was
found that the germline J_H sequence was strongly
cleaved at the deletion recombination site. Strong
cleavage was also observed at ATGGG repeats in the
germline switch region. We are currently attempting to
identify DNA binding activities that might be involved
in these recombinations.

INTRODUCTION

Immunoglobulin (Ig) variable region genes are generated
by site-specific DNA recombination during the differentiation

[*]This work was supported by grants from ACS (IM-366)
and NIH (AI-18790).

of B lymphocytes. In the heavy-chain genes, three DNA segments V, D, and J are involved in the recombination process.
Thus two recombination events, V-D and D-J joinings are
necessary to generate a complete heavy-chain gene (1). These
recombinations are mediated by two consensus sequences
CACTGTG and GGTTTTTGT which are separated by either a 12 or
23 base pair spacer (2,3).

Another DNA rearrangenment which takes place in the
heavy-chain gene is a class switch recombination (3-5). This
recombination replaces the $C\mu$ gene with another constant
region. DNA recombination sites for the class switch reside
in the 5' region of each constant gene where repetitive
sequences are usually found (3-5).

DNA elements responsible for the enhanced expression of
Ig genes have been found in the intron separating the J_H
region and the $C\mu$ exons (6-8). This element conveniently
lies 5' to the switch recombination site, and is therefore
unaffected by Ig class switching. Recently, four groups
including ours have reported enhancer deletions in Ig heavy-
chain genes (9-12). In all cases the deletion did not affect
the ability of the Ig genes to be expressed at a normal level.

In order to study the enhancer deletion at the DNA level,
we characterized two examples of the enhancer deletion by DNA
cloning and nucleotide sequencing. Enhancer deletions were
further analyzed by the in vitro cleavage assay with a mouse
fetal liver extract (13) using germline J_H and switch region
DNAs as substrates.

RESULTS

Enhancer deletion in the functional allele of a IgA
myeloma. In order to study the 3'-J_H deletion found in
myeloma M467, we cloned the region containing the deletion
into a phage vector (9). The cloned DNA was analyzed by
restriction enzyme mapping and DNA sequencing (9). As can
be seen in the restriction enzyme map in Figure 1A, the M467
allele contains three types of DNA rearrangements, a functional V-D-J joining, an isotype switch to $C\alpha$, and a ~3 kb
deletion between J_{H3} and the $C\alpha$ switch region. Nucleotide
sequence analysis revealed that the V-D-J joining is in
phase and the V-region sequence is derived from one of the
members of the V_{H-558} gene family (9). This is in agreement
with the previous demonstration that the M467 myeloma
secretes anti-flagellin IgA(κ) antibodies (16). We also
analyzed the other J_H allele of M467 and found that this
allele is a non-functional D-J intermediate.

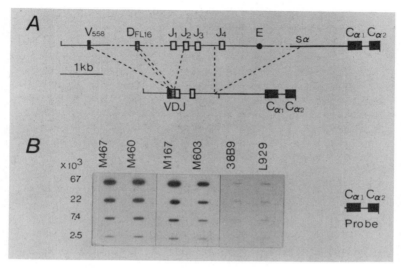

Figure 1 (A). Restriction enzyme map of the functional heavy-chain allele of myeloma MOPC467. This 3.5 kb EcoRI fragment contains three types of DNA rearrangements, a functional VDJ joining, a Ig class switch to Cα, and a 3 kb deletion between J_{H3} and the Sα region (9). (B) Analysis of the steady-state mRNA in MOPC467. Cytoplasmic mRNA was obtained from the indicated number of cells (x 10^3) by the procedure of White and Bancroft (14) and slot-blotted (BRL Hybri-Slot^{TM}) onto nitrocellulose. All samples were hybridized with the nick-translated (15) 800 bp XhoI/EcoR1 Cα probe.

Steady-state analysis of Ig mRNA expression in M467. Once we verified that the enhancer was deleted from a functional heavy-chain allele, we proceeded to analyze the cytoplasmic mRNA content of the M467 plasmacytoma. We found that the level of expression of α mRNA was not significantly affected by the loss of the enhancer. The level of α-mRNA in M467 is almost identical to three other IgA(κ) producing tumors (16) M460D, M167, and M603 (Figure 1B). As negative controls, we also isolated cytoplasmic mRNA from a pre-B cell line 38B9 and a fibroblast cell line L929, neither of which should have mRNA hybridizable to the Cα probe. As expected no hybridizable RNA was detected in these two cell lines (Figure 1B). As an internal control, we looked at κ mRNA expression and found that the level of the α message is similar in all the α/κ myelomas analyzed (9).

Enhancer deletion is not restricted to the functional
JH allele. Some myelomas were found to contain the enhancer
deletion in the non-functional J_H allele. For example,
myeloma M195 showed a D_{Q52}-J_H fragment whose size (~15 kb in
an Eco R1 digest) could not be explained by a simple D-J recombination. This aberrant allele was cloned and characterized by restriction enzyme mapping and DNA sequencing (9).
In addition to the D-J recombination, the non-functional
allele of M195 was found to contain two more recombination
events, a 1.5 kb deletion spanning from J_{H4} to the switch
region of $C\mu$, and a 15:12 c-myc translocation.

Nucleotide sequences around the deletion recombination
sites. In order to study the molecular mechanism of the
enhancer recombination, the two pairs of germline sequences
containing the deletion site were compared (9). Although
both deletions, one identified in M467 and the other in
M195, began around J_{H4} and ended in a switch region, no
characteristic sequences were evident around the recombination sites. Furthermore, there were no sequence homologies
between the 5' and 3' ends of the deletion. It is interesting to note that both in M467 and M195 extra nucleotides
were inserted at the site of the deletion; TAGGA in M467
(Figure 3) and A in M195 (9). These extra nucleotides
appear to have been inserted enzymatically during the
recombination process in a manner analogous to V-D-J joining
in both Ig heavy- and T-cell receptor β-chain genes (17-19).

In vitro cleavage of germline sequences corresponding to
the deletion recombination sites. Recently, we have identified in nuclear extracts of mouse fetal liver and chicken
bursa an endonucleolytic activity which cleaves Ig DNA in
the vicinity of the recombination sites (13). This activity
cleaves DNA at dinucleotide pairs TG/CA and is rapidly
inactivated by heat and proteinase K treatment (Figure 2,
ref. 13). In order to test the possibility that the same
endonucleolytic activity is responsible for the enhancer
deletion occurring in the J_H region, we cleaved the germline
J_H DNA in vitro with the mouse fetal liver nuclear extract
(13). As shown in Figure 3, a strong cleavage was detected
in the trinucleotide CCA at the deletion site. Since the
cleavage occurred one base downstream to the recombination
site, an exonucleolytic activity could have removed one
nucleotiotide after the cleavage prior to the joining.

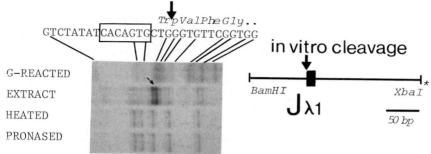

Figure 2. In vitro cleavage of the mouse Ig $J_{\lambda 1}$ gene.   A 300
bp BamHI-XbaI fragment containing the $J_{\lambda 1}$ sequence was end-
labelled (15) and reacted with fetal liver nuclear extracts
for 10 minutes at 37 °C (Extract).   Strong cleavage took
place in the first codon of the $J_{\lambda 1}$ gene.   However, no cleavage was observed, if the extracts were incubated at 65 °C for
3 minutes (Heated) or treated with proteinase K (Pronased)
prior to the reaction.   The reaction products were separated
on a 6% polyacrylamide gel containing urea alongside a Maxam
and Gilbert G-reacted sample (20).

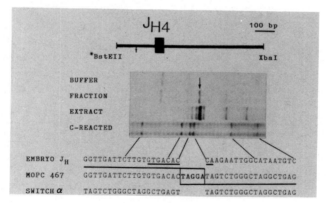

Figure 3. In vitro cleavage at the enhancer-deletion
breakpoint with the TG/CA endonucleolytic activity.   The
germline J_H fragment (BstEII-XbaI), end-labelled with ^{32}P at
the XbaI end, was reacted with either the crude nuclear
extract (EXTRACT) of fetal liver cells, DE52 fractionated
(13) extract (FRACTION), or reaction buffer alone (BUFFER).
Note that a strong cleavage (arrow) is found 3' of the
inverted heptamer (underlined) where the enhancer deletion
took place.   The boxed sequence TAGGA appears to have
been added enzymatically during the recombination process.

The recombination breakpoint in the switchα region has been identified in a 20 bp sequence (CTGGGCTAGGCTGAGT⬇TAGT) which is tandemly repeated 6 times over a 120 bp stretch (9). The germline switchα region was also analyzed for the in vitro cleavage.   Unlike the J_H breakpoint, we did not find the dinucleotide TG/CA at the deletion site and no in vitro cleavages were detected on the corresponding germline Sα sequences (data not shown).   However, very strong cleavages were observed in the ATGGG repeats located ~50 bp from the deletion site (Figure 4).

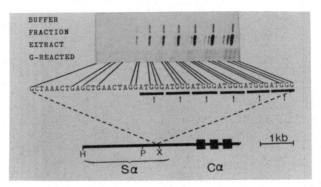

Figure 4.   In vitro cleavage of the Sα switch region with the endonucleolytic activity of mouse fetal liver nuclear extract. Strong cleaveages (indicated by arrows) were observed at ATGGG repeats (underlined) in the Sα region (9).

Examination of the Sα region for DNA binding properties. Due to the strong cleavage found at the ATGGG repeats in the Sα region we attempted to find if there were proteins binding to this region.   We have used a protein blotting assay (21, 22) which allows the detection and rough estimation of molecular weight of DNA binding proteins.   First the crude nuclear extract derived from myeloma J558L was separated on an SDS-PAGE gel and transfered electrophoretically to nitrocellulose (22).   The filter was placed in a DecaProbe^{TM} apparatus and each lane was probed with a 500 bp Sα probe in the presence of increasing amounts of specific and nonspecific competitor DNAs (Figure 5).   A couple of DNA binding proteins of an apparent molecular weight of ~90 kd appear to be competed for at the highest concentration of plasmid containing Sα sequences.   These specific binding proteins do not bind to the non-specific plasmid probe (right lane, Figure 5).

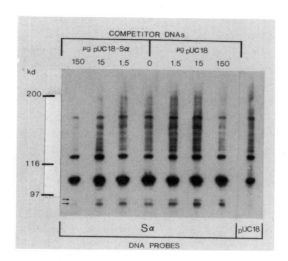

Figure 5. Detection of switchα DNA binding proteins by the
protein blotting procedure (21,22). Nuclear extracts were
prepared from myeloma J558L as previously described (13).
50 μg of the nuclear extract was separated on a 7 % SDS
acrylamide gel and transfered electrophoretically to a nitrocellulose filter (22). The filter was placed in a Hoefer
Deca-Probe^{TM} apparatus, pre-hybridized with low-fat milk
(22) to prevent non-specific binding, and hybridized (15) to
a nick-translated Sα probe (PstI-XbaI, see Figure 4) in the
presence of unlabelled plasmid DNA as a competitor. The
pUC18-Sα plasmid contains the entire 2.7 kb (HindIII-XbaI)
switch region. The competitor DNA was digested with
restriction endonuclease HaeIII, extracted with phenol and
precipitated with ethanol before use.

We compared the J558L nuclear extract with other
nuclear extracts from other lymphoid-cell lines and found it
was present in myeloma M467 but not in the T-cell (EL4) or
pre-B cell (38B9) nuclear extracts. We also analyzed
nuclear extracts from normal tissues such as adult Balb/c
spleen and liver and could only detect a ~66 kd non-specific
DNA binding protein which is found in all nuclear extracts
analyzed (not shown).

Discussion

We characterized two examples of enhancer deletions identified in murine plasmacytomas by DNA cloning and nucleotide sequencing (9). One of the clones that we analyzed is a functional α heavy-chain gene isolated from the anti-flagellin IgA myeloma MOPC467, and the other is a non-functional J_H allele of MOPC195 which contains a <u>myc</u> oncogene translocation (9). Although both deletions begin within the J_H region and end in a switch region, no characteristic or homologous sequences were evident around the recombination sites (9).

Our RNA analysis (Figure 1) revealed that the M467 heavy chain gene containing the enhancer deletion is expressed at a normal level. Similar observations were reported by other groups in different Ig heavy-chain genes (10-12). An intriguing question is why the enhancer deletion does not affect the level of transcription. It is possible that although the Ig heavy-chain enhancer is required for the transcriptional activation of the heavy-chain gene, it is not necessary to maintain its transcriptional activity (12). Alternatively, there might be other DNA elements, located in the vicinity of the Ig genes, which could replace the function of the deleted enhancer.

Recently, we have identified and isolated an endonucleolytic activity which cleaves Ig joining (J) sequences in the vicinity of the recombination sites (Figure 2, ref. 13). This activity can be isolated from nuclear extracts of cells in mouse fetal liver or chicken embryo bursa. These tissues are sites where B-cell progenitors first appear during embryonic development. The activity is known to cleave Ig recombination sequences at the dinucleotide TG/CA (13). We examined J_H germline DNA for <u>in vitro</u> cleavage with the extract because the 5' deletion site is close to J_{H4}. As shown in Figure 3, the endonuclease does indeed cleave germline DNA in the region where the M467 deletion took place. Immediately 5' to the cleavage site we found a heptamer GTGACAC which is identical, although inverted, to the consensus sequence CACTGTG used for normal V-(D)-J recombinations. This might indicate that the cleavage activity which is normally used for the V-(D)-J joining was responsible for the cleavage at the 5' deletion site. The switch region sequence was also examined for the cleavage, since the breakpoint of the enhancer deletion occurred in the $S\alpha$ switch region. <u>In vitro</u> cleavage analysis revealed that the endonuclease cleaved $S\alpha$ DNA about 50 bp from the recombination site, probably recognizing the TG dinucleotide in the A$\underline{\text{TG}}$GG repeat unit (Figure 4). We speculate that cleavage occurred at one of these repeats and exonucleolytic trimming produced the observed recombinant. It is interesting to note that other switch regions, including Sμ (3) Sγ , Sϵ (23) also contain TG or TGG in their repeat units. We postulate that the enhancer deletion is an aberrant event related to the V-(D)-J joining and/or to the switch recombination possibly mediated by the TG/CA endonucleolytic activity. It appears that the deletion between J_H and S_α occurs during the process of normal Ig gene rearrangement possibly due to the open chromatin structure of the region and its access to the recombination machinery.

Both in Ig and TcR genes two conserved sequences are commonly found adjacent to the recombination sites of the germline V-(D)-J gene segments. One is a heptamer (CACTGTG) and the other is a nonamer (GGTTTTTGT) (1,2). The spacer separating the two consensus sequences is always either 12 bp or 23 bp long. These structural features appear to be required for the recognition of the DNA by the recombinase. In the case of isotype swithing, repetitive sequences in the 5' region of the Ig C_H genes seem to play a key role in the recombination process. Although, a large amount of data has been accumulated at the DNA level, almost nothing is known about the proteins which are involved in these recombinations. We are currently attempting to identify DNA-binding proteins which might specifically interact with the recombination sites (Figure 5). These proteins, if isolated, will give us a new insight into the molecular mechanism of Ig somatic gene rearrangements.

ACKNOWLEDGEMENTS

We are greatfull to Drs. M. Potter, M. Weigert, F.W. Alt, V. Oi and R. Mishell for the tumor cells and cell lines. We also thank Dr. M.E. Koshland for her helpfull suggestions.

REFERENCES

1. Sakano H, Kurosawa Y, Weigert M, Tonegawa S (1981). Identification and nucleotide sequence of a diversity DNA segment (D) of immunoglobulin heavy-chain genes. Nature 290:562.

2. Early P, Huang H, Davis M, Calame K, Hood L (1980). An immunoglobulin heavy chain variable region gene is generated from three segments of DNA: VH, D and JH. Cell 19:981.
3. Sakano H, Maki R, Kurosawa Y, Roeder W, Tonegawa S (1980). Two types of somatic recombination are necessary for the generation of complete immunoglobulin heavy-chain genes. Nature 286:676.
4. Kataoka T, Kawakami T, Takahashi N, Honjo T (1980). Rearrangement of immunoglobulin Y_1-chain gene and mechanism for heavy-chain class switch. Proc Natn Acad Sci USA 77:919.
5. Davis MM, Kim ST, Hood LE (1980). DNA sequences mediating class switching in α-immunoglobulins. Science 209:1360.
6. Banerji J, Olson L, Schaffner W (1983). A lymphocyte-specific cellular enhancer is located downstream of the joining region in immunoglobulin heavy chain genes. Cell 33:729.
7. Gillies SD, Morrison SL, Oi VT, Tonegawa S (1983). A tissue-specific transcription enhancer element is located in the major intron of a rearranged immunoglobulin heavy chain gene. Cell 33:717.
8. Neuberger MS (1983). Expression and regulation of an immunoglobulin heavy chain gene transfected into lymphoid cells. EMBO J 2:1373.
9. Aguilera RJ, Hope TJ, Hitoshi S (1985). Characterization of immunoglobulin enhancer deletions in murine plasmacytomas. EMBO J 4:3689.
10. Klein S, Sablitzky F, Radbruch A (1984). Deletion of the IgH enhancer does not reduce immunoglobulin heavy chain production of a hybridoma IgD class switch variant. EMBO J 3:2473.
11. Wabl MR, Burrows, PD (1984). Expression of immunoglobulin heavy chain at a high level in the absence of a proposed immunoglobulin enhancer element in cis. Proc Natl Acad Sci USA 81:2452.
12. Zaller DM, Eckhardt LA (1985). Deletion of a B-cell-specific enhancer affects transfected, but not endogenous, immunoglogulin heavy-chain gene expression. Proc Natn Acad Sci USA 82:5088.
13. Hope TJ, Aguilera RJ, Minie ME, Sakano H (1986). Endonucleolytic activity that cleaves immunoglobulin recombination sequences. Science 231:1141.
14. White BA, Bancroft FC (1982). Cytoplasmic dot hybridization. J Biol Chem 15:8569.

15. Maniatis T, Fritsch EF, Sambrook J (eds) (1982). "Molecular Cloning. A Laboratory Manual," Cold Spring Harbor Laboratory Press, New York.
16. Potter M (1970). Mouse IgA myeloma proteins that bind polysaccharide antigens of enterobacterial origin. Federation Proc. 29:85.
17. Alt FW, Baltimore D (1982). Joining of immunoglobulin heavy chain gene segments: implications form a chromosome with evidence of three D-JH fusions. Proc Natn Acad Sci USA 79:4118.
18. Hagiya M, Davis DD, Takahashi T, Okuda K, Raschke WC, Sakano H (1986). Two types of immunoglobulin-negative Abelson murine leukemia virus-transformed cells: Implications for B-lymphocyte differentiation. Proc Natn Acad Sci USA 83:145.
19. Siu G, Kronenberg M, Strauss E, Haars R, Mak TW, Hood L (1984). The structure, rearrangement and expression of Dβ gene segments of the murine T-cell antigen receptor. Nature 311:344.
20. Maxam AM, Gilbert W (1980). Sequencing end-labelled DNA with base-specific chemical cleavages. Methods Enzymol 65:499.
21. Bowen B, Steinberg J, Laemmli UK, Weintraub H (1980). The detection of DNA-binding proteins by protein blotting. Nucleic Acids Res 8:1.
22. Miskimins WK, Roberts MP, McClelland A, Ruddle FH (1985). Use of a protein-blotting procedure and a specific DNA probe to identify nuclear proteins that recognize the promoter region of the transferrin receptor gene. Proc Natl Acad Sci USA 82:6741.
23. Nikaido T, Yamawaki-Kataoka Y, Honjo T (1982). Nucleotide sequences of switch regions of immunoglobulin Cϵ and Cγ genes and their comparison. J Biol Chem 257:7322.

Transcriptional Control Mechanisms, pages 83–101
© **1987 Alan R. Liss, Inc.**

NUCLEAR FACTORS INTERACTING WITH THE IMMUNOGLOBULIN HEAVY AND κ ENHANCERS

Ranjan Sen and David Baltimore

Whitehead Institute for Biomedical Research
Cambridge, MA 02142
and
Massachusetts Institute of Technology
Department of Biology
Cambridge, MA 02139

INTRODUCTION

The expression of immunoglobulin genes is regulated by tissue specific promoters and enhancers at both the heavy and the κ light chain loci (for a recent review see 1). Furthermore, at least in the μ heavy chain, the structural region of the gene also appears to confer some tissue specificity of expression (2). Although cis-acting regulatory sequences have been identified in a number of cellular genes, the mechanism of action of these sequences remains largely unknown. It is generally believed that tissue specific and developmentally regulated expression of these genes must be mediated by the action of specific trans-acting factors. The immunoglobulin genes represent a class of well characterized specifically expressed genes which lend themselves well to further analysis.

We have used an electrophoretic mobility shift assay to identify nuclear factors that interact with the enhancers associated with the immunoglobulin heavy and κ light chain loci (3). Briefly, this assay consists of incubating end-labelled DNA templates with nuclear extracts derived from tissue culture cell lines followed by electrophoretic analysis in low ionic strength polyacrylamide gels. In principle, if the DNA fragment interacts with proteins in the nuclear extract, its mobility in the gel should be retarded relative to the free fragment (4,5). We have used

competition experiments to show specificity of the
complexes observed and methylation interference
experiments to accurately define the binding sites of
the factors identified. Finally we have made extracts
from a wide variety of tissue culture lines to
determine the tissue distribution of factors.
     We have earlier used this protocol to identify a
ubiquitous factor (NFA1) which interacts with a highly
conserved octameric sequence (ATTTGCAT) found in the
promoters of all V_H and V_L genes sequenced to date
(6). Results from the immunoglobulin enhancer have led
to some interesting conclusions. First, we find that
multiple factors appear to interact with both the μ and
the κ enhancers. Secondly, these tissue specific
enhancers appear to be capable of interacting with
tissue specific as well as tissue non-specific
factors. Thirdly, both κ and μ enhancers interact with
at least one common factor and at least one factor
which is specific to each.

RESULTS

     The heavy chain enhancer has been located on a 700
bp XbaI/EcoRI fragment from the J_H-C_μ intron and
30-50% of this activity may be found in the 300 bp
PvuII/EcoRI fragment (μ300) shown in Fig. 1A. The
black boxes labelled μE3 and μE4 represent clusters of
residues that were found to be protected in vivo
against methylation by dimethyl sulfate (DMS) in a B
cell-specific manner (7,8). They were proposed to
represent binding sites for a B-cell-specific enhancer
binding protein. The open circle represents an
octa-nucleotide sequence element found in all
immunoglobulin promoters and at this position in the μ
enhancer. The μ300 was further subdivided by cleaving
with AluI, DdeI and HinfI to generate the fragments
labelled μ70, μ50, μ60-1 and μ60-2. When each of these
fragments was analyzed in a binding reaction, μ70 and
μ50 appeared to give specific complexes (3). The
technique of methylation interference was used to
locate the binding sites of these putative factors more
accurately and the results obtained are summarized in
Fig. 1B. The open circles above the G residues
indicate those positions that were found to be
protected against methylation by DMS in vivo, whereas

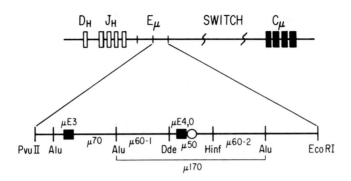

Figure 1: Factors interacting with the 300 bp PvuII-
        EcoRI fragment (μ300) of the μ enhancer.
        A. Schematic representation of μ300. To
facilitate identification of factors interacting with
this sequence, AluI, DdeI and HinfI were used to
generate smaller fragments labelled μ50, μ70, μ60-1 and
μ60-2. Binding and competition analysis using each
subfragment showed that specific nucleoprotein
complexes could be detected on μ50 and μ70 (3).

the encircled G's show the residues whose methylation
significantly affects the binding of the protein in the
in vitro assay. On the coding strand of the μ50
fragment (upper in Fig. 1B) modification at only one G
significantly alters the binding of the protein and on
the non-coding strand modification of 3A's show effects
on the binding and the G downstream shows only a
partial effect (represented by the dotted circle).
None of the other G residues associated with the μE4
site (Fig. 1A) appear to interfere with binding. This
suggests that in vitro we are only seeing the
interaction of NF-A1 with this fragment through its
recognition sequence, the octanucleotide ATTTGCAT.
        The complex detected on the fragment μ70 is
affected by methylation at any one of 3 G's (encircled)
on the coding strand and 2 G's on the non-coding strand
(Fig. 1B). The pattern of methylation interference

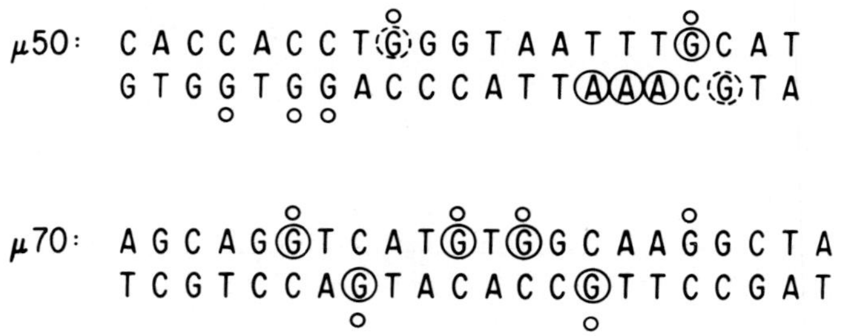

μ50: C A C C A C C T G G G T A A T T T G C A T
      G T G G T G G A C C C A T T A A A C G T A

μ70: A G C A G G T C A T G T G G C A A G G C T A
      T C G T C C A G T A C A C C G T T C C G A T

Figure 1B. Summary of methylation interference (in vitro) and methylation protection experiments (in vivo) used to precisely define protein binding sites within the μ enhancer. Briefly, end-labelled DNA was partially methylated using dimethyl sulfate (DMS) and then used in a preparative binding reaction. Following polyacrylamide gel electrophoresis, the nucleoprotein complex and the free DNA fragment were eluted from the gel, treated with piperidine (to achieve strand cleavage) and analyzed on sequencing gels. Those G residues whose modification seriously impairs DNA-protein interactions will be specifically depleted in the complex, thus allowing identification of critical residues. In the figure, the relevant regions of μ50 and μ70 are shown with the coding strand on top. The open circles above or below the sequence indicate those residues that were found to be protected against methylation by DMS in vivo. The encircled G's are the ones whose methylation interferes with protein-DNA interactions in vitro (dotted circles represent partial interference). The encircled A's represent those residues which are specifically depleted in the complex. The mechanism by which modification of A's by DMS leads to interference in protein DNA interaction under these conditions is not clear.

closely resembles the pattern of methylation protection observed in vivo thus suggesting strongly that the factor detected in vitro is the same that is bound to this region of the chromosomal DNA in B cells. We shall refer to this factor as NF-μE3.

In contrast to the observation that the in vivo protections are seen in a B cell specific manner, neither of the proteins detected in vitro appear to be tissue-specific (3). This implies firstly, that non-B cell specific factors may interact with the μ enhancer and secondly that perhaps some aspect of the chromatin organization prevents access of these factors to the enhancer in non-B cells.

In the course of competition experiments (using small fragments of non-radioactive DNA) to define the specificity of the nucleoprotein complex on the fragment μ70, we noticed that μ50 was unable to compete for the complex even when present in 60- to 100-fold molar excess relative to the labelled fragment. This was unexpected because μ50 carries a sequence very homologous to the binding site of μ70 (μE3) which, at least in vivo, appeared to interact with a nuclear factor. Since μ50 also contains an NF-A1 binding site, this lack of competition could be due to the competitor fragment being sequestered by NF-A1 and thus being unavailable for competing off factor NF-μE3. To eliminate this possibility, we separated the NF-A1 and NF-μE3 activities by a simple chromatographic step and repeated the competition experiments. The two factors were well resolved by heparin-agarose chromatography, the bulk of NF-A1 eluting at 100 mM KCl and NF-μE3 eluting at 0.45 M KCl. Even in a fraction significantly depleted of the relatively abundant octamer binding protein (NF-A1), μ170 (a fragment that contains μE4) failed to compete for the binding of NF-μE3 to its cognate DNA sequence (Fig. 1C). A DNA fragment from the κ enhancer (κ2) also has a binding site for NF-μE3 (3). This fragment generated a DNA-protein complex when incubated with the 0.45 M salt fraction of the heparin-agarose column (lane 2), which could be competed away by unlabelled μ300 fragment (lane 3) and μ70 fragment (lane 8) but not with equivalent amounts of μ170 fragment (lane 7), μ60 fragments (lanes 5,6), or μ400 fragment (lane 4). Thus the non-competition of the complex formed on μE3 by NF-μE3 is not due to presence of the octanucleotide sequence and its

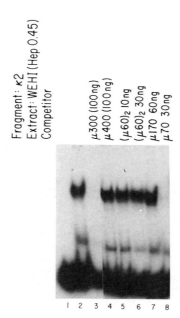

Figure 1C. Binding sites μE3 and μE4 are not equivalent. WEHI 231 nuclear extracts were fractionated on heparin agarose and the fraction eluting at 0.45 M KCl was found to contain the activity that interacts with either site μE3 in the μ enhancer or the site κE3 in the κ enhancer. Binding and competition analysis were carried out using the κ2 fragment (containing site κE3) and the 0.45 M fraction which was significantly depleted in octamer binding protein NF-A1 (3). Lane 2: No competitor DNA added; lanes 3-8 competitor DNA's added as shown above each lane. The fragments μ300 (contains sites μE3 and μE4) (μ60)₂, μ170 (contains site μE4) and μ70 are defined in Fig. 1A. μ400 is a 400 bp fragment just 5' to μ300 in the genomic sequence from XbaI to the PvuII (it contains one of the sites defined by Ephrussi et al. (7) to be protected against DMS in vivo).

interactions with NF-A1, but possibly because this
site, in spite of its strong homology, is not
equivalent to the site μE4. In this context when a
site having the identical sequence to μE4 within the
consensus region (CAGGTGG) was isolated from the kappa
enhancer and checked for its ability to bind to
proteins present in the nuclear extract we were unable
to detect any complex formation (data not shown).
Although the site μE4 itself (in the absence of the
octamer) has not been analyzed in a binding assay, the
result with the κ fragment supports the hypothesis that
this sequence has a much lower affinity for interaction
with a nuclear factor than any of the others studied so
far.

A cluster of in vivo protections against dimethyl
sulfate (which comprises the site μE1) lies in a 45 bp
HinfI-PvuII fragment just upstream of the PvuII site in
Fig. 1A. When this fragment was labelled and analyzed
for binding it showed the ability to form a discrete
nucleoprotein complex (Fig. 2, lane 2). The
specificity of this complex was shown by competition
analysis (lanes 3-10). Thus the complex was completely
eliminated when the binding reaction was carried out in
the presence of 10 ng of the unlabelled HinfI/PvuII
fragment (Fig. 2A, lanes 3,4) or unlabelled μ170
fragment (Fig. 2A, lanes 7,8). However similar amounts
of either μ70 (lanes 5,6) or the SV40 (lanes 9,10)
enhancer could not efficiently compete for complex
formation. The inability of μ70, a fragment which
bears the binding site μE3, to compete binding of this
fragment implied that the sites μE1 and μE3 were not
equivalent. Conversely, unlabelled HinfI/PvuII
fragment was unable to compete away the binding of
factor NF-μ3 to its cognate sequence μE3 (Fig. 2, lanes
11-14). The ability of both fragments containing μE1
and μE3 to bind to factors, but the lack of
cross-competition between them, suggests that different
proteins interact with these two sites (for a complete
description of site μE1 see ref (9)). Thus three of
four sites that were identified by Ephrussi et al. (7)
as the binding sites of a tissue-specific enhancer
binding protein, appear to be non-equivalent and at
least two of these appear to interact with distinct
nuclear factors. Furthermore, all the factors
identified as interacting with the μ enhancer could be
detected in all tissues and are therefore not B

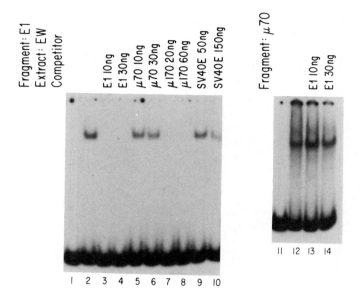

Figure 2: Binding and competition analysis using the
        HinfI-PvuII fragment.
        Competition experiments to show that
nucleoprotein complex formation on this fragment is
sequence-specific and different from the complex formed
on μ70 (containing μE3). End-labelled HinfI-PvuII
fragment was incubated with 12-15 μg of EW nuclear
extract in the presence or absence of competitor DNA's
as noted above each lane and analyzed by electrophoresis through low ionic strength polyacrylamide gels.
Competitor fragment E1: refers to the HinfI-PvuII
fragment. SV40E refers to a 170 bp fragment containing
both the 72 bp repeats of SV40. Nucleoprotein complex
on μ70 (μE3 site) is not competed out by the HinfI-
PvuII fragment (referred to in the Fig. as E1). Lane
12: no competitor added; Lane 13,14: 10 and 30 ng of E1
fragment were added to the binding reaction prior to
addition of protein. Lane 1, 11: free fragments.

cell-specific. There is also a tissue-specific
component of the octamer binding factor NF-A1 which
could potentially regulate the tissue specificity of
the heavy chain enhancer (10).

Analysis of the κ enhancer

Figure 3A shows a schematic representation of the
κ enhancer and the restriction enzyme sites used to
subdivide the enhancer into various fragments for
binding analysis. The black boxes show the sites which
are homologous to the consensus sequence derived by
Church et al. (8) on the basis of methylation

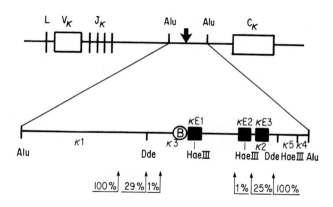

Figure 3: Dissection and binding analysis of the κ
enhancer.
A. Schematic representation of the 475 bp
AluI-AluI DNA fragment containing the κ enhancer.
Subfragments labelled κ1 through κ5 were derived by
cutting with DdeI and HaeIII and used as probes in
binding assays. κ4 and κ5 did not yield discernible
nucleoprotein complex whereas fragments κ2 and κ3 were
positive. Black boxes represent sequences homologous
to the consensus sequence for a B cell-specific
enhancer binding protein derived by Church et al.
(8). The lowest line summarizes the results of
deletion analysis of the κ enhancer carried out by
Queen and Stafford (11).

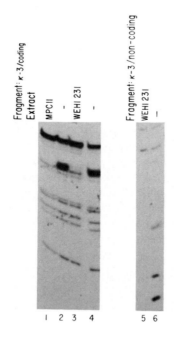

Figure 3B. Localization of binding sites of the factor interacting with fragment κ3 using methylation interference experiments. End-labelled κ3 fragment was partially methylated using dimethyl sulfate and then used for binding reactions in extracts derived from the mouse myeloma MPC11 (lanes 1,2) or the mouse B cell line WEHI 231 (lanes 3-6). Bands representing the nucleoprotein complex and the free DNA fragment were excised from preparative low ionic strength poly-acrylamide gels, the DNA eluted and treated with piper-idine and then analyzed by electrophoresis through sequencing gels. Coding strand analysis: G ladders corresponding to nucleoprotein complexes formed in MPC11 extracts (lane 1) or WEHI 231 extracts (lane 3). G ladders corresponding to the free DNA fragment present after binding in MPC11 extracts (lane 2) or WEHI 231 extracts (lane 4). Non-coding strand analysis in WEHI 231 extracts: G ladder derived from nucleo-protein complex (lane 5), G ladder derived from free DNA (lane 6). Asterisks indicate the position of G residues which are specifically depleted in the nucleoprotein complex.

protection experiments of the μ enhancer in vivo. When
the fragments generated by cutting with HaeIII and DdeI
were examined for their ability to bind to nuclear
factors, two fragments (κ4 and κ5) appeared to be
negative whereas κ2 and κ3 formed complexes (3). κ1 is
too large a fragment to be reliably assayed and has not
been further dissected yet.

Competition analysis using various fragments of
the μ enhancer and footprinting by methylation
interference showed that the protein binding to the
fragment κ2 was the same as that interacting with the
fragment μ70 (3). Further, the factors binding to
these sequences co-fractionated through 2 sequential
chromatographic steps suggesting that they are, in
fact, the same protein. Examination of the sequence
near the binding site shows that they are identical
between μ70 and κ2. Thus there is at least one protein
that appears to be able to interact with the μ enhancer
as well as the κ enhancer.

The fragment κ3 also forms a specific complex,
which was only competed off by either the κ enhancer or
the SV40 enhancer, but not by 2 fragments of the μ
enhancer or the κ promoter. Methylation interference
experiments were carried out to define the binding site
of this factor on κ3 accurately (Fig. 3B). On the
coding strand, complex formation is seriously impeded
by the prior methylation of any one of 3 consecutive
residues that lie towards one end of the κ3 fragment
(site B in Fig. 3A). On the non-coding strand, there
are 3 G residues indicated by the asterisks which
appear to block binding of the protein completely. The
complete binding site is shown in Fig. 3C. The circled
G's represent those residues whose methylation
interfered with nucleoprotein complex formation.
Definition of the binding site also served to explain
the specific competition observed with the SV40
enhancer, since the 11 bp stretch -- GGGGACTTTCC -- is
identically repeated in this enhancer.

Interestingly, deletion mapping of the κ enhancer
has shown that sequences within the κ3 fragment are
extremely important for enhancer function (11).

The tissue range of this factor was examined by
carrying out binding analysis with κ3 using extracts
from a variety of cell lines. Nucleoprotein complex
formation was detected in a mouse B cell line (WEHI
231) but not in five other non-B cell lines (COS, 3T3,
MEL, HeLa and PCC4) (3). The factor therefore appeared

κ-3: C A G A G G G G A C T T T C C G A G A G G
G T C T C C C C T G A A A G G C T C T C C

Figure 3C. Summary of in vitro methylation interference experiments to define the binding site on the κ3 fragment. The relevant section of the κ3 fragment is shown with the coding strand on top and the non-coding strand at the bottom. Circled G residues are those whose modification seriously impairs protein DNA interaction.

to be restricted in expression to lymphoid cells. We then examined extracts made from cells at various stages of B cell differentiation. Nucleoprotein complex formation was evident in the Abelson virus transformed pre-B cell line PD (Fig. 4, lane 5); 2 mouse B cell lines (WEHI 231 and AJ9, Fig. 4, lanes 6,7); a human B cell line (EW, Fig. 4, lane 8); 2 mouse myeloma lines (MPC 11 and SP2-0, Fig. 4, lanes 11,12) and 2 human myeloma lines (KR12 and 8226, Fig. 4, lanes 9,10). However, it was not apparent in a very early pre-B cell line (HAFTL, Fig. 4, lane 2) and 2 standard mouse pre-B cell lines (70Z and 38B9, Fig. 4, lanes 3,4). Thus, this factor appears not only to be tissue-specific but also stage-specific within the B cell lineage, being evident only in lines that have matured to the B cell stage or beyond. For these reasons we refer to the binding site as the B site (Fig. 3A) and the factor as NF-κB.

There appears to be one discrepancy to this generalization. The cell line PD (which is NF-κB positive) was derived by Abelson murine leukemia virus induced transformation of adult bone marrow cells (12) and undergoes κ light chain rearrangement in culture (13). Thus it appears to represent an earlier stage in the B cell differentiation pathway than 70Z which has stably rearranged one of its κ alleles (14) and can make κ protein when stimulated with the B cell mitogen LPS (15). However 70Z is NF-κB negative. We think

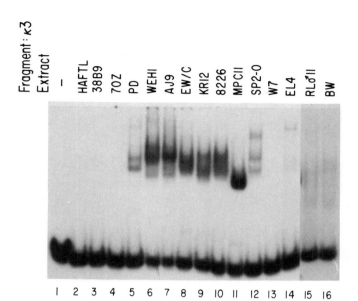

Figure 4. Lymphoid stage specificity of NF-κB, the
         factor interacting with the κ3 fragment.
     End-labelled κ3 fragment was used for binding
assays in extracts derived from a variety of B and T
lymphoid cells (shown above each lane). HAFTL (20):
38B9, 70Z and PD are mouse pre-B cell lines; WEHI 231
and AJ9 are B cell lines; EW is a human B cell line;
KR12 and 8226 are human myelomas; MPC 11 and SP2-0 are
mouse myelomas; W7, EL4, RLÓ11 and BW are mouse T cell
lines. Note that the mobility of the complex formed
differs slightly between extracts derived from mouse
cells (AJ9, WEHI 231) or human cells (EW, KR12, 8226).
The significantly altered mobility in MPC 11 extracts
is probably due to proteolysis, since many other
fragments containing binding sites show higher mobility
complexes in this extract.

that these may not be contradictory because (i) the 70Z
cell line does not have a DNAse 1 hypersensitive site

associated with the κ enhancer prior to induction with
LPS (16) whereas PD does (Sen and Baltimore,
unpublished results) and (ii) the κ enhancer appears to
be active after transfection of genes into PD whereas
it is not in 70Z (Speck and Baltimore, unpublished
results).

    To check whether LPS stimulation of 70Z would
induce NF-κB we induced the cells with 10 μg/ml of LPS
and carried out binding analysis using fragment κ3 in
extracts derived from these cells (Fig. 5A). Nuclear
extracts derived from 70Z cells 20 hr after induction
contain NF-κB (lanes 4,5) but not cells prior to
induction (Fig. 5A, lanes 2,3). Similarly, the low
level of NF-κB present in PD (Fig. 4, lane 5 or Fig.
5B, lane 3) can be raised to much higher levels after
induction with LPS for 20 hr (Fig. 5B, lanes 5,6).
Thus although pre-B cells appear to have little or no
NF-κB, this activity appears to be inducible by
treatment of the cells with LPS. Again, presence of
this factor correlates strongly with κ gene expression,
because 70Z/3 is transcriptionally inactive at the κ
locus prior to induction but accumulates fairly high
levels of κ mRNA even within 4 hrs post-induction
(17,18).

DISCUSSION

    We have detected interaction of multiple factors
with the Ig μ and κ enhancers (summarized in Table 1).
    At this stage our analysis has not directly
addressed the question of function of these factors;
however some insights may be derived from the data
presented. Enhancers are defined partly by their
ability to function in either orientation relative to
a promoter. As shown in Fig. 6, all the E series of
binding sites (first suggested by Ephrussi et al. (7)
to be homologous) contain an element of dyad symmetry
defined by the dinucleotide CA/TG separated by either 2
or 3 base pairs which may be part of the explanation
for orientation independence. The 2 sequences that
appear to bind the same factor (μE3 and κE3) have an
identical 2 base pair spacer of TG between their
symmetric residues. By contrast, sites μE1 or μE4 that
do not compete for binding to μE3, either have a 3 bp
spacer or a slightly different (GG) dinucleotide

Table 1

Factor	Binding Sites(s)	Tissue Distribution
1) NF-A	octamer sequence (ATTTGCAT) in V(H), V(L) promoters and μ enhancer	ubiquitous (B cell specific component)
2) NF-μE3	E3 site in μ and κ enhancer	ubiquitous
3) NF-μE1	E1 site in μ enhancer	ubiquitous
4) NF-κB	B site in κ enhancer	κ-producing B cells only

E1, E2, E3 etc., refers to the E homology identified by Ephrussi et al. (7). NF-μE1 has been identified by Weinberger et al. (9).

spacer. The inability of the closely homologous sequence μE4 to compete for μE3 binding implies that the factors may have almost a restriction enzyme-like specificity. Perhaps some of the enhancer binding proteins belong to families of related proteins with slightly differing binding specificites.

Secondly, there is an apparent discrepancy between the in vivo binding data (7,8) and the in vitro data we have presented. The protections observed by Ephrussi et al. (7) and Church et al. (8) on the immunoglobulin heavy chain enhancer occurred in a B cell specific manner and were absent in fibroblasts. We find that all the proteins interacting with the μ enhancer in vitro (i.e. NF-μE1, NF-μE3 and NF-A1) are not restricted to expression in B cells only. This implies that the mere presence of a factor in a cell is not enough to have it bound to its recognition sequences in vivo. Perhaps some aspect of the chromatin structure is "activated" to make some genes accessible to binding by ubiquitous factors. It is possible that tissue-specific binding proteins may play the role of activation and therefore 'open' the DNA to allow interaction with non-specific transcription enhancing

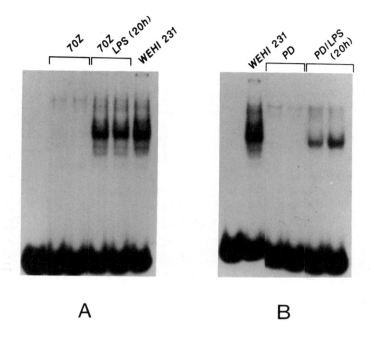

Figure 5: Inducibility of NF-κB in pre-B cells
          stimulated with lipopolysaccharide.
     A. κ3 fragment was end-labelled and used for
binding assays in extracts derived from unstimulated
7OZ cells (lanes 2,3) or from 7OZ cells that had been
treated for 20 hr with lipopolysaccharide (LPS) at
10-15 μg/ml (lanes 4,5). Lane 6: characteristic
nucleoprotein complex generated using this fragment and
extracts derived from B cells (WEHI 231).
     B. NF-κB induction in the Abelson murine leukemia
virus transformed pre B cell PD by LPS. Lane 2:
complex generated in WEHI 231 extracts; Lanes 3,4:
complex generated in unstimulated PD cell extracts;
Lanes 5,6: complex generated in 20 hr LPS stimulated PD
cell extracts.

proteins. In vitro, with naked DNA, such higher level
interactions would not be evident and the factors would
all appear to be equivalent DNA binding proteins.

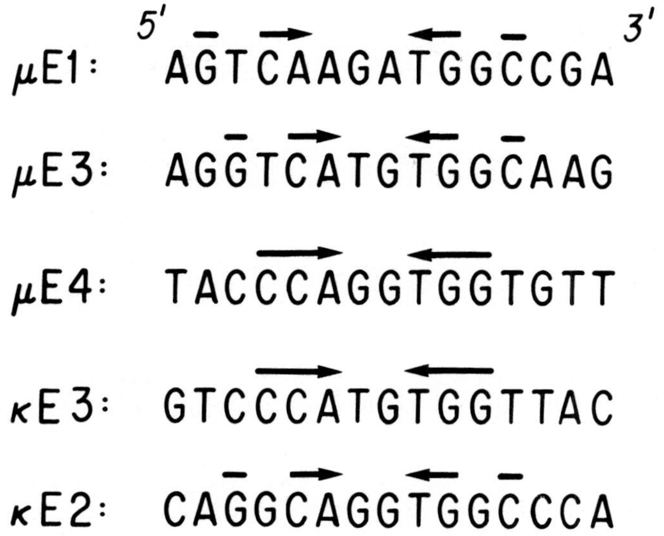

μE1:   5' $\overline{\text{AGT}}$ $\overrightarrow{\text{CAAGAT}}$ $\overleftarrow{\text{GGCC}}$ $\overline{\text{GA}}$ 3'

μE3: $\overline{\text{AG}}$ $\overrightarrow{\text{GTCATG}}$ $\overleftarrow{\text{TGGC}}$ $\overline{\text{AAG}}$

μE4:   TAC $\overrightarrow{\text{CCAGGT}}$ $\overleftarrow{\text{GGTG}}$ TT

κE3:   GTC $\overrightarrow{\text{CCATGT}}$ $\overleftarrow{\text{GGTT}}$ AC

κE2: $\overline{\text{CA}}$ $\overrightarrow{\text{GGCAGG}}$ $\overleftarrow{\text{TGGC}}$ $\overline{\text{CCA}}$

Figure 6: Comparison of 'E' domains from the μ enhancer and the κ enhancer. μE1-μE4 were defined by Ephrussi and Church in the μ enhancer on the basis of methylation protection experiments in vivo (7). κE1-κE3 were identified within the κ enhancer as being homologous to the consensus sequence derived by comparing μE1-μE4 (8).   The arrows over the sequence point out a mini dyad axis of symmetry within each domain.   In the fragments used by us to dissect the enhancers μE3 is completely present within the μ70 fragment, μE4 within the μ50 and μ170 fragments and κE3 within the κ2 fragment.

Finally, examination of the results of deletion analysis of the κ enhancer suggests that the B cell-specific site by itself is not sufficient for enhancer function.   Thus the 5' Alu1-HaeIII fragment of the enhancer (which contains fragments κ1 and κ3) is insufficient for function (19) in spite of containing the complete B binding site.

ACKNOWLEDGEMENTS

We would like to thank David Weaver, Lou Staudt and Nancy Speck for their help in generating the "library" of extracts described, Judah Weinberger and Phillip Sharp for communicating their results prior to publication and Ginger Pierce for quick and efficient typing of the manuscript. R.S. was a recipient of a fellowship from the Damon Runyon-Walter Winchell Cancer Fund. This work was supported by a grant from the American Cancer Society awarded to D.B.

REFERENCES

1. Calame KL, (1985). Mechanisms that regulate immun-oglobulin gene expression. Ann Rev Immunol 3:159.
2. Grosschedl R and Baltimore D (1985). Cell type specificity of immunoglobulin gene expression is regulated by at least 3 DNA sequence elements. Cell 41:885.
3. Sen R and Baltimore D (1986). Multiple nuclear factors interact with the immunoglobulin enhancer sequences. Cell, in press.
4. Garner MM and Revzin A (1981). A gel electrophoresis method for quantifying the binding of proteins to specific DNA regions: application to components of the E. coli lactose operon regulatory system. Nuc Acid Res 9:3047.
5. Fried MG and Crothers DM (1981). Equilibria and kinetics of lac repressor operator interactions by polyacrylamide gel electrophoresis. Nuc Acid Res 9:6505.
6. Singh H, Sen R, Baltimore D and Sharp PA (1986). A nuclear factor that binds a conserved sequence motif in transcriptional control elements of immunoglobulin genes. Nature 319:154.
7. Ephrussi A, Church GM, Tonegawa S and Gilbert W (1985). B lineage-specific interactions of an immunoglobulin enhancer with cellular factors in vivo. Science 227:134.
8. Church GM, Ephrussi A, Gilbert W and Tonegawa S (1985). Cell type specific contacts to immunoglobulin enhancer in nucleii. Nature 313:798.

9. Weinberger J, Baltimore D and Sharp PA (1986). Distinct factors bind to apparently homologous sequences in the immunoglobulin heavy chain enhancer. Nature, in press.

10. Staudt L, Singh H, Sen R, Wirth T, Sharp PA and Baltimore D (1986). A lymphoid-specific protein binding to the octamer motif of immunoglobulin genes. Nature, in press.

11. Queen C and Stafford J (1984). Fine mapping of an immunoglobulin gene activator. Mol Cell Biol 4:1042.

12. Rosenberg N and Baltimore D (1976). A quantitative assay for transformation of bone marrow cells by Abelson murine leukemia virus. J Exp Med 143:1453.

13. Lewis S, Rosenberg N, Alt FA and Baltimore D (1982). Continuing kappa gene rearrangement in a cell line transformed by Abelson murine leukemia virus. Cell 30:807.

14. Maki R, Kearney J, Paige CJ and Tonegawa S (1980). Immunoglobulin gene rearrangement in immature B cells. Science 209:1366.

15. Paige C, Kincade R and Ralph P (1978). Murine B cell leukemia line with inducible surface immunoglobulin expression. J Immunol 121:641.

16. Parslow TG and Granner DK (1982). Chromatin changes accompany immunoglobulin kappa gene activation: a potential control region within the gene. Nature 299:449.

17. Nelson KJ, Kelley DE and Perry RP (1985). Inducible transcription of the unrearranged κ constant region locus is a common feature of pre-B cells and does not require DNA or protein synthesis. Proc Nat Acad Sci USA 82:5305.

18. Wall R, Briskin M, Carter C, Govan H, Taylor A and Kincade P (1986). A labile inhibitor blocks immunoglobulin κ-light-chain-gene transcription in a pre-B leukemic cell line. Proc Nat Acad Sci USA 83:295.

19. Picard D and Schaffner W (1984). A lymphocyte specific enhancer in the mouse immunoglobulin kappa gene. Nature 307:80.

20. Pierce JH and Aaronson SA (1982). BALB and HARVEY murine sarcoma virus transformation of a novel lymphoid progenitor cell. J Exp Med 156:873.

Transcriptional Control Mechanisms, pages 103–112
© 1987 Alan R. Liss, Inc.

MITOCHONDRIAL TRANSCRIPTION FACTOR BINDS NOVEL CONTROL ELEMENTS OF BOTH MAJOR PROMOTERS OF HUMAN mtDNA[1]

Robert P. Fisher, James E. Hixson[2], and David A. Clayton

Department of Pathology, Stanford University School of Medicine, Stanford, California 94305

ABSTRACT   A mitochondrial transcription factor (mtTF), required for selective transcription initiation at both major promoters of human mtDNA (HSP and LSP), has been partially purified and its DNA binding specificity studied using DNase I footprint analysis. This factor binds strongly between 13 and 35 bp upstream of the LSP transcriptional start site, and weakly between 13 and 35 bp upstream of the HSP start site. Footprint analyses of mutant LSP templates containing clustered point substitutions indicate that mtTF binding is necessary, but not sufficient, for normal transcriptional selectivity. A comparison between the sequences protected at the two promoters reveals significant homology only when the orientations, relative to the major directions of transcription, are opposite, suggesting that mtTF can function (i.e., activate transcription) in either orientation.

[1]This work was supported by grant GM-33088-15 from the National Institute of General Medical Sciences. R.P.F. is a Medical Scientist Training Program trainee of the National Institute of General Medical Sciences (GM-07365-10) and J.E.H. was a Postdoctoral Fellow of the American Cancer Society, Inc. (PF-2288).

[2]Present address: Genetics Department, Southwest Foundation for Biomedical Research, San Antonio, TX 78284.

INTRODUCTION

The displacement loop (D-loop) region of the mammalian mitochondrial genome has evolved as a specialized control region for both transcription and replication. Biochemical analyses of human mitochondrial transcription, therefore, have been aimed at understanding the nature and complexity of protein-DNA interactions within this unique regulatory domain.

Extensive mapping of both in vivo and in vitro transcripts (1-4), together with fractionation of transcriptional extracts, form the basis for a model of transcription initiation illustrated in Figure 1. Each strand of the circular genome (heavy and light, or H- and L-strands) is transcribed from a single major promoter (HSP and LSP, respectively) situated in the D-loop region. Approximately 150 base pairs (bp) separate the start sites for H- and L-strand transcription, but deletion mapping has demonstrated that the promoters lie in close proximity to their respective start sites, and thus do not overlap, either spatially or functionally. As few as 16 bp upstream and seven bp downstream of the HSP start site are required for accurate initiation, but efficient promoter selection requires sequences further upstream; a similar situation seems to obtain at the LSP (3). In vitro transcriptional analyses of templates bearing point substitutions within the HSP and the LSP confirmed the importance of a weak consensus sequence encompassing the transcriptional start sites, but also provided additional evidence of upstream sequence requirements (5). Finally, both promoters appear to function "bidirectionally;" low levels of transcription initiate on the opposite strand at about the same positions, but at approximately 2% of the efficiency of transcription in the major direction. These minor "promoters"-- hsp and lsp-- respond to deletion and substitution much as the major promoters-- LSP and HSP, respectively-- do (4).

The transcriptional machinery of human mitochondria has been dissected into (at least) two components: an intrinsically nonselective (or weakly selective) RNA polymerase and a transcription factor (mtTF) that confers selectivity for both the HSP and LSP (6). Additional factor requirements have yet to be identified. In this report, we attempt to integrate our knowledge of the DNA sequences required for transcription with this simple picture of the transcriptional apparatus; we show that the

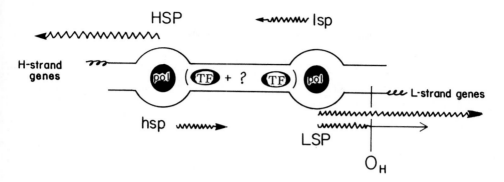

FIGURE 1. Transcriptional control in the D-loop region of human mtDNA. Sequence elements and proteins involved in transcription initiation events are represented schematically. HSP and LSP are the major promoters; lsp and hsp are their respective associated minor "promoters" for transcription of the opposite strand. A potential switch to H-strand DNA synthesis downstream of the LSP, at O_H, is indicated by the vertical line drawn through the L-strand transcript (see ref. 2). The location of transcription factor (TF) binding (this report) is indicated, but the recognition sequence (if any) for RNA polymerase (pol) has not been identified. We have placed it near the start sites to reflect the fact that specific nucleotides in this region are required for accurate initiation (see text and ref. 5).

mitochondrial transcription factor (mtTF) interacts specifically with novel control elements of the major mitochondrial promoters.

RESULTS

When an extract of human KB cell mitochondria, capable of selective initiation at both the HSP and the LSP, is chromatographed on phosphocellulose, a nonspecific RNA polymerase activity elutes at ~.45M KCl, while mtTF is retained on the column up to ~.65M KCl. Transcriptional selectivity can be reconstituted <u>in vitro</u> by the simultaneous addition of mtTF and mtRNA polymerase to a template containing either the HSP or the LSP (or both)

(6). That mtTF functions by binding to DNA is demonstrated by template competition experiments in which mtTF, when preincubated with template in the absence of polymerase, is able to sequester the DNA in a preinitiation complex (R.P. Fisher, J.N. Topper and D.A. Clayton, manuscript in preparation).

Figure 2A shows the DNase I "footprint" obtained when increasing amounts of phosphocellulose-purified mtTF are preincubated with an end-labelled DNA fragment, containing the LSP, prior to nucleolytic cleavage. A strongly protected region extends from position -13 to -35 with respect to the transcriptional start site. Deletion mapping studies have previously suggested that sequences between -28 and -56 were required for efficient promoter selection; deletion to -28 severely reduced transcription levels (3). Point substitutions have been introduced into the LSP sequence by bisulfite treatment, and the mutant templates assayed for efficiency of initiation in an unfractionated mitochondrial transcription extract (5). Transitions (C to T on noncoding strand) at positions -2, -5 and -10 combined to abolish selective transcription entirely, but footprinting of this mutant gave an essentially wild-type protection pattern (data not shown). This result implies that TF-binding is not sufficient to ensure proper promoter recognition. Since determinants of specific initiation lie outside of the region recognized by mtTF, and are not essential to that recognition, it seems likely that another component of the transcription machinery-- perhaps the RNA polymerase-- possesses promoter-specificity.

To ascertain whether mtTF binding to the -13 to -35 region was necessary for wild-type selectivity, we performed footprint analysis on a series of templates bearing G-to-A transitions on the noncoding strand of the LSP, as shown in Figure 3. These mutants were likely candidates to exhibit defective TF-binding for two reasons: they have changes located predominantly (but not exclusively) in the upstream region, with some substitutions within the footprinted sequence; and several of these templates are "down" mutants which support initiation at significantly reduced levels (20-40% of wild type). For each mutant analyzed in Figure 3, a rough estimate of transcriptional efficiency (relative to wild type, defined as 1.0) was obtained by liquid scintillation counting of excised gel bands containing full-length runoff transcripts (data not shown). These estimates are

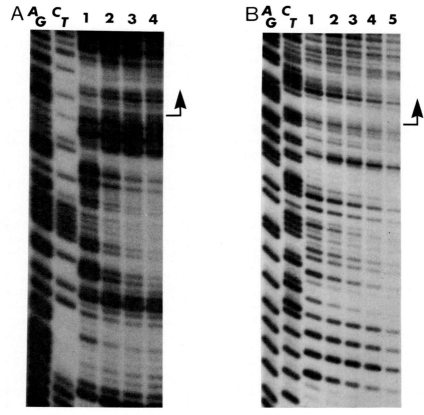

FIGURE 2.   Footprinting of the major promoters of
mtDNA with mtTF.   (A) A restriction fragment containing a
wild-type LSP sequence was uniquely end-labelled with ^{32}P
and incubated with increasing amounts of
phosphocellulose-purified mtTF prior to DNase I digestion
(1 μg/ml for 30 s at room temperature). Lanes: 1, no
mtTF added; 2, 1 μl mtTF; 3, 2 μl; 4, 4 μl; AG
and G, purine- and guanine-specific chemical cleavage
reactions, respectively. The transcriptional start site is
indicated by the arrow at right. (B) Same as (A) except
that labelled fragment contains the wild-type HSP sequence,
and total DNA concentration was one-fourth that in (A).
Lanes: 1, no mtTF added; 2, 1 μl mtTF; 3, 2 μl; 4,
4 μl; 5, 8 μl; AG and CT, purine- and
pyrimidine-specific chemical cleavage reactions,
respectively.

indicated below the corresponding panels of Figure 3. The positions mutated in each clone are indicated by asterisks in the G-specific chemical sequence ladders electrophoresed alongside the footprinting reactions. Mutant template 9 (panel B) has five transitions upstream of the LSP start site, but is transcribed with the same efficiency as is wild type. As expected, footprinting of this mutant appears essentially normal. Mutants 13 and 11, assayed in panels C and D, respectively, show partial defects in footprinting, which may explain their 60-70% reductions in transcription levels. However, both of these clones have a G-to-A transition four bp upstream of the start site which may contribute to the transcriptional defect. Mutant 10, on the other hand, has a wild-type sequence downstream of position -13, yet displays a severe loss of transcriptional capacity. It has seven substitutions-- six within the footprinted region-- and shows virtually no footprinting at mtTF concentrations sufficient to protect completely the unaltered site.

The data presented in Figure 3 suggest a degree of flexibility in the specificity of the mtTF-DNA interaction. Mutant templates containing two to four substitutions in the footprinted region give nearly normal protection patterns (panels B through D). Only when six positions (out of approximately 22 bp in the footprinted region) are altered is a dramatic defect in binding specificity seen (clone 10, panel E). An even more striking example of this flexibility is seen when footprint analysis is carried out with phosphocellulose-purified mtTF and labelled fragments containing the HSP (Figure 2B). A weakly protected region extends from approximately 13 to 35 bp upstream of the transcriptional start site. At the higher TF-to-DNA ratios needed to footprint the HSP, a significant amount of nonspecific "protection" is seen, making mutant analysis difficult; however, several similarities between "wild-type" LSP and HSP footprints are evident, and support the notion that they represent functionally analogous interactions. Most strikingly, the footprinted regions of the two promoters are positioned identically with respect to their transcriptional start sites. Also, in both cases, a cluster of strongly enhanced cleavages appears at the downstream boundary of the footprint.

However, when the noncoding strand sequences of the two promoters are compared (Figure 4A), no sequence homology is apparent in the footprinted region. The

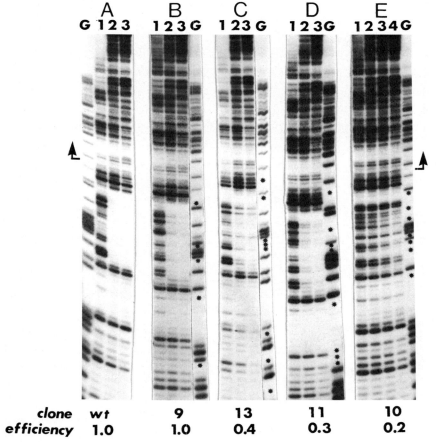

clone	wt	9	13	11	10
efficiency	1.0	1.0	0.4	0.3	0.2

FIGURE 3. Footprint analyses of LSP point substitution mutants. Each panel shows a different mutant clone assayed for footprinting ability as in Figure 2A. An estimate of transcriptional efficiency is presented for each template beneath the appropriate panel, expressed as a fraction of the level seen with a wild-type template (defined as 1.0; a wild-type footprint is shown in panel A). Lanes (panels A through D): 1, no mtTF added; 2, 2 μl mtTF; 3, 4 μl. In panel E only, lanes: 1, no mtTF; 2, 1 μl mtTF; 3, 2 μl; 4, 4 μl. Asterisks in G-specific sequence ladders (lanes G) indicate, for each mutant, the positions mutated to A, while arrows at left and right denote the transcriptional start sites and the direction of major transcription.

A

```
        440                              420                      ┌─────►
A A T G T G T T│A G T T G G G G G G T G A C T G T T A A A A G│T G C A T A C C G C C A A A A G A T A
C A C A C A C C G│C T G C T A A C C C C A T A C C C C G A A C│C A A C C A A A C C C C A A A G A C A
            530                              550                      └─────►
```

B

```
5'GTTAGTTGGGGGGGTGACTGTTAAAAGTG 3'  LSP noncoding

   │││  │││  ││││        ││││   ││  │

5'GTTGGTTCGGGGTATGGGGTTAGCAGCG 3'   HSP coding
```

FIGURE 4. Sequence homology of mtTF binding sites. (A) shows the noncoding strand sequences of the LSP (top) and HSP (bottom) upstream regions, aligned at the transcription start sites (arrows). The protected bases in the noncoding strand footprints are boxed. In (B) the complementary (coding strand) sequence of the HSP (bottom) has been compared to the noncoding strand of the LSP (top). The sequence alignment maximizes matching, but is also identical to the alignment obtained when the footprinted region is precisely inverted (if the one base "overhang" at the upstream boundary of the LSP footprint is ignored). The LSP sequence presented in B has been extended to include 5'-GTT-3', even though this sequence is unprotected, because its complement-- 5'-AAC-3'-- is protected in coding strand footprints (data not shown).

sequences have been aligned by the major transcriptional start sites (arrows), and the footprinted regions on both noncoding strands are boxed. In Figure 4B, the coding strand sequence of the HSP footprint region is compared with the noncoding LSP sequence. A limited but striking homology is apparent, consisting in two conserved "domains"-- one with a 10/12 match, and the other a 7/10 match-- flanking a central nonconserved hexanucleotide. It seems likely that these conserved elements form the essential requirements for sequence-specific mtTF-binding, while the nonconserved bases may play a role in determining the relative affinities of the factor for the two sites. Most interesting, though, is the implication that the transcription factor-- presumably oriented within its binding site by the orientation of its recognition sequence-- can activate transcription in either direction.

DISCUSSION

In this report, we have focused on a single protein-DNA interaction, placing it on the pathway leading to selective transcription initiation in human mitochondria. We have identified novel regulatory elements-- transcription factor binding sequences-- present in both major promoters of human mtDNA; these signals are positioned analogously with respect to the transcriptional start sites, but are in opposite orientations, suggesting that directionality of major transcription may be determined by other factors. Moreover, mtTF can apparently activate transcription even when it binds the target sequence relatively weakly (e.g., at the HSP), albeit at lower levels than at the wild-type LSP. The involvement of additional, perhaps HSP-specific factors, in enhancing this level cannot be ruled out at present.

Characterization of the mtTF-DNA interaction has sharpened the picture of transcription initiation presented in Figure 1. However, it has also served to re-emphasize areas of uncertainty in this model. Since TF-binding is confined to the upstream region-- not actually required for initiaton accuracy-- any function of the start site consensus sequence (3) remains unknown. Another critical question, raised above, is left unanswered: how is a >95% preponderance of transcription in one direction maintained, when mtTF itself appears to function bidirectionally? An attractive hypothesis-- one that would not require an additional factor or sequence element for promoter recognition-- is that the start site consensus sequence binds and orients RNA polymerase molecules so that they initiate transcription preferentially on one strand. However, attempts to demonstrate any sequence-specific binding by partially purified mtRNA polymerase have been unsuccessful. The observed low levels of transcription on the "other" strand may reflect leakiness in this mechanism. Alternatively, mtTF, bound upstream of the start site, may represent a block to polymerase elongation in that direction.

In any case, these data indicate a greater complexity for mitochondrial promoters than was previously recognized. The proper juxtaposition of separate promoter elements is clearly required to constitute a functional promoter that is responsive to mtTF and has a distinct polarity. Promoter elements that can function bidirectionally have been described previously, most recently by McKnight and coworkers (7). The relatively simple mitochondrial system may afford unique insight into how such elements function.

REFERENCES

1.   Yoza BK, Bogenhagen DF (1984). Identification and in vitro capping of a primary transcript of human mitochondrial DNA. J Biol Chem 259:3909.

2.   Chang DD, Clayton DA (1985). Priming of human mitochondrial DNA replication occurs at the light-strand promoter. Proc Natl Acad Sci USA 82:351.

3.   Chang DD, Clayton DA (1984). Precise identification of individual promoters for transcription of each strand of human mitochondrial DNA. Cell 36:351.

4.   Chang DD, Hixson JE, Clayton DA (1986). Minor transcription events indicate that both human mitochondrial promoters function bidirectionally. Mol Cell Biol 6:294.

5.   Hixson JE, Clayton DA (1985). Initiation of transcription from each of the two human mitochondrial promoters requires unique nucleotides at the transcriptional start sites. Proc Natl Acad Sci USA 82:2660.

6.   Fisher RP, Clayton DA (1985). A transcription factor required for promoter recognition by human mitochondrial RNA polymerase. J Biol Chem 260:11330.

7.   Graves BJ, Johnson PF, McKnight SL (1986). Homologous recognition of a promoter domain common to the MSV LTR and the HSV tk gene. Cell 44:565.

Transcriptional Control Mechanisms, pages 113–122
© 1987 Alan R. Liss, Inc.

THE ACTION OF THE SEQUENCES AND FACTORS THAT DIRECT RIBOSOMAL RNA TRANSCRIPTION AND PROCESSING OF THE PRIMARY rRNA TRANSCRIPT[1]

Barbara Sollner-Webb, Jolene Windle, Sheryl Henderson, John Tower, Val Culotta, Susan Kass, and Nessly Craig[2]

Department of Biological Chemistry, The Johns Hopkins University School of Medicine, Baltimore, Maryland 21205

ABSTRACT We report that two factors are necessary and sufficient to catalyze rDNA transcription. One is a promoter-specific, rDNA-binding protein and the other is a subform of RNA polymerase I that is specifically activated to participate in accurate rRNA initiation. Many, but not all, instances of rDNA transcriptional regulation appear to be mediated by an altered availability of this activated polymerase. The promoters that direct rDNA transcription are large complex control regions, the various domains of which are best seen under stringent reaction conditions. Recent experiments demonstrate that an upstream domain of the Xenopus promoter directs transcriptional initiation in a stereospecific manner, while the large central segment of the Xenopus promoter (the enhancer cognate region) may be deleted with only minimal adverse effects. Moreover, we show that a novel upstream domain of the mouse rDNA promoter that is essential in vivo acts to provide a transcriptional terminator function. Finally, the primary mouse rRNA processing event that we described earlier is directed by the rRNA sequences that reside within an ~150 bp region 3' to the processing site. The primary human

— — — — — — —

[1] This work was supported by grants from the National Institutes of Health and the March of Dimes.
[2] Department of Biological Science, The University of Maryland Baltimore County, Catonsville, Maryland.

rRNA transcript is similarly processed and the human sequences also direct faithful processing by the mouse factors, indicating that this rRNA processing is well-conserved across mammalian species.[3]

THE rDNA TRANSCRIPTION FACTORS

We have approached the question of how rDNA transcription is accomplished and regulated by studying the cellular factors as well as the nucleotide sequences that are involved. The protein components necessary for mouse rDNA transcription have been resolved and substantially purified by chromatographic fractionation of a mouse S-100 cell extract, and two factors appear necessary and sufficient to catalyze transcription (2). One component (factor 'D'), which has been enriched ~10,000-fold by DNA cellulose and Heparin or phosphocellulose chromatography, binds stably and specifically to the 5' region of the mouse rDNA core promoter (residues ~—35 to ~—15; Fig. 1A, B). It is to this D/DNA complex that the second rDNA transcription component (factor 'C') can then bind.

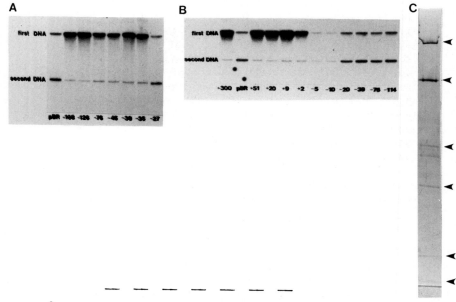

3   For a more complete review of the rDNA transcription field through Fall, 1985, the reader is referred to (1).

Factor C is a form of RNA polymerase I that is specifically activated to participate in accurate rDNA transcription. It has been purified by chromatography on DNA cellulose and gradient elution from DEAE and Heparin resins followed by glycerol gradient sedimentation; this results in a preparation that is ~90% pure in the polymerase I polypeptides (Fig. 1C). Indeed, through all the chromatographic fractionations that we have examined, this transcription factor and non-specific polymerase I activity co-purify. The simplest explanation would be that factor C is simply bulk polymerase I, but this is not correct, for these two activities can be substantially resolved using a fractionation in which the transcription complexes are sedimented. The resultant pelleted fraction contains >90% of this transcription factor but only <10% of bulk polymerase I activity, while the supernatant contains the converse (2). We therefore conclude that this second mammalian rDNA transcription factor is a polymerase I subform that is in some manner modified to allow specific rDNA transcriptional initiation in combination with factor D (Tower and Sollner-Webb, in preparation).[4]

Many instances of in vivo rDNA transcriptional regulation are mirrored by the activity of extracts made from the treated cells (1, 4-7). For instance, extracts from both cyclohexamide treated cells and stationary phase cells have markedly reduced transcriptional capacity. This inactivity is due to their lack of the activated polymerase subclass; they retain normal levels of factor D and bulk polymerase I (6). However, this regulation is not the ubiquitous method of rDNA transcriptional control, for in at least one other instance the decreased transcriptional activity is due to reduced availability of factor D (Tower and Sollner-Webb, in preparation).

THE rDNA PROMOTER

Our earlier work has shown that the sequences that

---

[4] While this conclusion is in contrast to a recent report (3), we find their data can be fully explained by fractionation-induced activation of inhibitory components. On the other hand, our conclusion is fully consistent with results on Acanthamoeba rDNA transcription factors (4).

promote transcription of the rRNA gene extend over a large
region of ~170 bp (8-10). This promoter region contains a
small core domain that overlaps the initiation site and is
essential for efficient transcription and several adjacent
domains in the 5' flanking region that greatly augment the
initiation process. These upstream domains are not
detected under the most efficient transcription conditions,
but become increasingly crucial as reaction conditions are
made more stringent by any of a number of alterations.
This is perhaps best illustrated by analyzing the
transcriptional efficiency of 5' and 3' deletions of mouse
rDNA under systematically altered in vitro reaction
conditions (9). As the transcription reactions are made
more stringent, the 5' border of the apparent promoter
region is seen to move progressively upstream from
residue -35 to -39, -45, -100, and to -140, while the
apparent 3' border of the promoter region remains at
residue ~+5.

A similar effect is observed with the Xenopus rDNA
promoter. When transcribed in injected oocytes under
conditions that yield maximal expression, only a small
core domain (residue ~-7 to +6) is detected (8), while
under conditions that yield lower transcription levels, the
requirement for a promoter domain that extends upstream to
residue ~-150 is apparent (10). This effect is a concerted
transition: by merely lowering the number of injected rDNA
promoters approximately two fold, there is a dramatic
change in the apparent 5' border of the promoter. At ≥ 2
pmols of injected promoter per oocyte a 5'Δ-7 mutant
displays maximal promoter activity, while at ≤ 1 pmol of
injected promoter per oocyte, 5' deletions that extend to
or beyond residue ~-150 are virtually inactive (Fig. 2).

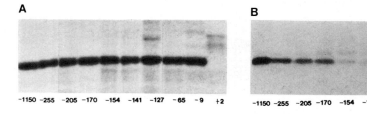

A

-1150 -255 -205 -170 -154 -141 -127 -65 -9 +2

B

-1150 -255 -205 -170 -154 -141 -127 -65 -9

FIGURE 2.   The Xenopus rDNA promoter boundaries. The
indicated 5' deletions of Xenopus rDNA were transcribed
upon injection into oocytes at 200 μg/ml (A) or 100 μg/ml
(B) plasmid concentration.

To define the internal organization of the Xenopus rDNA promoter, we have constructed 'linker scanner' mutants across this region. When linker scanners that do not alter spacing are injected at low concentration into Xenopus oocytes, the crucial domains are seen to be the most upstream (residue ~−140 to −128) and downstream (residue ~−36 to +10) segments of the promoter (Fig. 3A; Windle and Sollner-Webb, in preparation). Similar results are obtained when the transcription is conducted in vitro, using either Xenopus or mouse cell extracts (data not shown).

By transcribing these mutants and other linker scanner mutants that have controlled spacing changes in the mouse cell extract, one obtains the striking result that the site of initiation is determined by sequences ~130 base pairs upstream of this position. All linker mutants between residues −120 and −28 that change the spacing by ~−4 or ~+4 residues (+ or − approximately half a helix turn) cause the initiation site to move downstream by 4 residues (Fig. 3B).

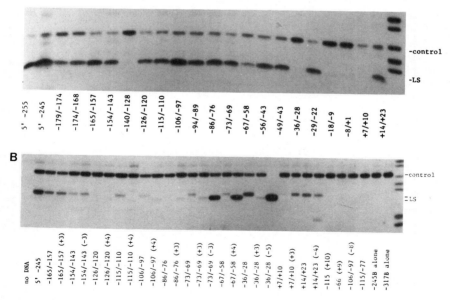

FIGURE 3.   Internal organization of the Xenopus promoter. The indicated linker-scanner mutants were transcribed in injected oocytes (A) or in the mouse cell extract (B) along with a control gene. The spacing alterations are shown in parentheses.

Spacing changes of ~10 residues markedly reduce the level
of transcription but do not change the start site, while
alterations outside of the promoter region have no effect.
We conclude that sequences between ~−36 and ~+10 are
important for initiation and specify potential start sites,
while sequences between −140 and −120 specify on which face
of the helix transcription will initiate and determine the
preferred number of helix turns to the initiation point
(Windle and Sollner-Webb, in preparation). In light of the
crucial role of the −140 to −120 region of the Xenopus rDNA
it is, perhaps, not surprising that this sequence is highly
conserved (13 out of 17) in a similar region of the mouse
rDNA promoter.

THE XENOPUS rDNA ENHANCER

As can be seen in Fig. 2B lane 1, sequences yet
further upstream in the Xenopus rDNA also act to augment
the level of transcription in the injected oocyte. This is
an effect of the rDNA enhancer (11), a segment of the rDNA
spacer that contains multiple duplications of the promoter
region from residue ~−114 to ~−72. While enhancer action
has previously been demonstrated only under co-injection
(competition) conditions, where it provides a relatively
position and orientation independent competitive advantage
to an adjacent gene (12, 13), these sequences also elevate
the level of transcription from genes injected singly. To
begin to describe how the rDNA enhancer acts, we first
showed that its effect is polymerase-specific: when cloned
adjacent to a polymerase II or polymerase III promoter, the
rDNA enhancer has no detectable effect on the transcription
of that gene (data not shown). Two lines of evidence
indicate that the pol I enhancer actually acts in
conjunction with the upstream domain of the rDNA promoter
(Windle and Sollner-Webb, in preparation). First, the
action of the rDNA enhancer is only observed at those rDNA
concentrations where the upstream domain of the promoter is
detected (Fig. 2). Second, when the rDNA enhancer segment
is cloned adjacent to various of the 5' deletion mutants,
it augments 5'Δ−154 (and less extensive deletions which
have complete promoters), but it has a much smaller effect
when cloned adjacent to 5'Δ−127 or more extensive promoter
deletions (Fig. 4). It is particularly striking that
5'Δ−127, which lacks the far upstream promoter domain but
retains the entire enhancer cognate sequence, appears

largely unenhanceable.

An obvious model to explain rDNA enhancer action is that a transcription component binds to the enhancer-cognate sequence within the promoter (residues -114 to -72) and that the additional presence of this duplicated sequence in the upstream spacer enables more rapid or higher level of binding of this important factor (11, 13). In light of this, we constructed a mutant promoter in which virtually the entire enhancer-cognate region (residue -115 to -77) is replaced by an equal length segment of bacterial DNA. Surprisingly, in injected oocytes this promoter initiates transcription at control levels and it competes for essential transcription factors as effectively as does a complete promoter, indicating that this sequence is not crucial for promoter function. We anticipate that further study of this mutant, in particular whether or not it is enhanceable, should result in a better understanding of how the rDNA enhancer sequence exerts its effect.

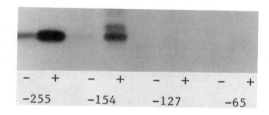

FIGURE 4. Promoter requirement for enhancement.   The indicated Xenopus 5' deletions, cloned either in the absence (-) or presence (+) of the rDNA enhancer, were transcribed in injected oocytes.

A TRANSCRIPTIONAL TERMINATOR IS A PROMOTER ELEMENT

Recent studies with the mouse rRNA gene have shown there to be another, novel promoter domain. When the transcriptional capacity of 5' deletion mutants is assessed in transiently transfected rodent cells, the results shown in Fig. 5A are obtained. 5' deletions extending to residue -168 initiate efficiently, while deletions to or beyond residue -163 exhibit only minimal levels of expression. A marked effect of the ~-168 region is also observed in vitro when closed circular templates are transcribed in the mouse cell extract (Fig. 5B). In this instance, as the transcription falls off in the more extensive deletions, the amount of RNA that reads into the promoter region from upstream increases markedly. This read-in RNA, detected as an S1 divergence band ('div.') in Fig. 5B, results from normally initiated rDNA transcripts that have read completely around the plasmid, for it is virtually absent in the initiation deficient 5'Δ-5 mutant. This suggests that the ~-168 region acts as a transcriptional terminator (14), and that the amount of read-in RNA relative to the amount of +1 RNA provides a measure of the efficiency of termination (Henderson and Sollner-Webb, in preparation).

Formation of 3' RNA ends by this upstream promoter domain can also be detected by other analyses. S1 mapping using a 3' end-labeled probe reveals spacer transcripts that end at residue ~-185. Moreover, when this region is positioned downstream of the rDNA promoter and the closed circular plasmid is used as an in vitro transcription template, a discrete transcript is obtained which extends from the initiation site to residue ~-185 of the downstream termination region (Fig. 5C). It should be noted that the sequence of this -168 region exhibits a striking 13 out of 15 homology with a 15 base pair repeated sequence ~600 bp 3' of the 28S coding region that has been reported to have terminator function (15). While the in vivo role of the terminator within the rDNA promoter may be to eliminate transcripts that initiate fortuitously within the 'nontranscribed spacer', another possibility should also be considered. This terminator may be part of an elegant system to deliver polymerase to the rDNA promoter and thereby converts what would be a three dimensional search by polymerase for a promoter into a directed one dimensional process.

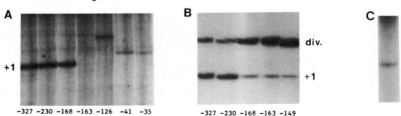

FIGURE 5. **The terminator domain of the mouse rDNA promoter.** The indicated 5' deletions were transcribed upon transient transfection **(A)** or in vitro **(B)**. In **(C)** the terminator region (-230 to -39) was cloned downstream from an rDNA promoter and causes a formation of a discrete length transcript.

THE PRIMARY rRNA PROCESSING EVENT

Once the rRNA transcript is synthesized, it is processed to yield the mature rRNA species. We earlier demonstrated that the primary mouse rRNA transcript is rapidly and efficiently cleaved at residue +650, both in vivo and in vitro (16). More recently, we have found that an analogous processing takes place in human rRNA (Kass, Craig, and Sollner-Webb, in preparation). Comparison of the mouse and human rDNA reveals an ~80% sequence identity in the 200 base pairs 3' to the processing site but no significant homology in the surrounding regions. To directly determine which sequences specify the processing, 5' and 3' deletion mutants of this mouse rDNA region have been constructed and used as in vitro transcription templates. The 5' border of the sequences required for processing maps precisely at the cleavage site, which is also the 5' border of the conserved region (Fig. 6A). The 3' border of the regions is less discrete; the processing efficiency gradually falls off as increasing regions of the mouse-human conserved sequences are deleted. While the proximal ~40 nucleotides are sufficient to specify rRNA cleavage and allow the rapid destruction of the upstream RNA fragment, the bulk of the conserved sequence augments processing efficiency. Notably, this rRNA processing appears more highly conserved in evolution than does rDNA

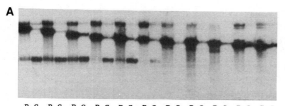
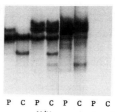

FIGURE 6. The rDNA processing signal. **A.** The indicated 5' deletion mutants of the mouse rRNA processing region, cloned adjacent to a rDNA promoter, were transcribed during a radioactive pulse (P) and then allowed to process during a cold chase (C). **B.** The indicated templates were transcribed in the mouse S-100 extract: M/M = mouse promoter with mouse processing region; M/H = mouse promoter with human processing region; H/H = human promoter with human processing region.

transcription itself. In contrast to rRNA initiation which does not function across the mouse-human species barrier, a chimeric gene containing the human processing region adjacent to a mouse rDNA promoter directs faithful processing when reacted in a mouse cell extract (Fig. 6B). Experiments are currently underway to identify the protein factors that catalyze this primary rRNA processing event.

REFERENCES

1. Sollner-Webb, B. and Tower, J. (1986) Ann. Rev. Biochem. 55, 801-30.
2. Tower, J., Culotta, V. and Sollner-Webb, B. (1986) Manuscript submitted.
3. Buttgereit, D., Pflugfelder, G. and Grummt, I. (1985) Nuc. Acids Res. 13, 8165-80.
4. Paule, M., Iida, C., Perna, P., Harris, G., Knoll, D. and D'Alessio, J. (1984) Nuc. Acids Res. 12, 8161-80.
5. Grummt, I. (1981) (1981) Proc. Natl. Acad. Sci. U.S.A. 78, 727-31.
6. Tower, J., Culotta, V. and Sollner-Webb, B. (1985) J. Cell Biochem. 9B, 206.
7. Gokal, P., Cavanaugh, A. and Thompson, E.A. (1986) J. Biol. Chem. 261, 2536-41.
8. Sollner-Webb, B., Wilkinson, J., Roan, J., and Reeder, R. (1983) Cell 35, 199-206.
9. Miller, K., Tower, J. and Sollner-Webb, B. (1985) Mol. Cell. Biol., 5, 554-562.
10. Windle, J. and Sollner-Webb, B. (1986) Mol. Cell. Biol. 6, 1228-34.
11. Reeder, R. (1984) Cell 38, 349-51.
12. Moss, T. (1983) Nature 302, 223-28.
13. Reeder, R., Roan, J. and Dunaway, M. (1983) Cell 35, 449-56.
14. For convenience we will use the term 'transcriptional terminator', although our experiments do not yet distinguish pausing from true termination.
15. Grummt, I., Maier, U., Ohrlein, A., Hassouna, N., Bachelleria, J.-P. (1985) Cell 43, 801-10.
16. Miller, K. and Sollner-Webb, B. (1981) Cell 27, 165-174.

Transcriptional Control Mechanisms, pages 123–133
© **1987 Alan R. Liss, Inc.**

NEGATIVE REGULATION OF RAT INSULIN 1 GENE EXPRESSION[1]

Uri Nir, Michael D. Walker & William J. Rutter

Hormone Research Institute
and
Department of Biochemistry & Biophysics
University of California
San Francisco, CA 94143

ABSTRACT Two cis-acting elements, the enhancer and the promoter, independently contribute to the cell-specific expression of the rat insulin 1 gene. The activities of these elements are presumably mediated by trans-acting factors. We have performed intracellular competition experiments suggesting the presence of negative factor(s) that repress the enhancer activity in cells not expressing the insulin gene. In these experiments fibroblast cells (COS-7) were transfected with two plasmids: a test plasmid containing the chloramphenicol acetyltransferase (CAT) gene under the control of the thymidine kinase promoter and the insulin enhancer; and a competitor plasmid containing insulin enhancer sequences and the SV40 virus origin of replication to permit its replication in the recipient cells. The presence of the competitor plasmid led to a 5 fold increase in CAT RNA level as compared with the level detected when insulin enhancer was absent from either the competitor or the test plasmid. Efficient derepression required additional sequences downstream from those essential for enhancer activity. On the other hand, the insulin enhancer itself (without these additional downstream sequences) was repressed by the adenovirus Ela gene product. We propose that the activity of the enhancer is modulated by a negative trans-acting factor(s) which is active in cells not expressing insulin but is overridden by the dominant positive trans-acting factor(s) present in insulin producing cells.

INTRODUCTION

We have recently demonstrated that 5' flanking DNA sequences of the rat insulin 1 (rINS1) gene contain two distinct DNA elements which control cell specific gene expression (7). The properties of these elements are consistent with the hypothesis that differentiated cells contain positive trans-acting proteins (differentiators) which interact with the specific cis-acting sequences to stimulate the expression of the associated genes. This idea is also consistent with the observations and concepts of other workers (8-12).

On the other hand, "extinction" of the differentiated phenotype following fusion of differentiated cells with fibroblasts is frequently observed (13). These results are consistent with a role for repressor-like molecules in non-expressing cells. Furthermore the adenovirus E1a protein has been shown to reduce the activity of several enhancers (1-3). We therefore designed competition experiments to obtain evidence for the existence of such negative regulators by titrating them out with high copy number plasmid (6). To maximize the ratio of competitor to test DNA, we have developed an intracellular amplification competition assay in which a replicating plasmid containing the target rINS1 DNA 5' flanking sequences is co-introduced with a test plasmid containing a reporter gene whose expression is controlled by the same target DNA sequences. Such a competition using insulin gene 5' flanking DNA led to activation of the insulin enhancer activity in a cell type where it is normally totally inactive.

RESULTS

The Rat Insulin I Gene Enhancer Could Be Activated in Non-Pancreatic Cells

To achieve a high molar excess of competitor over test sequences, the competing rat insulin I gene 5' DNA flanking sequences were introduced into a vector, pSV, that carries the SV40 early promoter region including the transcriptional enhancer and a functional origin of replication. This plasmid was transferred to COS-7 cells together with a non-replicating test plasmid pTE1rINS1. The test plasmid contains the rat insulin 1 enhancer linked to the Herpes Simplex Virus thymidine kinase promoter TK promoter which drives the bacterial chloramphenicol acetyltransferase (CAT) gene (Fig. 1). The presence of insulin enhancer sequences in the replicating plasmid led to a 5 fold increase in CAT RNA levels (Fig. 2, lanes 1,2). The TK promoter preceded by no enhancer or by an inactive insulin enhancer was not activated under these conditions (4). We have verified that the rInsl sequences did not affect the replication level of the pSV plasmid using quantitative Southern blot analysis of extrachromosomal DNA preparations. The cellular levels of both pSV and pSVrINS1 were estimated to be about $5\text{-}10\times10^4$ copies per cell (data not shown).

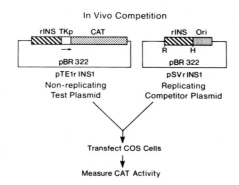

FIGURE 1. Competition by in vivo amplification. The non-replicating plasmids are based on the plasmid pTE1 (7) into which 5' flanking DNA sequences of the rInsl gene were introduced. The arrow show the direction of the CAT gene transcription under the control of the TK promoter. The replicating plasmids were derived from the vector pSV into which rInsl gene 5' DNA flanking sequences were introduced. H, HindIII; R, EcoRI.

The Moloney murine sarcoma virus (MSV) enhancer while driving the TK promoter was not affected by the presence of the replicating rInsl sequences. In contrast, insulin enhancer sequences, -410 to +51 or -249 to -103 were derepressed and led to significant elevation in CAT activity (4). However, when the latter sequences (-249 to -103) were present 3' to the CAT gene, no derepression was observed (4) This is consistent with the observed lack of enhancer activity of this fragment when placed 3' of the CAT gene in

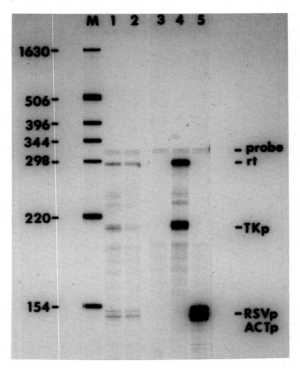

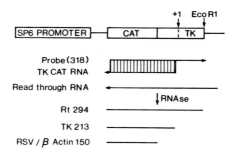

FIGURE 2.   Analysis of CAT RNA from transfected cells.
Total cellular RNA was extracted from cells transfected with
the following plasmid combinations:
1)   5 µg pTElrINS1 (-333 to +51) DNA cotransferred with 5
      µg pSVrINS1 (-370 to +51) DNA. 1 µg rat β-actin CAT
      DNA was included as an internal control. 30 µg RNA from
      the transfected COS-7 cells were analyzed.

2)   pTElrINS1 cotransferred with pSV and 1 µg rat β-actin CAT. 30 µg RNA from the COS-7 cells were analyzed.
3)   Analysis of 30 µg RNA from non-transfected COS-7 cells.
4)   15 µg RNA from HIT (insulin-producing hamster islet cells) cells transfected with 5 µg pTE1.
5)   5 µg RNA from HIT cell transfected with 5 µg RSV CAT DNA (18).
M)   Size marker produced by HinfI digestion of pBR322.
Length of the fragments is given in nucleotides. rt, read-through transcripts; TKp, initiation of RNA synthesized from TK promoter; RSVp, initiation of RNA synthesized from RSV promoter; ACTp, initiation of RNA synthesized from rat β-actin promoter. RNA samples were analyzed with uniformly labeled RNA probe as illustrated in the lower part of the figure. RNase digestion products were separated on 5% denaturing acrylamide urea gels. The expected lengths of the protected bands are shown.

transfected HIT cells (Edlund et al., unpublished results) and eliminates the possibility that the derepression resulted from homologous recombination between insulin sequences, leading to transfer of the SV40 origin of replication to the test plasmid. Significantly, the amylase 5' flanking sequences (containing an enhancer active in pancreatic exocrine cells (14)) were also capable of derepressing the insulin enhancer sequences (4). The amylase DNA sequences contain no striking homologies with the insulin enhancer region. Thus a strictly conserved nucleotide sequence is not required for efficient derepression. The result also supports the conclusion that homologous recombination cannot explain our results.

Derepression Required Sequences 3' to Essential Enhancer Sequences

In order to map the DNA sequences required for the enhancer derepression, progressively shorter fragments of the 5' flanking sequences of rat insulin 1 gene were introduced into the pSV vector and cotransfected with the test plasmid pTElrINS1 (-333 to +51). The smallest fragment tested which retained full ability to derepress, mapped from -219 to +51 (Figure 3, construct 3). Deletion of sequences from -219 to -159 (construct 4) or from +51 to -103 (construct 6) caused dramatic reductions in derepression activity. However, a replicating plasmid lacking sequences downstream from -103 was capable of derepressing an insulin enhancer fragment on the test plasmid which also lacked these sequences (4). This raises the possibility that multiple DNA domains are involved in derepression.

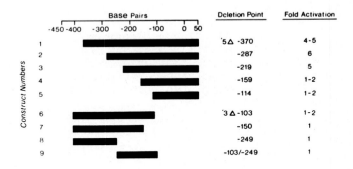

FIGURE 3. Mapping by deletion analysis of sequences capable of derepressing the insulin enhancer. DNA fragments spanning the regions indicated by black bars were introduced into the pSV replicating vector and were cotransferred with pTE1rINS1 plasmid to COS-7 cells. CAT activities were measured and compared to that seen using pSV. The values shown represent the mean of 8 independent experiments. The fold activation varied by up to 30% of the mean value.

Ela Represses the Activity of the Insulin Enhancer

The adenovirus Ela protein has been shown to repress the activity of a number of enhancers. We therefore co-transfected cells with plasmids containing the insulin enhancer and different viral promoters upstream of the CAT gene and plasmids containing the Ela coding region. The expression of CAT activity was then measured. Two different Ela containing plasmids were used. One plasmid, JN20 (5), which gives rise to a 13S form of Ela RNA generates a protein which both represses enhancers and activates promoters. The other, PJOAC (5), generates a 12S RNA which can also repress enhancers but has very little effect on promoters. The presence of the 12S plasmid led to a 5-10 fold reduction in activity of the insulin enhancer but to a very small effect on the TK promoter.

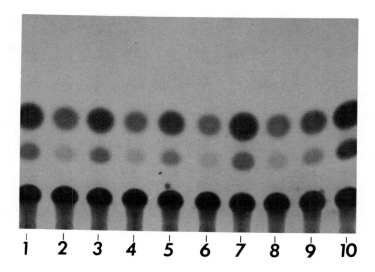

FIGURE 4.   CAT activity following transfection of HIT cells with the following CAT containing plasmids:

Lanes 1 & 2:   Insulin enhancer (-287 to +51) upstream of TK promoter

Lanes 3 & 4:   Insulin flanking sequences (-410 to +51) driving CAT

Lanes 5-7:   Insulin enhancer (-410 to -103) upstream of SV40 promoter.

Lanes 8-10:   TK promoter upstream of CAT.

In addition the following plasmids were co-transfected:

Lanes 1,3,5,8: pBR322 containing an insulin cDNA insert

Lanes 2,4,6,9: 12S Ela (PJOAC) plasmid

Lanes 7 & 10: 13S Ela (JN20) plasmid

All transfections were performed using 2.5 µg of each plasmid.

DISCUSSION

In the present study, we have tested for repressing trans-acting factors interacting with the rat insulin I gene enhancer by competition experiments. We used intracellular amplification to achieve a high ratio of competitor to test plasmid in transfected cells without toxic effects. A sufficiently high ratio is important since standard competition protocols using non-replicating plasmids in our system produced insignificant effects on the insulin enhancer (unpublished data). Our results suggest that the rat insulin enhancer is repressed in non-pancreatic cells: when the rat insulin 5' flanking DNA sequences were present in high copy number the putative repressor was titrated out. This led to derepression and activation of the enhancer (Fig. 2). The enhancer activity observed (5 fold) was substantially lower than that found in insulin-producing cells (40 fold) (7).

The simplest explanation for these results is that the insulin enhancer is controlled by both positive and negative regulators. Thus the high degree of transcriptional control of insulin gene expression seen in vivo may result from the activity of positive regulators in insulin-producing cells and negative regulators in non-producing cells. Titration of the repressor(s) in non-expressing cells led to some activation of tissue specific enhancers, presumably via interaction with positive factors. These positive factors might be non-specific enhancer activator proteins (19) or specific trans-activator molecules (differentiators) present at low concentrations. Full activity of the enhancer was only seen in the appropriate cell type where the differentiator is presumably present at high concentration (Fig. 5).

Wasylyk et al. (19) have recently shown that a central fragment of the immunoglobulin heavy chain enhancer (115 bp long) has significant activity in both lymphoid and non-lymphoid cells in contrast to the well documented lymphoid specificity of longer enhancer fragments. These results are consistent with the above model. An analogous situation in which gene expression is controlled by the equilibrium between positive and negative factors has been described for Xenopus 5S genes (15).

Since the rat amylase gene 5' flanking sequences could activate the rIns1 gene enhancer, the function of the putative repressor may be to control a subset of cell-specific enhancers. It therefore may be also present in insulin-producing cells, where its activity is overridden by positive acting differentiators by virtue of their relative concentration or affinity (Fig. 5).

The precise nature of such repressor molecules is not yet clear. In order to approach the question of how enhancer repression may occur, we chose to look at the effects of a well-studied transcriptional regulatory protein, the adenovirus Ela protein. This protein has been shown to reduce the activity of both viral and cellular enhancers (1-3) and to increase the activity of promoters. Our experi-

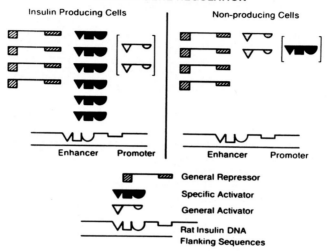

INSULIN GENE REGULATION

FIGURE 5.   Model for transcriptional regulation of the insulin gene.   Insulin producing cells contain <u>sequence specific</u> positive regulators, negative regulators with less stringent sequence specificity and possibly <u>nonspecific</u> enhancer activators.   Non-producing cells contain repressors, low concentrations of sequence specific activators and/or non-specific activators.

ments clearly show that the activity of the insulin enhancer is repressed. We are currently testing whether repression by Ela requires similar enhancer sequences as those for repression in COS cells.

   The binding region of the repressor is not congruent with that of the differentiator: whereas full enhancer activity required sequences from -103 to -333 (7), efficient derepression required sequences downstream from -103 but not upstream of -219.   Thus the repressor binding domain may overlap with elements of both enhancer and promoter.   DNase I protection experiments have demonstrated that both pancreatic and non-pancreatic cells contain proteins which bind between nucleotides -110 and -210, within the binding regions of repressor and differentiator (16; E. Fodor and W.J. Rutter, unpublished results). From these experiments and those of others, it seems that the cell specific regulation of transcription involves a number of cooperating and overlapping regulatory molecules which can act upon a relatively small DNA domain. The study of mechanism will be greatly aided by identification of the trans-activators and characterization of their properties.

ACKNOWLEDGEMENTS

We thank T. Edlund, O. Karlsson, D. Standring, N. Hay and H. Edenberg for valuable discussions, U. Nudel, N. Hay and N.C. Jones for gifts of plasmids, J. Turner for assistance with plasmid preparations and L. Spector for preparation of the manuscript. This research was supported by grants from the NIH (GM 28520) and the March of Dimes to W.J.R. U.N. is supported by a fellowship from the Juveniles Diabetes Foundation.

REFERENCES

1. Borrelli E, Hen R, Chambon P (1984). Adenovirus-2 E1A products repress enhancer-induced stimulation of transcription. Nature 31: 608.
2. Velcich A, Ziff E (1985). Adenovirus E1A proteins repress transcription from the SV40 early promoter. Cell 40:705.
3. Hen R, Borrelli E, Chambon P. (1985). Repression of the immunoglobulin heavy chain enhancer by the adenovirus-2 E1A products. Science 230:1391.
4. Nir U, Walker MD, Rutter WJ (1986). Regulation of rat insulin 1 gene expression: evidence for negative regulation in non-pancreatic cells. Proc Nat Acad Sci USA 84:in press.
5. Jones NC, Richter JD, Weeks DL, Smith LB (1983). Regulation of adenovirus transcription by an E1A gene in microinjected Xenopus laevis oocytes. Mol Cell Biol 3:2131.
6. Scholer HR, Gruss P (1984). Specific interaction between enhancer containing molecules and cellular components. Cell 36:403.
7. Edlund T, Walker MD, Barr PJ, Rutter WJ (1985). Cell

specific expression of the rat insulin gene: evidence for role of two distinct 5' flanking elements. Science 230:912.

8. Payvar F, DeFranco D, Firestone, GL, Edgar B, Wrange O, Okret S, Gustafsson J-A, Yamamoto KR (1983). Sequence specific binding of the glucocorticoid receptor to MTV DNA at sites within and upstream of the transcribed region. Cell 35:381.

9. Sassone-Corsi P, Wildeman A, Chambon P (1985). A trans-acting factor is responsible for the Simian Virus 40 enhancer activity in vitro. Nature 313: 458.

10. Dynan WS, Tjian R (1985). Control of eukaryotic mRNA synthesis by sequence specific DNA binding proteins. Nature 316:774.

11. Ephrussi A, Church GM, Tonegawa S, Gilbert W (1985). B lineage specific interactions of an immunoglobulin enhancer with cellular factors in vivo. Science 227:134.

12. Piette J, Kryszke MH, Yaniv M (1985). Specific interaction of cellular factors with the B enhancer of polyoma virus. EMBO J 4:2675.

13. Killary AM, Fournier REK (1984). A genetic analysis of extinction: trans-dominant loci regulate expression of liver specific traits in hepatoma hybrid cells. Cell 38:523.

14. Boulet AM, Erwin C, Rutter WJ (1986). Cell specific enhancers in the rat exocrine pancreas. Proc Nat Acad Sci USA, in press.

15. Brown DD, Schlissel MS (1985). A positive transcription factor controls the differential expression of two 5S genes. Cell 42:759.

16. Ohlson H, Edlund T (1986). Sequence-specific interactions of nuclear factors with the insulin gene enhancer. Cell, in press.

17. Melloul D, Aloni B, Calvo J, Yaffe D, Nudel U (1984). Developmentally regulated expression of chimeric genes containing muscle actin DNA sequences in transfected myogenic cells. EMBO J 3:983.

18. Gorman CM, Merlino GT, Willingham, MC, Pastan TR, Howard BH (1982). The Rous sarcoma virus long terminal repeat is a strong promoter when introduced into a variety of eukaryotic cells by DNA mediated transfection. Proc Nat Acad Sci USA 79:6777.

19. Wasylyk C, Wasylyk B. (1986). The immunoglobulin heavy chain B-lymphocyte enhancer efficiently stimulates transcription in non-lymphoid cells. EMBO J 5:553.

Transcriptional Control Mechanisms, pages 135–146
© **1987 Alan R. Liss, Inc.**

AN EPIGENETICALLY CONTROLLED B-CELL SPECIFIC
FACTOR ACTIVATES THE MOUSE IMMUNOGLOBULIN
KAPPA GENE PROMOTER

T. Venkat Gopal

Clinical Hematology Branch
National Heart, Lung, and Blood Institute, NIH
Bethesda, Maryland 20852

ABSTRACT   The mouse immunoglobulin (Ig) Kappa (K) gene
promoter contributes to its B-cell specific expression. A
genetic approach has been designed to define the role of
B-cell specific trans-acting regulatory factor(s)
governing Ig K gene expression. Stable transformants of
3T3 cells containing the bacterial neomycin resistance
gene linked to the mouse Ig Kgene pomoter and a "neutral"
enhancer derived from Harvey Murine Sarcoma virus were
isolated. The neomycin resistance gene in the 3T3 test
clones was activated by fusion only with mouse and human
lymphoid cells of B-cell origin, but not with non B-cells.
Activation of the neomycin resistance gene in the 3T3 test
clone following fusion with functionally enucleated human
or mouse lymphoid cells indicates epigenetic control (an
inheritable alteration in phenotype without transfer of
nuclear genetic material) of Ig K gene promoter
activation.

INTRODUCTION

Developmental gene regulation in various cell types
of eukaryotic organisms appears to operate at the level of
initiation of transcription. Techniques of molecular
cloning of developmentally regulated genes, _in vitro_
manipulation and transfer of these cloned genes into
various cell types are used to study the molecular events
responsible for cell type specific gene expression. These
studies have revealed that promoters are essential for
initiation of transcription, and are often involved in

cell specific expression of these genes. Additional
sequences which enhance the transcriptional activity of
these promoters, and therefore are referred to as
enhancers or activators, have also been shown to be
necessary for tissue-specific gene expression (1,2,3, and
references therein). Tissue specificity of cis-acting
regulatory sequences is thought to be due to interaction
with cell-specific trans-acting regulatory factors. The
most common approach to the identification and
purification of such tissue-specific trans-acting
regulatory factors, has involved DNA-protein binding
studies with tissue-specific regulatory sequences and
nuclear extract from appropriate cell types (4,5). These
studies have led to the identification of protein-binding
domains within the regulatory regions of these sequences.

I have developed an alternative genetic approach to
study the action of trans-acting regulatory factors, using
the B-cell specific expression of rearranged
immunoglobulin (Ig) genes as a model system. Both the
promoter and the intron between the joining (J) and
constant (C) region enhancer sequences contribute to
tissue-specific expression of rearranged Ig heavy and
light chain genes transfected into B-cells (1,3, and
references therein). Cells of B-cell origin would seem
therefore to contain trans-acting regulatory factors
specific for Ig gene promoter and enhancer sequences
respectively (6, and references therein). This report
deals with the tissue-specificity of trans-acting
regulatory factor(s) for the mouse immunoglobulin kappa
(K) gene promoter, which I will refer to as Promoter
Competence Factor, PCF. These studies also indicate that
Ig K gene PCF is epigenetically controlled.

RESULTS

Figure 1 shows the features of the test vector pG1
that was designed to probe for the presence of, and test
the tissue-specificity of trans-acting factor(s) for the
mouse Ig K gene promoter. This vector contains a
selectable gene trasncriptional unit- the bacterial
neomycin (Neo) resistance gene linked to the B-cell-
specific mouse Ig K gene promoter (1) and a tissue
non-specific enhancer derived from the Harvey Murine
Sarcoma virus (HaMuSV) LTR. The vector also contains the
herpes simplex virus thymidine kinase (HSVTK) gene. The

only component of this vector that is B-cell specific for
its function is the Ig K gene promoter. We have shown
previously that the Neo gene transcriptional unit in pG1
is expressed in mouse myeloma and mouse fibroblast L
cells, but not in 3T3 cells (1).

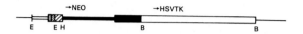

Figure 1.   General map of the test vector pG1.   The
important features of the vector are: 1) the 75 bp repeat
of HaMuSV enhancer region (dotted); 2) the Ig K gene
promoter consisting of 250 bp of the flanking region
immediately 5' to the Kappa gene V-region coding sequence
(hatched); 3) coding sequences of the bacterial neomycin
(Neo) resistance gene (dark line); 4) SV40 t splice and
poly A addition sequences (heavy dark); and 5) 3.4 kb
herpes simplex virus thymidine kinase (HSVTK) gene (open
box).   The direction of the arrows indicates the
transcriptional orientation.   The rest of the vector
consisted of pBR322 sequences.

The test vector was introduced into thymidine kinase
deficient mouse 3T3 cells using standard $CaPO_4$
coprecipitation method and selected in HAT.   Clones were
isolated from HAT medium, propagated, and 2 x 10^6 cells of
each clone were tested for their ability to grow in medium
containing G418.   Eight G418^s/HAT^r clones were fused with
J558 mouse myeloma cells and selected in HAT + G418 to
identify 3T3 clones containing the Neo gene
transcriptional unit in an inactive form and which can be
activated by a PCF from B-cells.   Four out of eight 3T3
clones tested were activated by fusion with mouse myeloma
cells.   One clone, 414G (test clone), was chosen for
fusion with various other lymphoid and non-lymphoid cell
types to determine the tissue-specificity of the K gene
PCF.

Activation of the Neo Gene in the 3T3 Test Clone by Cell Fusion.

The tissue specificity of the Ig K gene PCF was tested by fusion of the test clone with a variety of cell types followed by selection in G418. Table 1 shows the phenotypes of various cell lines used in these fusions, and the activation frequency of the Neo gene. Both mouse and human lymphoid cells of B-cell origin were effective in activating the Neo gene in 414G cells although the activation frequency varied in fusions with different cell types. Fusion of the test clone with mouse and human non B-cells did not result in hybrids capable of growing in G418. Cell fusions, monitored microscopically for the formation of heterokaryons, showed that all the cell lines formed significant number of heterokaryons upon fusion with 414G cells. Therefore, the activation of the Neo gene in 414G cells must be due to either transfer of genes coding for B-cell specific trans-acting regulatory factor(s) or the factor(s) itself for the mouse Ig K gene promoter from mouse and human a B-cells. Such B-cell specific factors must be absent in other cell types that did not activate the Neo gene in 414G cells by fusion.

TABLE 1

Cell Type	Activation	Cell Line Frequency
J558	Mouse Myeloma	48×10^{-6}
p3x63Ag8	Mouse Myeloma	4×10^{-6}
CCL156	Human Peripheral lymphocyte IgM secreting	0.5×10^{-6}
C-NABL	Burkitt Lymphoma IgM, K	15×10^{-6}
BHM-23	Burkitt Lymphoma IgM, K	8×10^{-6}
Daudi	Burkitt Lymphoma IgM, K	20×10^{-6}
HeLa	Human cervical carcinoma	No activation
HT1080	Human fibrosarcoma	No activation
HUT78	Human T-cells	No activation
MOLT3	Human T-cells	No activation

HL60	Human Promyelocytic	No activation
	leukemia cells	
3T3	Mouse fibroblast	No activation

Activation frequency was estimated from the number of G418^r hybrids obtained divided by the number of 414G cells used in various fusions. Spontaneous activation of Neo gene in 414G cells was $<2 \times 10^{-7}$. Each fusion was carried out in triplicate.

Southern blot analysis of DNA from 414G and the Burkitt lymphoma activated hybrid clones showed that lymphoid cell mediated activation of the Neo gene transcriptional unit in 414G cells was not due to its amplification (data not shown). S1 nuclease analysis of RNA from human lymphoid cell activated hybrid clones showed that the Neo gene transcripts utilized the correct transcription start site of Ig Kappa gene promoter (data not shown).

Human Sequences in the Burkitt Lymphoma Cell Activated Hybrid Clones.

Activated hybrid clones derived from fusions of 414G cells with the Burkitt lymphoma cells, C-NABL and BHM-23, consisted of two morphologically distinct phenotypes. Colonies morphologically similar to 414G cells appeared about 3 weeks after fusion and selection in G418 (Type I morphology). The second type of hybrid colonies which were larger than the 3T3 test clone and therefore presumed to be hybrids, appeared about five to six weeks after fusion (Type II morphology). Representative samples of both types of colonies were isolated and analyzed for the presence of human chromosomes or human DNA sequences by C-banding and hybridization to human "Alu" repetitive sequence probe, respectively. Alu repeat sequence is a highly repeated (300,000 per haploid) sequence which is interspersed throughout the human genome. The Alu probe does not cross hybridize to mouse DNA. Therefore hybridization of DNA from the activated hybrids with the Alu probe can serve as a sensitive method to detect the presence of sequences derived from the human genome in these hybrids. Activated hybrid clones of type I

morphology did not contain any human chromosome or
sequences derived from the human genome (Figure 2A).   On
the other hand, activated hybrid clones that were
morphologically different from 414G cells contained human
chromosomes (data not shown), and were positive for the
presence of human "Alu" sequences (Figure 2B).

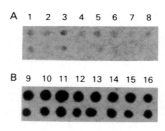

Figure 2.   Detection of sequences derived from the human
genome with the Alu probe in Burkitt lymphoma cell
activated hybrid clones using 5.0 ug and 2.5 ug of DNA
(top and bottom row respectively of each panel).
A.   Hybrids of type I morphology (414G cell like).
    1. 414G   2 through 8- various hybrid clones.
B.   Hybrids of type II morphology.   9-Positive control -
    DNA from a secondary transformant of NIH 3T3 cell
    transformed with the human colon carcinoma DNA.   10
    through 16 - various hybrid clones.

Epigenetic Control of the Ig Kappa Gene PCF.

     Derivation of Burkitt lymphoma cell activated hybrid
clones without the presence of any detectable human
sequences suggested that the Ig K gene promoter activation
could be an epigenetic phenomenon.   Even in the event that
the human PCF gene is not proximal to any member of the
highly dispersed alu family it is highly unlikely that all
the activated hybrid clones with type I morphology
retained the putative PCF gene alone and no other flanking
or distant sequences.   Therefore, it is possible that the
activated clones with type I morphology were derived by
the transfer of only the cytoplasmic contents of lymphoid
cells into the 3T3 test clone.   This may occur through the

unstable or abortive fusion events which arise during fusion of Burkitt lymphoma cells with 414G cells using the lectin coating method and PEG as the fusogen (7). To test this hypothesis, C-NABL and Raji Burkitt lymphoma cells were treated with mitomycin C (100 ug/ml at 37° for 15 minutes in Hepes buffered saline) to achieve functional enucleation prior to fusion with 414G cells (8). Mitomycin C covalently cross-links DNA causing depolymerization and thus brings about the inactivation of functional genes. Cells actively growing in G418 cells were obtained in all these fusions between 414G cells and functionally enucleated human B-cells lines. These activated cells resembled 414G cells morphologically (Type I). Also, these activated hybrid clones did not contain any human DNA sequences as tested by hybridization with the human repetitive sequence probe (data not shown). Fusion of mitomycin treated mouse myeloma cells J558 with the test clone also gave rise to activated hybrid clones that morphologically resembled 414G cells.

Ig Kappa Gene PCF in the Activated Hybrid Clones.

Epigenetic activation of the Ig K gene promoter in the 3T3 test clone could be due either to its stable activation by the putative PCF transferred from the myeloma or lymphoid cell cytoplasm or to activation of the endogenous PCF gene in the mouse fibroblast nuclei. The former would involve a single-hit activation mechanism whereas the latter requires continuous presence of the PCF gene product in the activated clones. To discern between these two possibilities, test vectors containing the bacterial chloramphenicol acetyltransferase (CAT) gene linked to the Ig K gene promoter with the HaMuSV enhancer (1; Figure 3) were introduced into the test clone and the activated hybrid clones.

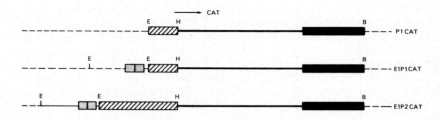

Figure 3.   CAT gene test vectors.   Two forms of the Ig K
gene promoter (hatched) containing either 225 bp (P1) or
625 bp (P2) of 5' - V region flanking sequences were used.
The vector P1 CAT did not contain any HaMuSV enhancer
(dotted) element.

    The results of such an assay are shown in Figure 4.
These test vectors will be activated only in the presence
of the Ig K gene PCF.

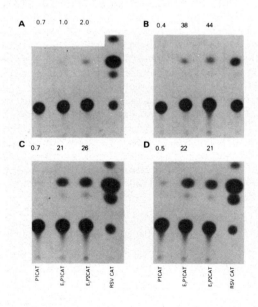

Figure 4. Assay for the presence of Ig K gene specific PCF in the parental 3T3 cell clone 414G and in the activated hybrid clones by a transient expression system using CAT vectors. Transfections and CAT assays were carried out as described previously (9) with minor modifications. Twenty five micrograms of each cell extract was assayed for 1 hour at 37°. A - 414G; B - C-NABL; C - GN-5; D - GN-12. C and D are activated hybrid clones obtained from fusion between 414G and C-NABL cells. The vectors used are denoted at the bottom of the CAT assays. The number on top of each lane of CAT assay figure shows the relative CAT activity of each vector normalized with respect to the expression of pRSVCAT in these cell lines.

   Both of the promoters P1 and P2 were activated in Burkitt lymphoma cells as well as in the Burkitt lymphoma activated hybrid clones (Figure 4B, C, and D) with the HaMuSV in tandem. None of the CAT vectors expressed in 414G cells (Figure 4A). These results imply that maintenance of the Neo gene transcription in the activated hybrids must be due to the continuous presence of the Ig K gene PCF, as opposed to stable activation of the Ig K gene promoter.

                              DISCUSSION

   The data presented in this report show that a genetic approach can be used to define the role of trans-acting regulatory factors governing tissue-specific function of the mouse Ig K gene promoter. This approach uses a heterologous system to assay for the presence of an Ig K gene promoter specific trans-acting factor found only in B-cells. This system presents an opportunity to directly clone by DNA transfer methods the genes coding for trans-acting factors necessary for Ig gene expression. The specificity of the system is demonstrated by the activation of the mouse Ig K gene promoter in the 3T3 test clone after fusion with only B-cells but not with any one of six non B-cells tested.

   Recently, DNA-protein binding studies have shown that a nuclear factor from human B-cells and human HeLa cells binds to both the mouse heavy and K light chain gene promoters and also to the mouse heavy chain J-C intron region enhancer (6). These results imply the presence of sequence specific, but not cell type specific, DNA binding proteins. Fusion of HeLa cells with the 3T3 test clone does not bring about the activation of the mouse Ig K promoter. These studies of Ig K gene promoter activation by cell fusion and of DNA-protein binding suggest ubiquitous and sequence specific DNA binding proteins may be interacting with other cell-specific regulatory factors such as the putative Ig K gene PCF. Such interactions may be responsible for the tissue-specific expression Ig genes.

Activation of the Ig K gene promoter is analogous to epigenetic activation of phenylalanine hydroxylase gene in mouse erythroleukemia cells by the cytoplast of rat hepatoma cells (10). The activation of phenylalanine hydroxylase gene also appears to be due to continuous synthesis of a positive regulatory factor rather than a single-hit activation mechanism. We have previously proposed that self-perpetuating regulatory factor(s) are involved in controlling the activity of the phenylalanine hydroxylase structural gene. Similar mechanisms may be operative in the tissue-specific activation of the Ig K gene promoter. The Ig K gene PCF transferred from the myeloma or human lymphoid cell cytoplasm into the 3T3 test clone is present even after several passages of the activated clones, despite the inevitable dilution of the transferred factor during subsequent cell divisions of the activated cells.

Stable activation and repression of differentiated functions in somatic cell cybrids (10,11,12 and references therein) suggest that genes for various differentiated functions are controlled by cytoplasmic regulatory factors. This could be through either autoregulation of trans-acting activators and repressors or through the formation of stable transcription complexes involving such regulatory factors (13). Activation and repression of differentiated functions in somatic cell hybrids, heterokaryons, and cybrids would argue for the cotinuous production of these putative regulatory factors in the cytoplasm through every cell division. The minimal hypothesis for the epigenetic activation of the Ig K gene promoter would be that either the PCF itself or a trans-acting regulatory factor for the PCF has to be maintained continuously in the cytoplasm of B-cells, perhaps through autoregulation or other as yet unconceived of mechanisms. To further understand developmental gene regulation at the molecular level it is necessary to purify tissue-specific trans-acting regulatory factors or to clone genes coding for such factors. The system described in this report, when paired with cDNA subtraction or gene transfer, provides a promising approach to achieve this goal.

Acknowledgments

This work was carried out in the Clinical Hematolgoy Branch of the NHLBI.   I am thankful to Dr. Arthur Nienhuis for support and encouragement, and to Ms. Anne Baur for excellent technical assistance.   I also thank Ms. Rhonda Mays for preparation of the manuscript.

REFERENCES

(1)   Gopal TV, Shimada T, Baur A, Nienhuis AW (1985). Contribution of promoter to tissue-specific expression of the mouse immunoglobulin kappa gene. Science, 229:1102.
(2)   Edlund T, Walker MD, Barr PJ, Rutter WJ (1985). Cell-specific expression of the rat insulin gene: Evidence for role of two distinct 5' flanking elements.   Science, 230:912.
(3)   Grosschedl R, Baltimore D (1985). Cell-type specificity of immunoglobulin gene expression is regulated by at least three DNA sequence elements. Cell, 41:885.
(4)   Emerson BM, Lewis CD, Felsenfeld G (1985). Interaction of specific nuclear factors with the nuclease hypersensitive region of the chicken adult β-globin gene. Cell, 41:21.
(5)   Levens D, Howley PM (1985).   Novel method for identifying sequence-specific binding proteins. Mol. Cell. Biol, 5:2307.
(6)   Singh H, Sen R, Baltimore D, Sharp PA (1986).   A nuclear factor that binds to a conserved sequence motif in transcriptional control elements of immunoglobulin genes. Nature, 319:154.
(7)   Gopalakrishnan TV (1984).   DNA-mediated restoration of phenylalanine hydroxylase gene expression in enzyme-deficient derivatives of enzyme-constitutive mouse cell hybrids. Somatic Cell and Mol. Genetics, 10:3.
(8)   Gopalakrishnan TV, Littlefield JW (1983).   RNA from rat hepatoma cells can activate phenylalanine

hydroxylase gene of mouse erythroleukemia cells. Somatic Cell Genetics, 9:121.

(9) Gopal TV (1985). Gene transfer method for transient gene expression, stable transformation, and cotransformation of suspension cell cultures. Mol. Cell. Biol, 5:1188.

(10) Gopalakrishnan TV, Anderson WF (1979). Epigenetic activation of phenylalanine hydroxylase in mouse erythroleukemia cells by the cytoplast of rat hepatoma cells. Proc. Natl. Acad. Sci. USA, 74:3932.

(11) Gopalakrishnan TV, Thompson EB, Anderson WF (1977). Extinction of hemoglobin inducibility in Friend erythroleukemia cells by fusion with cytoplasm of enucleated mouse neuroblastoma or fibroblast cells. Proc. Natl. Acad. Sci., 74:1642.

(12) Iwakura Y, Nozaki M, Asano M, Yoshida MC, Tsukada Y, Hibi N, Ochiai A, Tahara E, Tosu M, Sekiguchi T (1985). Pleiotropic phenotypic expression in cybrids derived from mouse teratocarcinoma cells fused with rat myoblast cytoplasts. Cell, 43:777.

(13) Brown DD (1985). The role of stable complexes that repress and activate eukaryotic genes. Cell, 37:359.

Transcriptional Control Mechanisms, pages 147–157
© **1987 Alan R. Liss, Inc.**

INDIRECT CONTROL OF SPORULATION BY THE MATING TYPE LOCUS
IN YEAST

Aaron P. Mitchell

Department of Biochemistry and Biophysics
University of California at San Francisco
San Francisco, CA 94143

ABSTRACT   Yeast cells sporulate (undergo meiosis and
form spores) in response to starvation only if they
express both alleles of the mating type locus: MATa
and MATα. The simultaneous expression of MATa and
MATα gives rise to a negative regulator, a1–α2, that
represses the set of haploid-specific genes. One
haploid-specific gene, RME1, encodes an inhibitor of
meiosis. Thus a1–α2 promotes sporulation through
repression of an inhibitor of sporulation.

INTRODUCTION

There are three cell types in the yeast Saccharomyces
cerevisiae: a cells, α cells, and a/α cells (reviewed in
refs. 1 and 2). a and α cells, which are typically
haploid, are specialized for mating. Each produces a
peptide mating pheromone that is recognized by a receptor
on the surface of cells of the opposite cell type.
Exposure to mating pheromone causes cell division arrest
and elaboration of cell surface agglutinins by the target
cell. Ultimately, one a cell fuses with one α cell to
produce the third type of cell, the a/α diploid. a/α
cells do not mate and do not produce either pheromone.
However, a/α cells are uniquely capable of sporulation
(meiosis and spore formation) to yield haploid a and α
cells (reviewed in ref. 3). Whereas nutritional
limitation induces sporulation in a/α cells, it causes
only cell division arrest in a and α cells. Thus cells of

a given type are specialized to change ploidy through
either mating or meiosis in response to specific
environmental cues.

     Cell type is determined by a single genetic locus,
the mating type locus (MAT). The two MAT alleles, MATa
and MATα, encode regulatory proteins that govern
expression of sets of unlinked target genes (4,5). The
specific sets of target genes that a cell expresses
determine which specialized properties it will have.
These target genes fall into three groups, based on their
regulation: a-specific genes, α-specific genes, and
haploid-specific genes (see Figure 1). The α cell type
depends on the two regulators encoded by MATα, α1 and α2.
α1 turns the α-specific genes on (6); α2 turns the a-
specific genes off (7,8); neither product affects the
haploid-specific genes, which are on in α cells (9). An a
cell type derives essentially from the absence of the MATα
products: α-specific genes are off because α1 is absent
and a-specific genes are on because α2 is absent; again,
the haploid-specific genes are on. In the a/α cell, all
three groups of genes are off. a-specific genes are
repressed by α2 and, in addition, the α- and haploid-
specific genes are silent because of a unique regulatory
activity, a1-α2, that arises from simultaneous expression
of both MAT alleles. a1-α2 represses MATα1, thus ensuring
that α-specific genes are off (10,11). a1-α2 also
represses the haploid-specific genes and, in doing so,
permits a/α cells to enter meiosis and sporulate. The
question I will focus on is how repression by a1-α2 acts
to promote sporulation. I first briefly review what is
known about a1-α2 and the regulation of haploid-specific
genes.

          REPRESSION OF HAPLOID-SPECIFIC GENES BY a1-α2

     Repression of haploid-specific genes in a/α cells
depends on the a1 and α2 products and on a particular DNA
sequence found upstream of the target gene's transcript
initiation site. It is not known how the combination of
a1 and α2 exerts repression, but the simplest type of
model is that a1 protein modifies the activity of α2. The
α2 protein (in the absence of a1) is a repressor of the
set of a-specific genes. Johnson and Herskowitz have

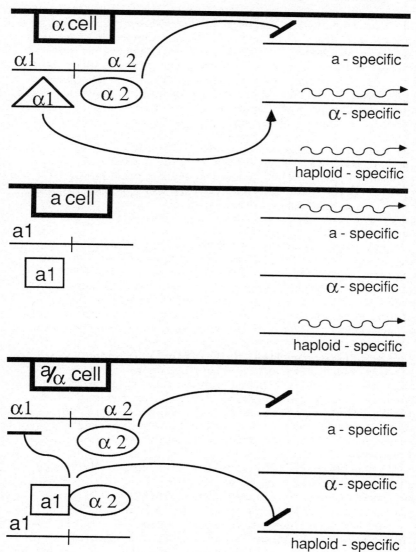

Figure 1. The α1–α2 hypothesis. In α cells, the MATα1
product is a positive regulator of α–specific genes and
the MATα2 product is a repressor of **a**–specific genes.   In
a cells, the **a**l product has no known role.   In **a**/α cells,
al–α2 is a negative regulator of MATα1 and of haploid–
specific genes.

shown that an α2-β-galactosidase hybrid protein binds <u>in vitro</u> to a 32 base pair sequence found upstream of several **a**-specific genes (8). Analysis of both deletions and hybrid promoters demonstrates that the binding site is necessary and sufficient to impose repression by α2 on a target gene (7,8,12). Thus α2 can both bind to a specific sequence and repress transcription in the vicinity of that sequence. Perhaps, then, **a**l protein alters the binding specificity of α2 protein such that a different target sequence is recognized. One particular model (13) is that **a**l and α2 physically associate to form a mixed oligomer that binds to the **a**l-α2 target sequence (see below). However, there is no direct evidence for the interaction of these two proteins. Therefore, the term "**a**l-α2" is used to describe this negative regulator without any mechanistic connotations.

Sites that render genes repressible by **a**l-α2 have been identified at the <u>MATαl</u> and <u>HO</u> genes. Both of these genes are repressed over 100-fold in **a**/α cells (9,10,11). Siliciano and Tatchell (14) found that repression of <u>MATαl</u> depends on a 28 bp sequence located between the <u>MATαl</u> upstream activation site and TATA box. Deletions of this sequence result in constitutive expression; introduction of this site into a heterologous promoter places it under negative control by **a**l-α2. Therefore, the site is both necessary and sufficient for repression by **a**l-α2. Sequences required for <u>MATαl</u> expression are completely separable from the repression site. This finding indicates that negative control by **a**l-α2 is exerted directly on the target gene and not, for example, through repression of a transcription factor necessary for <u>MATαl</u> expression.

If haploid-specific genes share a common mechanism for their repression by **a**l-α2, then a similar repression site should exist near other genes of this set. This prediction is borne out by studies of Miller et al. on the <u>HO</u> gene (12). Evidence that **a**l-α2 repression is exerted at several sites upstream of this gene led them to identify 10 repeats of a 20 bp sequence that is defined by 14 consensus nucleotides: TC(A/G)TGTNN(A/T)NANNTACATCA. This repeat is homologous to the repression site at <u>MATαl</u> and two examples tested rendered a heterologous promoter repressible in **a**/α cells. Two sites homologous to this sequence were also found upstream of the haploid-specific gene STE5 (ref. 12). Therefore, this sequence may be widely used as a site at which a1-α2 represses transcription.

CELL TYPE, RME1, AND CONTROL OF MEIOSIS

In standard strains of yeast, only a/α diploids can sporulate; haploids and a/a and α/α diploids cannot (3,15). Sporulation requires both a1 and α2 products (5,16) and is induced by starvation (which generates a signal unrelated to MAT control as far as we know; 17). Kassir and Simchen observed that some strains are able to sporulate regardless of cell type (16). This ability is due to a single recessive genetic lesion called the rme1-1 mutation (RME1, Regulator of Meiosis). Subsequent studies by Rine et al. led to a simple model to explain how the mating type locus governs meiosis (18). The model specifies that RME1 product is a negative regulator of meiosis; it is expressed in a/a and α/α cells and blocks initiation of meiosis. To explain how a/α cells can sporulate, it was proposed that RME1 activity or expression is negatively regulated in a/α cells by a1-α2. Thus a/α diploids are able to sporulate because they inhibit an inhibitor of meiosis. I cloned the RME1 gene and used it to test this hypothesis (17).

First, I asked whether RME1 activity was under control of the mating type locus. Although there was no way of knowing the level at which regulation would be exerted, I started with the simplest possibility and this turned out to be correct: RME1 is controlled by MAT at the level of gene expression (17). Haploid a and α cells have similar levels of the RME1 transcript, but a/α diploids have 10- to 20-fold lower levels. Diploids with mutations in either of the proposed negative regulators of RME1 -- MATa1 and MATα2 -- have the same elevated RME1 transcript levels as the haploid strains. Therefore, the mating type locus controls RME1 expression through repression by a1 and α2 products.

Second, if failure to express RME1 is sufficient to permit meiosis to occur, then high levels of RME1 product should prevent meiosis even in a/α cells. Overexpression of RME1 was achieved by placing the gene on a multi-copy plasmid (17). a/α cells carrying this plasmid produce

100-fold elevated levels of RME1 transcript (19). Such
cells do not undergo sporulation, as assayed by failure to
produce spores. In fact, these cells are blocked at a
very early point in meiosis as indicated by their failure
to produce intragenic recombinants. This behavior is the
same as that of **a/a** and α/α diploids, which naturally
express RME1. Therefore, the RME1 product is capable of
blocking meiosis in any cell type.

Finally, if RME1 product is responsible for blocking
sporulation in cells lacking either **a**1 or α2, then
inactivation of the RME1 gene by an insertion mutation
should permit these cells to sporulate. I inserted the
LEU2 gene into the middle of the RME1 transcribed region
and replaced chromosomal sequences with this mutated gene.
Northern blots confirmed that the insertion allele
produces no RME1 transcript. This mutation permits
diploids lacking **a**1-α2 to undergo intragenic recombination
and form spores (17). In fact, the mutation permits
derepession of SPO13, a sporulation-specific gene, in
a1⁻/α cells incubated in sporulation medium (A Mitchell,
HT Wang, and RE Esposito, unpublished results). It seems
likely that other sporulation-specific genes that are
essential for sporulation, such as SPS1 (A Percival-Smith
and J Segall, submitted), are also derepressed in rmel
mutant strains irrespective of **a**1-α2 activity. Thus RME1
is formally a negative regulator of sporulation-specific
genes. These observations demonstrate that repression of
RME1 in **a**/α cells is sufficient to explain their ability
to sporulate.

It was important to use a null rmel mutation for the
experiment described above because the original rmel-1
mutation confers only partial sporulation ability; this
property could indicate that rmel-1 does not completely
abolish RME1 activity (16-20). However, among strains
homozygous for the rmel insertion mutation, **a**/α cells
sporulate more rapidly and more efficiently than isogenic
strains lacking **a**1-α2 (17,19; unpublished results). Thus
a null rmel mutation has a leaky phenotype. This
observation suggests that **a**1-α2 promotes sporulation
through some route in addition to repression of RME1. One
simple model is that **a**1-α2 represses a second inhibitor of
meiosis (18). It is presence or absence of RME1 product
alone, though, that determines qualitatively whether cells
will enter meiosis.

a/α cells have several unique properties in addition to sporulation ability; does RME1 expression influence any of these phenotypes? In the cases examined so far, the answer is no (18,20; A Mitchell and S Lovett, unpublished results). The ability of **a** and α cells to mate is not reduced by either naturally occurring rme1 alleles or by the rme1 insertion mutation. **a**/α cells exhibit higher levels of intragenic recombination in response to ultraviolet irradiation and increased X-ray resistance compared to cells lacking **a**1-α2; these properties are not conferred to cells lacking **a**1-α2 by rme1 mutations. Thus every indication is that the targets of RME1 inhibition are specifically required for sporulation.

I have recently sequenced a 3 kilobase region of DNA that includes the RME1 gene. Within this region is an open reading frame that may represent the RME1 product. Upstream of the open reading frame are several sequences that resemble the previously characterized **a**1-α2 repression sites. Whether any or all of these sites are functional is currently under study. Assuming that at least one of the **a**1-α2 sites is functional, how can we account for the differences in repression ratios of the haploid-specific genes? For example, RME1 is repressed 20-fold in **a**/α cells while HO is repressed over 100-fold. A simple explanation may lie in the placement of the **a**1-α2 repression site relative to the upstream activation sequence: an **a**1-α2 repression site is less potent upstream of an activation sequence than it is between an activation sequence and TATA box. A case in point is MATα: MATα1 is repressed over 100-fold in **a**/α cells but the divergent MATα2 transcript is repressed only 10-fold. Expression of both genes relies on the same activation sequence and repression site (14,21). However, the activation site is upstream of the repression site relative to MATα1; the repression site is upstream of the activation site relative to MATα2. The importance of order of **a**1-α2 repression site and activation site has been confirmed with hybrid promoter constructions (14). Thus the **a**1-α2 repression site may be upstream of the activation site at RME1 in order to achieve a modest 20-fold repression ratio.

Control of meiosis by the mating type locus may be summarized as follows (see Figure 2). Haploid **a** and α

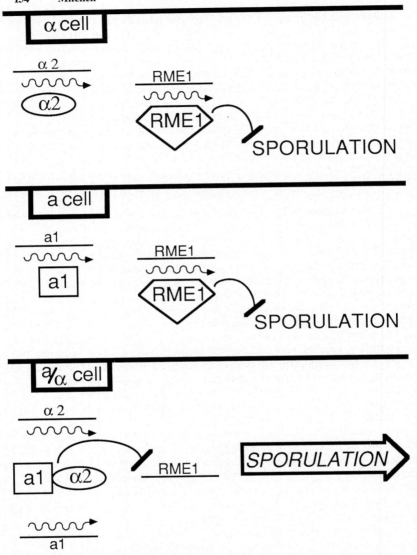

Figure 2. Control of sporulation by the mating type
locus. α cells do not have the negative regulator **a**1–α2;
therefore RME1 is expressed and the RME1 product blocks
sporulation. A similar situation exists in **a** cells.   **a**/α
cells possess **a**1–α2, which represses RME1 transcript
levels. Absence of RME1 product permits **a**/α cells to
sporulate.

cells possess either a1 or α2 but not both. Thus these cells produce RME1 transcript at high levels and the RME1 product inhibits entry into meiosis. However, a/α diploids possess both a1 and α2. These products act together in some way to repress RME1 and thus confer the ability to enter meiosis and sporulate. a1-α2 improves sporulation ability independently of repression of RME1; a simple explanation is that other haploid-specific gene products interfere with quantitative aspects of sporulation. Several unique properties of a/α cells unrelated to sporulation are not regulated by RME1; it is likely that these phenotypes depend on repression of other haploid-specific genes.

How does this study of RME1 expand our understanding of cell type determination in yeast? On the one hand, there is a common theme: the mating type locus governs the properties of each cell type through transcriptional regulation of cell type-specific genes. On the other hand, RME1 is different from many of the previously characterized cell type-specific genes. They are structural genes whose products play a direct role in cell phenotype; they encode such functions as mating pheromones and pheromone receptors. RME1 is a regulatory gene through which MAT controls an entire set of genes, the sporulation-specific genes, by proxy. Whether RME1 represses sporulation-specific genes directly or through a cascade of regulators is an open question. However, we now know that the mating type locus governs expression of genes through both direct and indirect mechanisms.

ACKNOWLEDGEMENTS

I am most grateful to Ira Herskowitz, Paul Sternberg, Susan Michaelis, Beth Shuster, Lorraine Marsh, Ira Clark and Paul Siliciano for their comments on this manuscript. I thank H. Tong Wang, Rochelle Easton Esposito, Tony Percival-Smith, Jacqueline Segall, Paul Siliciano and Kelly Tatchell for providing information prior to publication. This work was supported by N.I.H. research grant AI 18738 (to Ira Herskowitz) and by a postdoctoral fellowship from the Damon Runyon -- Walter Winchell Cancer Fund.

REFERENCES

1.   Herskowitz I, Oshima Y (1981). Control of cell type
      in Saccharomyces cerevisiae: Mating type and mating
      type interconversion. In Strathern JN, Jones EW,
      Broach JB (eds): "The Molecular Biology of the Yeast
      Saccharomyces: Life Cycle and Inheritance," Cold
      Spring Harbor, New York; Cold Spring Harbor
      Laboratory Press, pp 181-209.
2.   Sprague Jr GF, Blair LC, Thorner J (1983). Cell
      interactions and regulation of cell type in the yeast
      Saccharomyces cerevisiae. Ann Rev Microbiol 37:623-660.
3.   Esposito RE, Klapholz S (1981). Meiosis and
      ascospore development. In Strathern JN, Jones EW,
      Broach JB (eds): "The Molecular Biology of the Yeast
      Saccharomyces: Life Cycle and Inheritance," Cold
      Spring Harbor, New York; Cold Spring Harbor
      Laboratory Press, pp 211-287.
4.   MacKay V, Manney TR (1974). Mutations affecting
      sexual conjugation and related processes in
      Saccharomyces cerevisiae. II. Genetic analysis of
      nonmating mutants. Genetics 76:273-288.
5.   Strathern J, Hicks J, Herskowitz I (1981). Control
      of yeast cell type by the mating type locus: The α1-
      α2 hypothesis. J Mol Biol 147:357-372.
6.   Sprague Jr GF, Jensen R, Herskowitz I (1983).
      Control of yeast cell type by the mating type locus:
      Positive regulation of the α-specific STE3 gene by
      the MATα1 product. Cell 32:409-415.
7.   Wilson KL, Herskowitz I (1984). Negative regulation
      of STE6 gene expression by the α2 product of
      Saccharomyces cerevisiae. Mol Cell Biol 4:2420-2427.
8.   Johnson AD, Herskowitz I (1985). A repressor (MATα2
      product) and its operator control expression of a set
      of cell type specific genes in yeast. Cell 42:237-247.
9.   Jensen R, Sprague Jr GF, Herskowitz I (1983).
      Regulation of yeast mating-type interconversion:
      Feedback control of HO gene expression by the yeast
      mating type locus. Proc Natl Acad Sci 80:3035-3039.
10.   Klar AJS, Strathern JN, Broach JR, Hicks JB (1981).
      Regulation of transcription in expressed and
      unexpressed mating type cassettes of yeast. Nature
      289:239-244.
11.   Nasmyth KA, Tatchell K, Hall BD, Astell C, Smith M

(1981). A position effect in the control of transcription at the yeast mating type loci. Nature 289:244-250.

12. Miller AM, MacKay VL, Nasmyth KA (1985). Identification and comparison of two sequence elements that confer cell-type specific transcription in yeast. Nature 314:598-603.

13. Herskowitz I (1982). The MATα2 gene. Rec Adv Yeast Mol Biol 1:320-331.

14. Siliciano PG, Tatchell K (1986). Identification of the DNA sequences controlling the expression of the MATα locus of yeast. Proc Natl Acad Sci, in press.

15. Roman H, Philips MM, Sands SM (1955). Studies of polyploid Saccharomyces. I. Tetraploid segregation. Genetics 40:546-561.

16. Kassir Y, Simchen G (1976). Regulation of mating and meiosis in yeast by the mating-type region. Genetics 82:187-206.

17. Mitchell AP, Herskowitz I (1986). Activation of meiosis and sporulation by repression of the RME1 product of yeast. Nature 319:738-742.

18. Rine JD, Sprague Jr GF, Herskowitz I (1981). rme1 mutation of Saccharomyces cerevisiae: map position and bypass of mating type locus control of sporulation. Mol Cell Biol 1:958-960.

19. Mitchell AP (1986). Control of sporulation by the mating type locus and the RME1 gene product. In "Symposium on the Biochemistry and Molecular Biology of Industrial Yeasts," New York, New York; Academic Press, in press.

20. Hopper AK, Kirsch J, Hall BD (1975). Mating type and sporulation in yeast. II. Meiosis, recombination, and radiation sensitivity in an αα diploid with altered sporulation control. Genetics 80:61-76.

21. Siliciano PG, Tatchell K (1984). Transcription and regulatory signals at the mating type locus in yeast. Cell 37:969-978.

Transcriptional Control Mechanisms, pages 159–169
© 1987 Alan R. Liss, Inc.

THE BINDING OF THE MAJOR REGULATORY PROTEIN, α4, TO THE
PROMOTER-REGULATORY DOMAINS OF HERPES SIMPLEX VIRUS 1 GENES[1]

T.M. Kristie and B. Roizman

Marjorie B. Kovler Viral Oncology Laboratories, University of
Chicago, 910 East 58th Street, Chicago, Illinois 60637

ABSTRACT   Earlier studies have shown that functional α4
gene product is essential for the transition from α to β
and γ protein synthesis and that α4 is auto-regulatory.
We report that labeled DNA fragments containing promoter-
regulatory domains of 3 α (α0, α4, and α27) and a γ_2 genes
form stable complexes with proteins from infected cell
lysates as detected by a gel electrophoresis binding assay.
Monoclonal antibody to the α4 protein, but not to other
α proteins, reduced the electrophoretic mobility of the
labeled DNA-infected cell protein complex. One monoclonal
antibody, H950, blocked the binding of α4 complexes to
the promoter and regulatory domains of α genes. The nucleotide sequence of the α0 promoter domain protected from
exonuclease III digestion by α4 protein in the absence
of H950 monoclonal antibody but not in its presence is
presented. A second binding site for α4 protein was
identified within a 59bp sequence from the regulatory
domain of the α4 gene. Deletion clones of this fragment
localize sequence elements required for the α4 protein-DNA
complex formation.

INTRODUCTION

Herpes simplex virus 1 genes form at least 5 groups (α, β_1,
β_2, γ_1, and γ_2) whose expression is coordinately regulated and
sequentially ordered in a cascade fashion (1). The focus of our
interest is the function of α proteins and the regulation of α

[1]These studies were aided by grants from the National Cancer
Institute (CA08494 and CA19264), USPHS, and the American Cancer
Society (MV2T).

gene expression, the first viral genes to be transcribed after infection (2). The promoter-regulatory region of α genes contains at least two distinct elements which enable α gene expression. The first responds to both host and viral factors while the second responds to an α-specific trans-inducing factor (αTIF) packaged in the virion (2,3).

α gene expression is turned off by viral proteins made later in infection and probably also by the product of the α4 gene itself (3,4,5). The α4 protein is the major regulatory protein of HSV-1 inasmuch as temperature-sensitive mutants in this protein fail to make the transition from α to β and γ protein synthesis (3). The newly synthesized α4 protein has a mol. weight of 160,000 but the processed protein exists in at least 3 forms of higher apparent molecular weight (6). In its native state, α4 exists as a homodimer-phosphoprotein (7,8) and is reported to bind DNA (7,9), although one report states that the binding is dependent on the presence of host proteins (9).

We report that labeled HSV-1 DNA fragments containing promoter and regulatory domains of α, β, and γ genes specifically bind infected cell proteins as detected by retardation of their electrophoretic mobility in non-denaturing polyacrylamide gels (10-12). These complexes contain the α4 protein inasmuch as labeled DNA-infected cell protein complexes incubated with monoclonal antibody to the α4 protein retard the electrophoretic mobility of the DNA-protein complex whereas monoclonal antibodies to several other HSV α proteins had no effect. Two α4 protein binding sites were identified by protection from exonuclease III digestion and by specific deletion analysis.

MATERIALS AND METHODS

Preparation of cellular extracts. HeLa cells were mock infected or infected with 5 pfu of HSV-1(F) (13), ts502Δ305 (14), or tsHA1tk⁻ (15) for 10-12 hours. Whole cell extracts and nuclear extracts were prepared as described (16,17).

Cloning and preparation of DNA probes. The HSV-1 DNA fragments shown in Figure 1 were cloned, purified, and 5' end-labeled according to standard procedures. DNA binding assays and exonuclease III digestions were done as described in the legends to figures 2 and 4 respectively.

Monoclonal antibodies. All monoclonal antibodies used in these studies were a gift of Lenore Pereira and were, in part, described elsewhere (18,19).

RESULTS

   We have used the DNA-protein gel retardation assay to
identify viral-encoded DNA binding proteins specific for the α 0
promoter domain shown in Figure 1. The assay involved the
binding of labeled α0 P DNA in extracts of mock-infected
or HSV-1 infected cells in the presence of excess unlabeled
competitor DNA and visualizing the protein-α0 P DNA probe
complexes by the retardation of their electrophoretic mobility
in non-denaturing polyacrylamide gels (10-12). Complexes formed
in the presence of HSV-1 infected cell proteins, but not in the
presence of mock infected proteins, were likely to represent
sequence specific viral-encoded factors or cellular proteins
whose relative abundance or activity were altered by the viral
infection. The specificity of the   α0 P DNA-infected cell
protein complex was demonstrated by the ability of homologous
unlabeled α0 P DNA, but not heterologous DNA derived from
pUC9, to compete effectively for the infected cell protein(s).

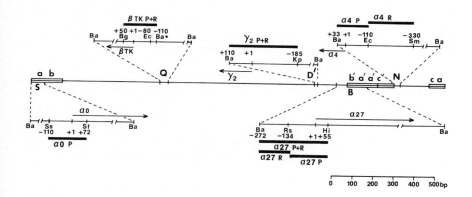

FIGURE 1.   Sequence arrangement and locations of HSV-1
DNA fragments used in these studies. The boxes represent the
terminal repeat sequences ab and ca and their internal inverted
repeats b'a'a'c' (20). Expanded scales show the restriction
patterns and the locations of the promoter (P), regulatory (R),
and coding sequences for the α genes 4, 0, and 27 (21), the βTK
gene (22), and a γ2 gene (15,23). Enzyme cleavage sites are
abbreviated as follows: Ba, BamHI; Ss, SstII; Sf, SfaNI; Bg,
BglII; Ec, EcoRI; Sa, SalI; Rs, RsaI; Hi, HinfI; Sm, SmaI.
* indicates the location of the synthetic BamH1 linker (22).

The identification of the protein in infected cell specific-protein- α0 P DNA complex was approached using temperature sensitive viral mutants defective in the transition from α to β or from β to Y₂ protein synthesis. We tested lysates of cells infected with ts502Δ305 (14) and tsHA1tk⁻ (15), maintained at the permissive (34°C) or the non-permissive (39.5°C) temperatures for the ability to form the specific infected cell protein-α 0 P DNA complex. ts502Δ305 carries a ts lesion in the α4 gene and infected cells incubated at 39.5°C accumulate predominantly α proteins. tsHA1tk⁻ carries a ts lesion in the major DNA binding protein (β8) and the infected cells do not synthesize viral DNA or Y₂ proteins at the non-permissive temperature. As lysates of cells infected with either ts mutant and maintained at 39.5°C were capable of forming the infected cell specific-α0 P DNA complex, the protein involved in this complex was available very early in the infectious cycle.

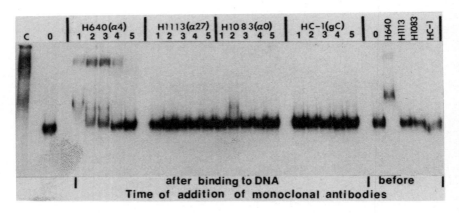

FIGURE 2. Autoradiographic images of α0 P DNA-infected cell protein complex in the presence of murine monoclonal antibodies against α4 (H640), α27 (H1113), α0 (H1083) and Y₂gC (HC-1). All protein-DNA binding assays were done by incubating labeled DNA probe (2.5 ng) with 1000 ng of protein extract in the presence or absence of 1000 ng of competitor nucleic acid poly(dI)·poly(dC) in 20 mM Tris (pH 7.6), 50 mM KCl, 0.05% Nonidet P40, 5% glycerol, 50 ug/ml bovine serum albumin, 10 mM β-mercaptoethanol, and 1 mM EDTA for 0.5 hour at 25°C. MOCK, mock infected cell extract; INFECTED, HSV-1(F) infected cell extract. The monoclonal antibodies were either added after protein-DNA binding or were preincubated with the protein extract for 0.5 hour prior to the incubation of the extract

with labeled DNA. Lane C, DNA-protein complexes formed in the
absence of poly(dI)·poly(dC); lane 0, no monoclonal antibody
added. Lanes 1-5 differ with respect to the amount of
monoclonal antibody added to the reaction as follows: 1, 500
ng; 2, 250 ng; 3, 100 ng; 4, 10 ng; 5, 1 ng. Preincubations
were done with 500 ng antibody/1000 ng protein extract.

To test the hypothesis that the complexes contained α
proteins, monoclonal antibody specific for α4 (H640), α27 (H1113),
α0 (H1083), or control monoclonal antibody (HC-1) directed
against glycoprotein C (γ2gC) were added to the binding reaction
either after or before complex formation. As shown in Figure 2,
monoclonal antibody H640 retarded the electrophoretic mobility
of the α0 P-infected cell protein complex when added either
before or after the binding reaction. In contrast, none of the
monoclonal antibodies used in these studies to α0, α27 or γ2gC
affected the mobility or inhibited the formation of this complex.
A panel of independently derived anti-α4 monoclonal antibodies was subsequently screened for the ability to inhibit
the formation of α4 protein-α0 DNA complex. Of 10 antibodies
tested, two (H942 and H950) were capable of substantially reducing the complex formation (Kristie and Roizman, PNAS, in press).
     To determine whether α4 protein was present in complexes
formed with other genes, the promoter and regulatory domains of
other α genes and representative βι and γ2 promoter-regulatory
domains shown in Figure 1, were reacted with infected cell
proteins. Monoclonal antibodies specific for α4, α0, α27
and γ2gC were used to demonstrate that the α4 protein was
present in complexes containing its own promoter and regulatory
domains as well as in those formed by the γ2 promoter-regulatory
region and promoter-regulatory domains of the α27 gene. In
contrast, complexes containing α4 protein and the βTK
promoter-regulatory domain were only detected in the presence
of monoclonal antibody. In no instance were we able to demonstrate
reactivity of the monoclonal antibodies used in these studies to
α27 or α0 with any of the complexes tested.
     Mapping of the binding site of the α4 protein to the α0
promoter DNA. As illustrated in Figures 3 and 4, DNA fragments
labeled at one 5' terminus and specifically bound to infected
cell proteins were digested with exonuclease III. The DNA
protected from digestion was sized on sequencing gels and the
nucleotide sequence bound by the protein complex emerged from
the comparison of the 3' boundaries of the protected coding and
non-coding strands. Exonuclease III has been used to detect
protein-DNA complexes in vitro and in vivo and is well suited

for mapping complexes in crude nuclear extracts (24,25). The
left panel of Figure 4 shows the results of exonuclease III
digestion of α 0 promoter DNA labeled at the 5' end of the non-
coding strand. This DNA was incubated without extract, with
mock-infected cell extract, with infected cell extract, or with
infected cell extract preincubated with anti-α4 antibody H950.
As previously mentioned, H950 significantly inhibited the binding
of α4 protein complex to the α0 promoter DNA. With increasing
time of digestion, a unique 69 nucleotide fragment becomes
progressively evident in the infected extract reaction. In
contrast, no unique DNA fragment was protected from exonucleolytic digestion when incubated without extract or with mock-
infected extract. DNA digested in the presence of infected cell
extract, preincubated with H950, showed a substantial reduction
in the protection of the 69 base fragment. A similar digestion
of α0 promoter DNA labeled at the 5' end of the coding strand
yielded a 163 base protected fragment as shown in the right
panel of Figure 4. The protected strands share the common 26
nucleotide sequence represented in Figure 3.

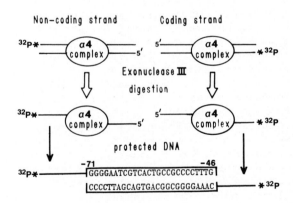

FIGURE 3. Schematic representation of the exonuclease III
protection experiment and the location of the α0 promoter DNA
sequences protected by the α4 protein complex. Non-coding or
coding strand refers to the strand 5' end-labeled. The overlapping nucleotide sequences (-71 to -46) of α0 P DNA protected
from digestion by α4 protein complex were numbered relative
to the transcription initiation site of the α0 mRNA (21).

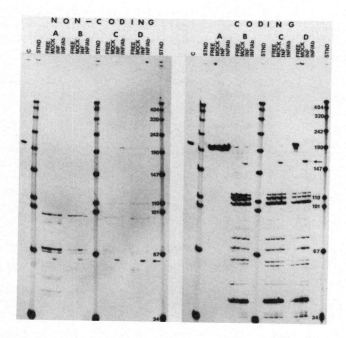

FIGURE 4. Autoradiogram of electrophoretically separated exonuclease III digests of α 0 P DNA in the absence or presence of available α4 protein. DNA-protein binding was done in the presence of 2000 ng competitor. After incubation, the mixture was adjusted to 2.5 mM MgCl$_2$ and digested with exonuclease III. C, nondigested α0 P; FREE, digestion after incubation without protein extract; MOCK, digestion after incubation with mock-infected cell extract; INF, digestion after incubation with infected cell extract; INF/Ab, digestion after incubation with infected cell extract that had been preincubated for 0.5 hour with monoclonal antibody H950. The left panel represents digests of α0 P DNA labeled at the 5' end of the non-coding strand. The exonuclease III digestions were: 100 units for 15 minutes (A); 100 units for 30 minutes (B); 100 units for 45 minutes (C); 200 units for 30 minutes (D). The right panel shows digests of α0 P DNA labeled at the 5' end of the coding strand. The exonuclease III digestions were for 15 minutes at the following enzyme concentrations: 1 unit (A); 25 units (B); 50 units (C); 100 units (D). pUC9 digested with HpaII and 5'end-labeled (STND) served as size markers. The arrows indicate the position of bands representing DNA protected by the α4 protein complex.

Location of the binding site of α 4 protein complex to its own regulatory domain. DNA-binding assays with the α4 promoter and regulatory domains indicated that the α4 protein is present in specific complexes formed with these DNA fragments. To locate the sequences required for this complex formation, the α4 regulatory domain, which extends from -110 to -330 relative to the α4 mRNA transcription initiation site, was cloned as 3 DNA fragments: 150bp (-110 to -260), 59bp (-135 to -194), and 70bp (-260 to -330). These DNA probes were tested for the formation of an α4 protein-DNA complex in infected cell extracts in the presence or absence of anti-α4 monoclonal antibody. In this manner, the α4 protein binding site was localized to the 59bp subclone (-135 to -194) of the α4 regulatory region. To further characterize this site, we constructed a series of deletions of this fragment and tested the deleted constructs for the formation of the α4 protein-DNA complex as described above. The results suggest that the α4 protein binding site is located within the right half of the 59bp fragment, proximal to the α4 promoter domain. It remains to be determined, however, whether the previously identified cis-acting functions of the 59bp clone were also conserved in the deleted fragments (26). Several of the constructed deletions progressively remove the sequence TGGGCGGGGC which has been shown to bind the SP1 factor in vitro (27). The removal of this binding site does not affect the binding of the α4 protein complex to these DNA fragments.

DISCUSSION

We report that the HSV-1 α4 protein is present in complexes formed by promoter and regulatory domains of selected α and γ2 genes with infected cell proteins. The binding of α4 protein to the promoter-regulatory domains of α and γ2 genes is sequence specific. The amounts of protein-β gene promoter-regulatory domain complexes containing α4 protein were small. We have mapped the DNA binding site of the α4 protein complex to the promoter domain of the α0 gene and identified a second binding site within the regulatory domain of the α4 gene.

The α4 protein appears to induce the expression of a variety of genes other than HSVβι or γ genes (28,29) and in this respect it is similar to the adenovirus E1A and pseudorabies virus immediate early proteins (30,31). Current evidence also suggests that α4 protein negatively regulates the transcription of α genes (3,4,5). An interesting feature of the α4 DNA binding sites is their proximity to various host factor binding sites such as the CCAAT (32) and SP1 (27). It is conceivable that the proximity of host components may be critical in both the positive and negative regulatory functions of the α 4 protein. As shown in the Results, however, the actual binding of α4 does not require conservation of the SP1 binding site.

The identification of the α4 binding site in the 59bp sequence of the α4 regulatory region is of particular interest in light of the results of earlier studies (26) showing that this sequence conferred a high constitutive level of expression when fused 5' of chimeric genes containing the α4 promoter (-110 to +33) or the β regulated promoter of the thymidine kinase gene, but did not confer the capacity to be regulated as an α gene. The higher constitutive level of expression of the chimeric genes may be due to juxtaposition of additional binding sites for host factors, e.g. SP1, rather than for α4 protein inasmuch as the elevated basal level of expression was independent of α4 protein. Moreover, the fusion of the 59bp fragment to the β regulated promoter chimera restored the capacity of this gene to be highly induced by α4 protein. The studies presented in this paper suggest that the 59bp fragment substituted for elements in the 2nd distal signal mapped by McKnight (22) by providing an α4 binding site.

In the case of the α genes, the α4 protein binds to both promoter and regulatory domains. It remains to be determined whether the binding of α4 interferes with the α <u>trans</u>-inducing factor and host regulatory factors specific for sequences in the α regulatory domains.

REFERENCES

1.   Honess RW, Roizman B (1974). Regulation of herpesvirus macromolecular synthesis. I. Cascade regulation of the synthesis of three groups of viral proteins. J Virol 14:8.
2.   Roizman B, Batterson W (1985). The replication of herpes-viruses (Chapter 25). In Fields B (ed): "Virology," p 497.
3.   McKnight JLC, Kristie TM, Silver S, Pellett PE, Mavromara-Nazos P, Campadelli-Fiume G, Arsenakis M, Roizman B (1986). Regulation of herpes simplex virus gene expression: The effect of genomic enviroments and its implications for model systems. In Botchan M, Grodzicker T, Sharp P (eds): "Cancer Cells 4, Control of gene expression and replication," Cold Spring Harbor Laboratories (in press).
4.   O'Hare P, Hayward GS (1985). Three trans-acting regulatory proteins of herpes simplex virus modulate immediate early-early gene expression in a pathway involving positive and negative feedback regulation. J Virol 56:723.

5. Preston CM (1979). Control of herpes simplex virus type 1 mRNA synthesis in cells infected with wildtype virus or the temperature-sensitive mutant tsK. J Virol 29:275.
6. Morse LS, Pereira L, Roizman B, Schaffer P (1978). Anatomy of HSV DNA. XI. Mapping of viral genes by analysis of polypeptides and functions specified by HSV-1 X HSV-2 recombinants. J Virol 26:389.
7. Wilcox KW, Kohn A, Sklyanskaya E, Roizman B (1980). Herpes simplex virus phosphoproteins. I. Phosphate cycles on and off some viral polypeptides and can alter their affinity for DNA. J Virol 33:167.
8. Metzler DW, Wilcox K (1985). Isolation of herpes simplex virus regulatory protein ICP4 as a homodimeric complex. J Virol 55:329.
9. Freeman MJ, Powell KL (1982). DNA-binding properties of a herpes simplex virus immediate early protein. J Virol 44:1084.
10. Fried M, Crothers D (1981). Equilibria and kinetics of lac repressor-operator interactions by polyacrylamide gel electrophoresis. Nucleic Acids Res 9:6505.
11. Garner MM, Revzin A (1981). A gel electrophoresis method for quantifying the binding of proteins to specific DNA regions. Nucleic Acids Res 9:3047.
12. Strauss F, Varshavsky A (1984). A protein binds to a satellite DNA repeat at three specific sites that would be brought into mutual proximity by DNA folding in the nucleosome. Cell 37:889.
13. Ejercito PM, Kieff ED, Roizman B (1968). Characterization of herpes simplex virus strains differing in their effect on the social behavior of infected cells. J Gen Virol 3:357.
14. Post LE, Mackem S, Roizman B (1981). The regulation of α genes of herpes simplex virus: expression of chimeric genes produced by fusion of thymidine kinase with α gene promoters. Cell 24:555.
15. Silver S, Roizman B (1985). γ_2-thymidine kinase chimeras are identically transcribed but regulated as γ_2 genes in herpes simplex virus genomes and as $\beta\iota$ genes in cell genomes. Mol Cell Biol 5:518.
16. Manley JL, Fire A, Cano A, Sharp PA, Gefter ML (1980). DNA-dependent transcription of adenovirus genes in a soluble whole-cell extract. Proc Natl Acad Sci USA 77:3855.
17. Dignam JD, Lebovitz RM, Roeder R (1983). Accurate transcription initiation by RNA polymerase II in a soluble extract from mammalian nuclei. Nucleic Acids Res 11:1475.
18. Ackermann M, Braun DK, Pereira L, Roizman B (1984). Characterization of α proteins 0, 4, and 27 with monoclonal

antibodies. J Virol 52:108.

19. Pereira L, Klassen T, Baringer JR (1980). Type-common and type-specific monoclonal antibody to herpes simplex virus type 1. Infect Immun 29:724.

20. Roizman B (1979). The organization of the herpes simplex virus genomes. Ann Rev Genet 13:25.

21. Mackem S, Roizman B (1982). Structural features of the herpes simplex virus α genes 4, 0, and 27 promoter-regulatory sequences which confer α regulation on chimeric thymidine kinase genes. J Virol 44:939.

22. McKnight SL, Kingsbury R (1982). Transcriptional control signals of a eukaryotic protein-coding gene. Science 217:216.

23. Hall LM, Draper KG, Frink RJ, Costa RH, Wagner EK (1982). Herpes simplex virus mRNA species mapping in EcoRI fragment I. J Virol 43:594.

24. Wu C (1984). Two protein-binding sites in chromatin implicated in the activation of heat-shock genes. Nature 309:229.

25. Wu C (1985). An exonuclease protection assay reveals heat-shock element and TATA box DNA-binding proteins in crude nuclear extracts. Nature 317:84.

26. Kristie TM, Roizman B (1984). Separation of sequences defining basal expression from those conferring α gene regulation within the regulatory domains of herpes simplex virus 1 α genes. Proc Natl Acad Sci USA 81:4065.

27. Jones KA, Tjian R (1985). Sp1 binds to promoter sequences and activates herpes simplex virus 'immediate-early' gene transcription in vitro. Nature 317:179.

28. Everett RD (1983). DNA sequence elements required for regulated expression of the HSV-1 glycoprotein D gene lie within 83 bp of the RNA capsites. Nucleic Acids Res 11:6647.

29. Schek N, Bachenheimer SL (1985). Degradation of cellular mRNAs induced by a virion-associated factor during herpes simplex virus infection of vero cells. J Virol 55:601.

30. Green MR, Treisman R, Maniatis T (1983). Transcriptional activation of cloned human β globin genes by viral immediate-early gene products. Cell 35:137.

31. Imperiale MJ, Feldman LT, Nevins JR (1983). Activation of gene expression by adenovirus and herpesvirus regulatory genes acting in trans and by a cis-acting adenovirus enhancer element. Cell 35:127.

32. Breathnach R, Chambon P (1981). Organization and expression of eucaryotic split genes coding for proteins. Ann Rev Biochem 50:349.

Transcriptional Control Mechanisms, pages 171–180
© 1987 Alan R. Liss, Inc.

VIROIDS CONTAIN SEQUENCES CHARACTERISTIC OF GROUP I INTRONS[1]

Gail Dinter-Gottlieb and Thomas R. Cech

Department of Chemistry and Biochemistry
University of Colorado, Boulder, Colorado 80309-0215

ABSTRACT Group I introns have been located in nuclear
rRNA genes, mitochondrial mRNA and rRNA genes, and
chloroplast tRNA genes. The hallmarks of this class
of introns are a sixteen nucleotide consensus sequence
and three sets of complementary bases. The viroids,
circular pathogenic plant RNAs, also contain the sixteen nucleotide sequence and the three sets of complementary bases. Pairing of the complementary sequences
would generate a viroid structure resembling a Group I
intron, which might be stabilized in vivo through
interactions with proteins. The Tetrahymena self-
splicing rRNA intron also contains sequence homologies
with regions of the potato spindle tuber viroid which
correlate with viroid infectivity.

INTRODUCTION

Many eukaryotic genes are interrupted by apparently
extraneous segments of DNA, termed introns. In order for the
gene to be properly expressed, the introns must be removed
from the RNA transcripts and the exons ligated in a reaction
known as RNA splicing. Following excision, the intron RNA
may exist in a lariat structure (1,2), as a linear molecule
(3,4,5), or as a circle (5,6).

One intron class, the Group I introns, is defined by
conserved sequence and structural features and is found
among mitochondrial messenger and ribosomal RNA genes,

[1] This work was supported by an NIH Postdoctoral Fellowship,
 #5 F32 GM 09831 to G.D.-G., and by NIH Grant GM 28039 and
 ACS Grant NP-374 to T.R.C.

chloroplast transfer RNA genes, and nuclear ribosomal RNA
genes, a surprisingly broad distribution (3,7-14). The
hallmark of this intron class is a sixteen nucleotide
phylogenetically conserved sequence (3), the Group I consensus sequence (Figure 1a). Twelve bases of this sequence,
termed Box 9L, comprise a region of cis-dominant mutations
in the splicing of the yeast mitochondrial cytochrome b
(cob) intron 4. Elsewhere in the intron is Box 2, five bases
of which are complementary to five bases of Box 9L. Another
pair of sequence elements, called A and B (or P and Q), is
less conserved as to sequence, but located upstream of Box
9L and rich in GC (15). A third pair of sequences, Box 9R
and 9R', although not conserved as to sequence, are conserved in location. 9R is located just downstream of 9L, and 9R'
is just upstream of A. The ability of 9R and 9R' to base
pair has been implicated in yeast cob mRNA splicing, since a
point mutation at the base of the helix destroys splicing,
while a double mutation, restoring pairing, also restores
splicing of the intron (16).

The pairing of the complementary boxes, and the order
in which they occur within the intron, 9R', A, B, 9L, 9R and
2, force the RNA into a characteristic structure (Figure
1a). The Box 9L: Box 2 pairing would be at the level of
tertiary structure in this folding.

The phenomenon of self-splicing has been described in
vitro for several Group I introns, the nuclear ribosomal RNA
intron of Tetrahymena thermophila (17,18,19), Neurospora
mitochondrial cob intron 1 (4) and the large rRNA and two
mRNA introns from yeast mitochondria (20,21). The sole
requirements for these reactions in vitro are GTP and a
divalent cation, usually magnesium, and in some cases
monovalent cations or polyamines. No enzymes or other
proteins are required. A series of transesterification
reactions occurs, as guanosine becomes covalently bonded to
the 5' end of the excised intron, followed by exon ligation.
In many cases, the 3' terminal G-OH of the intron can then
attack a specific bond near the 5' end of the intron, thus
creating a covalently closed circle and releasing a short
oligonucleotide. While both the Tetrahymena and the yeast
rRNA introns cyclize, no circle form has been found for the
Neurospora intron. Based on similarities in intron structure, it seems reasonable that all Group I introns splice
through a similar mechanism (22). In some cases, however,
proteins may be necessary for stabilization of the structure
and maintaining the base-pairing of the boxes in vivo.

Since viroids are single-stranded, circular RNAs,

similar in size to some of the circular introns, comparisons have been drawn between the two types of molecules. Diener (23) proposed an intron origin for viroids, and subsequently, noting the similarities between the RNAs based on homologies between the negative strand of the viroid and the U1 snRNP, proposed that the viroids might be escaped introns (24). As additional viroids have been sequenced, further comparisons between the viroids and the various intron types have become possible.

RESULTS

Viroids and Group I introns

The sixteen nucleotide Group I consensus sequence has been located in all of the viroids (25). In fact, this sequence represents the lower portion of the "central conserved region" of the viroids. As shown in Figure 2, this sequence is present even in viroids such as coconut cadang-cadang viroid (CCCV), which has only 11 % sequence homology with the potato spindle tuber viroid (PSTV) group (26).

The conserved sequence elements of Group I introns also occur in the viroids (25). Box 9L, a portion of the consensus sequence, is present in all viroids examined, although there is one base change in all viroids except for TASV and TPMV. The initial G has been changed to a U or A. A Box 2 region has been located, and in some cases it contains a single base change such that its ability to form five base pairs with Box 9L is preserved (Figure 2). Such compensatory base changes provide evidence that sequence elements are paired in folded RNA structures (27). Boxes 9R and 9R' are also present, as well as the GC-rich Boxes A and B. Significantly, the 5' to 3' order of the sequence elements, 9R', A, B, 9L, 9R and 2, is the same as in the Group I introns.

This order is important, because the pairing of the sequence elements determines the structure of the Group I intron (Figure 1a). When the boxes in PSTV are paired, a viroid structure is generated which is strikingly similar to that of Group I introns in the region of the boxes (Figure 1b). The calculated free energy of this molecule, approximately -100 kcal/mole, is not nearly as favorable as that calculated for the rod structure, -209.8 kcal/mole (M. Zuker, personal communication), yet it is possible that a structure such as this might be stabilized by proteins in vivo.

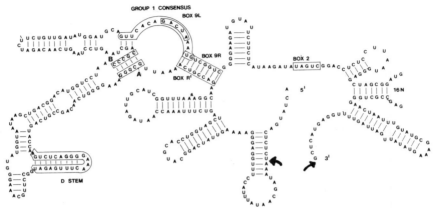

Figure 1a. The structure of a Group I intron (exemplified by the self-splicing intron of <u>Tetrahymena</u>)
generated by pairing Box 9R with Box 9R', and Box A
with Box B in the secondary structure, and Box 9L with
Box 2 at the level of tertiary structure (after Michel
and Dujon, 1983). Arrows denote splice sites.

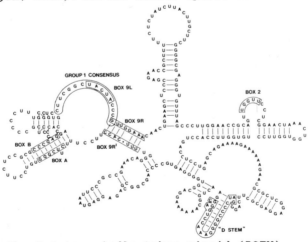

Figure 1b. Potato spindle tuber viroid (PSTV)
structure derived from pairing the conserved sequence
elements which are also found in Group I introns.

	GA (CGU)YUCAACGACUACANG	Box 9L: Box 2	Box 9R: Box 9R'	Box A: Box B
Tetrahymena IVS	GUUCACAGACUAAAUG	GACUA	UGUCGGUC	UGCGGG .....·.
PSTV	CUUCGGCUACUACCCG	CUGAU UACUA	ACUGCCAG GGUGGAAA	ACGCCC UUCGGG . ....
CEV	CUCUGGAUACUACCCG	UUGGU UACUA ...	CCCACUUU GGUGGAAA	ACGCCC CUGG ....
CSV	CUUUGGCUACUACCCG	UGGGU AACUA	CUUCCUCU GGUGGAAA	GGCC GGCC
TPMV	CUUCGGAGACUACCCG	UUGGU GACUA	CCCACUUU GGUGGAAA	CCGG CGGGU .....
TASV	CUCUGGAGACUACCCG	CUAGG GACUA	CCACUUUU CCUCCAAA	GCCCG CUUCUGG ....
CCCV	CUUGGGAGACUACCCG	CUGGU GACUA	CCCACUUU GGUGGAAA	GAAGGCC CUUCUGG ....
		CUGGU	CCUCCUCU	GAAGGCC

Figure 2. A series of conserved sequence elements which pair and delineate a characteristic structure for the Group I introns are also found in viroids, in the same 5' to 3' order.

All Group I introns terminate in a guanosine in the region following Box 2. In the circular form of a Group I intron, this G residue is between Box 2 and Box 9R'. Recently Visvader et al. (28) determined the site of processing of a cloned monomer of citrus exocortis viroid (CEV) and concluded that the three processing sites all followed a guanosine residue in the region corresponding to the viroid central conserved region(CCCGGG), which lies between Box 2 and Box 9R' (Figures 1a,b).

Viroids and the self-splicing intron of Tetrahymena

The homologies between Group I introns and viroids prompted a search for other similarities with the self-splicing intron of Tetrahymena thermophila. The two circular RNAs are similar in size; PSTV is 359 bases in length, and the Tetrahymena intron is 399 bases in its circular form. The similarities between the two classes of RNAs in their secondary and tertiary structures as Group I introns have been described. Surprisingly, a strong homology was found between the "d stem" of the Tetrahymena intron and the viroid central conserved region, comprising 15/19 bases (Figure 3a). When the viroid structure is written as a rod, this portion of the viroid is shown paired, in the central conserved region, with the 16 nucleotide consensus sequence found in all Group I introns. While the significance of this region is not yet known, experiments in our laboratory have shown that the "d stem" portion of the Tetrahymena intron can be deleted with little or no effect on the self-splicing activity of the molecule (G. Dinter-Gottlieb, L.A.H. Dokken and T.R. Cech, unpublished results).

Tetrahymena rRNA intron: AGUCUCAGGGGAAACUUUGAGA

Potato spindle tuber viroid: GAUCCCCGGGGAAACCUGGAGC

Vicia faba tRNA intron: AGCCUUGGUAUGGAAACAUAUUAAG

    Figure 3a. Homologies with a region of the Tetrahy-
mena intron termed the "d stem" are seen in viroids and the two plant chloroplast tRNA Group I introns which have been sequenced.

    The differences in symptom severity caused by mild, intermediate and severe strains of PSTV depend upon the nucleotide sequences in three regions of the molecule (29,30). Figure 3b reveals that homologies are found between the Tetrahymena linear intron and the mild PSTV strain.

         I. PSTV (49-54):   GAAAAG
             IVS (27-32):   GAAAAG

         II. PSTV (114-122):   UGGCAAUAAG
             IVS (179-188):   UGGUAAUAAG

        III. PSTV (303-314):   UAUCUUUCUUUG
             IVS (10-22):   UAUUUACCUUUG

    Figure 3b. Sequence variations in the "infectivity regions" of PSTV determine the extent of pathogenic effects caused by the viroid. Homologies with these regions exist in the Tetrahymena intron.

The region of homology in the IVS from nucleotides 10 to 22 contains the site of cyclization of the linear molecule (19). Since even minor changes in the sequence of PSTV may eliminate replication and infectivity, it is difficult to ascertain whether these similarities have any significance, but they further reinforce the relationship between the two molecules and emphasize the conservation of significant sequences across species.

DISCUSSION

The hypothesis that viroids might be related to introns has been further supported by numerous sequence and structural homologies between Group I introns and, specifically, with the self-splicing intron of Tetrahymena thermophila. In fact, based on the crucial sequence similarities, the viroids appear to be closely related to Group I introns. The question still remains as to whether viroids evolved from introns,or whether both evolved from a common ancestor molecule. The phylogenetic conservation of the consensus sequence of 16 nucleotides attests to its importance.

Fortunately, some of the pertinent questions can be approached on an experimental level. First, can viroids self-cleave and auto-cyclize?

Intermediates of viroid replication have been detected in vivo, comprising both plus and minus strand concatemers. The detection of minus strand PSTV linear concatemers has led to a model for viroid replication (31). An unknown plant polymerase might copy the circular plus strand via a rolling circle form of replication producing minus strand linear concatemers. These could be copied into plus strand concatemers, cleaved and cyclized via plant enzymes. Conversely, the minus strand concatemer might first be cleaved, then copied into plus strand monomers and cyclized. The enzymes for these reactions have not been identified, but an RNA ligase activity from wheat germ can cyclize the natural PSTV linear monomers (32).

Yet linear concatemers can be formed in a non-enzymatic fashion. The linear Tetrahymena intron can form concatemers in vitro. The 3' terminal G-OH attacks the junction between nucleotides 15 and 16 in another linear molecule, becoming covalently attached, and releasing an oligonucleotide of length 15, in a reaction analogous to the cyclization reaction. Dimer, trimer and larger linear molecules have been detected, as well as their circular forms. Under cyclization conditions, these will then form monomer circles (33). The cloning of PSTV cDNA (34) allows in vitro transcription of viroid molecules, and these might be assessed for such activity.

PSTV dimers derived from transcription of cloned viroid cDNA have been isolated and placed under splicing and cyclization conditions, in reactions analogous to those seen for the Tetrahymena self-splicing intron (6,19). Robertson et al have recently reported that a monomer length PSTV

containing a 2',3'cyclic phosphate was produced with low
efficiency under such conditions (35). If this reaction
proves to be relevant to the viroid replication cycle in
vivo, it indicates a different cleavage and ligation
mechanism than that seen for the Tetrahymena intron.

While the self-splicing intron of Tetrahymena pre-rRNA
requires no proteins for splicing or to maintain its structure, other introns with less stable base pairing or with
competing non-productive secondary structures may require
proteins to maintain the active structure. In vitro PSTV
oligomeric transcripts can be specifically cleaved and
circularized when incubated in cell extracts (H.L. Sanger,
personal communication). These reactions may be carried out
by enzymes in the extract, or cellular proteins may be
required for the formation of stable viroid RNPs, allowing
their subsequent self-splicing. The nuclear rRNA intron of
Neurospora crassa contains the six boxes, but they do not
pair to give the core structure in deproteinized RNA because
stronger alternative base pairing forces the region into a
rod structure. In vivo, however, the intron exists as a
ribonucleoprotein, and psoralen cross-linking studies indicate a structure that is consistent with pairing of the
conserved sequence elements (4,36).

Recent evidence indicates that PSTV in vivo is found in
a ribonucleoprotein complex (37,38). The intriguing possibility exists then that viroids, found as rods in the
deproteinized state, can assume other foldings when stabilized by proteins in the cell.

A drawback to the in vitro experiments concerns the
structure of the viroid transcript, which, once
deproteinized, will be in the rod form, and presumably
inactive. However, the existence of metastable forms might
contribute a population of active molecules with base-
pairing similar to Group I introns.

The natural linear viroid molecules arising from the
viroid circles upon storage are also of interest. When the
circular Tetrahymena intron is incubated in a Mg++ containing buffer, a unique bond is broken, at the same site at
which the circle was formed. This has been termed
autoreopening or site-specific hydrolysis (39). The reaction
rate increases rapidly with pH in the pH range 7.5-9.5 (39).
The viroid circles might be subjected to similar conditions,
to see if such a specific bond might be selected.

The discovery that the "central conserved region" of
the viroids is not unique to viroids, but that one portion
of it exists in all Group I introns as the Group I consensus

and another portion is found in some introns as the "d stem" (Figure 2), raises questions about the function of this region. The "d stem" in the Tetrahymena intron is located in the region bounded by the A and B boxes, while the homologous region in viroids is located upstream of box A and box 9R'. In the plant tRNA Group I introns it is located near the 5' end of the intron. This may be an example of the intermolecular RNA rearrangement recently proposed for viroid evolution (40). While this region is not necessary for self-splicing of the Tetrahymena intron, it may serve as a recognition site for structural proteins, or it may be essential for replication of the viroids.

While the in scripto analysis of similarities between the molecules provides interesting insights, the final comparisons must be made in the laboratory.

REFERENCES

1. Padgett RA, Konarska MM, Grabowski PJ, Hardy SF and Sharp PA (1984). Science 225:898.
2. Ruskin B, Krainer A, Maniatis T, and Green MR (1984). Cell 38:317.
3. Burke J and RajBhandary U (1982). Cell 31:509.
4. Garriga G and Lambowitz A (1984). Cell 39:631.
5. Grabowski P, Zaug AJ, and Cech TR (1981). Cell 23:467.
6. Halbreich A, Pajot P, Foucher M, Grandchamp C and Slonimski PP (1980). Cell 19:321.
7. Michel F and Dujon B (1983). The EMBO J 2:33.
8. Davies RW, Waring RB, Ray JA, Brown TA and Scazzocchio C (1982). Nature 300:719.
9. Anziano PQ, Hanson DK, Mahler HR and Perlman PS (1982). Cell 30:925.
10. De La Salle H, Jacq C and Slonimski PP (1982). Cell 28:721.
11. Netter P, Jacq C, Carignani G and Slonimski PP (1982). Cell 28:733.
12. Weiss-Brummer B, Rodel G, Schweyen RJ and Kaudewitz F (1982). Cell 29:527.
13. Steinmetz A, Gubbins E and Bogorad L (1982). Nuc. Acids Res.10:3027.
14. Bonnard G, Michel F, Weil J and Steinmetz A (1984). Mol.Gen.Genet. 194:330.
15. Waring RB, Ray JA, Edwards SW, Scazzocchio C and Davies R W (1985). Cell 40:371.
16. Weiss-Brummer B, Holl J, Schweyen RJ, Rodel G and

Kaudewitz F   (1983). Cell 33:195.
17. Cech TR, Zaug AJ and Grabowski PJ (1981). Cell 27:487.
18. Kruger K, Grabowski PJ, Zaug AJ, Sands J, Gottschling D and Cech TR (1982). Cell 31:147.
19. Zaug AJ, Grabowski PJ and Cech TR (1983). Nature 301:578.
20. Tabak HF, van der Horst G, Osinga KA and Arnberg AC (1984). Cell 39:623.
21. van der Horst G and Tabak HF (1985) Cell 40:759.
22. Cech TR, Tanner NK, Tinoco I, Weir BR, Zuker M and Perlman PS (1983). Proc. Natl. Acad. Sci. USA 80:3903.
23. Diener TO (1979). Science 205:859.
24. Diener TO (1981). Proc. Natl. Acad. Sci. USA 78:5014.
25. Dinter-Gottlieb G and Cech TR (1984) in "Abstracts of the Molecular Basis of Plant Disease Conference" University of California: Davis, CA.
26. Haseloff J, Mohamed NA and Symons RH (1982). Nature 299:316.
27. Noller HF and Woese CR (1981). Science 212:403.
28. Visvader JE, Forster AC and Symons RH (1985). Nuc. Acids Res. 13:5843.
29. Dickson E, Robertson HD, Niblett CL, Horst RK and Zaitlin M (1979). Nature 277:60.
30. Schnolzer M, Haas B, Ramm K, Hofmann H and Sanger H (1985). The EMBO Journal 4:2181.
31. Branch AD and Robertson HD (1983). Science 223:450.
32. Branch AD, Robertson HD, Greer C, Gegenheimer P, Peebles C and Abelson J (1982). Science 217:1148.
33. Zaug AJ and Cech TR (1985). Science 229:1060.
34. Cress D, Kiefer M and Owens RA (1983). Nuc. Acids Res. 11:6821.
35. Robertson HD, Rosen DL and Branch AD (1985). Virology 142:441.
36. Wollenzein PL, Cantor CR, Grant DM and Lambowitz AM (1983). Cell 32:397.
37. Schumacher J, Sanger HL and Riesner D (1983). The EMBO J 2:1549.
38. Wolff P, Gilz R, Schumacher J and Riesner D (1985). Nuc. Acids Res. 13:355.
39. Zaug AJ, Kent J and Cech TR (1984). Science 224:374.
40. Keese P and Symons RH (1985). Proc. Natl.Acad. Sci.USA 82:4582.

Transcriptional Control Mechanisms, pages 181–194
© 1987 Alan R. Liss, Inc.

AN ANALYSIS OF SPECIFIC RNA–FACTOR INTERACTIONS DURING PRE–mRNA SPLICING IN VITRO[1]

Albrecht Bindereif, Barbara Ruskin, and M.R. Green

Department of Molecular Biology and Biochemistry
Harvard University
Cambridge, Massachusetts 02138

ABSTRACT The ribonucleoprotein (RNP) structures of pre–mRNA and RNA processing products generated during in vitro splicing of an SP6/β–globin pre–mRNA were characterized in the crude nuclear extract and follow-ing partial purification by sucrose gradient sedimenta-tion. Biochemical components (splicing factors) specif-ically associate with the branch point and the 5' splice site of the pre–mRNA as determined by protection from RNase A digestion and from oligonucleotide-directed RNase H cleavage. The latter assay also allowed the detection of the same RNA–factor interac-tions in the IVS1–containing RNA processing products. The various RNP complexes of the pre–mRNA and RNA pro-cessing products were partially fractionated by sucrose gradient sedimentation and characterized by protection from RNase A digestion, protection of the lariat species from enzymatic debranching, and by immunopre-cipitation.

Taken together, the results of these experiments indi-cate that, early in the pre–mRNA splicing reaction, the branch point region and subsequently the 5' splice site associate with factor(s) to assemble a functional 60S pre–mRNA RNP complex, that these interactions are main-tained in the 60S lariat intermediate RNP complex, and that, following 3' splice site cleavage and exon–exon ligation, most of the Sm–determinants segregate with the excised IVS1 RNP complex. Evidence is presented that factors dissociate from the excised IVS1 RNP com-plex in an ordered manner. The implications of these results for understanding the role of RNP structure in the pre–mRNA splicing mechanism and in intron degrada-tion are discussed.

INTRODUCTION

Most structural genes in higher eucaryotes contain intervening sequences that are removed from the primary transcripts (pre-mRNAs) by splicing (1). In addition to being required for the correct alignment of protein coding sequences, RNA splicing also provides a mechanism to regulate gene expression; the differential splicing of a pre-mRNA can generate multiple mRNA products, which encode different proteins. This strategy has been exploited by DNA tumor viruses such as SV40 and adenovirus (2, 3) and is also found in some cellular genes.

In order to understand how splicing can provide a regulatory function, it is first necessary to analyze the basic biochemical mechanisms that occur during the pre-mRNA splicing reaction. The pre-mRNA primary sequences including the consensus sequence elements at the splice junctions (4) and at the branch point (5) are not sufficient to answer the critical question of what determines how the accuracy of splicing and correct splice-site selection are achieved. In addition, it is not known how pre-mRNAs with more than one intron are correctly spliced without deleting introns. These questions have led to the speculation that features other than the primary sequence of the pre-mRNA, such as the RNP structure, are required for the specificity of the splicing reaction (6-8). The development of efficient in vitro splicing systems (9-12) now allows a direct biochemical analysis of the RNP structure of a well-characterized eucaryotic primary transcript, the human β-globin pre-mRNA. Investigating the contribution of RNP structures, both in crude nuclear extract and after purification by sucrose gradient sedimentation, to splice site specificity should provide important insight into pre-mRNA splicing mechanisms. By a similar approach, it has recently been shown in yeast (13) and mammalian cell extracts with an adenovirus major late substrate (14, 15) that the pre-mRNA, splicing intermediates, and products are in the form of rapidly sedimentable RNP complexes.

[1] This work was supported by a Public Health Service grant from the National Institutes of Health to M.R.G., A.B. was supported by a postdoctoral fellowship from the Deutsche Forschungsgemeinschaft, and B.R. by a National Institutes of Health predoctoral training grant.

RESULTS

1. Characterization of Specific RNA-factor Interactions in the Crude Nuclear Extract

We have been studying the interaction of biochemical components with specific regions of the pre-mRNA and RNA processing products. Due to the diverse, complex structures RNA molecules can form, mapping specific RNA-factor interactions is more difficult than mapping specific DNA-factor interactions; secondary and tertiary structure can render the RNAs resistant to the same cleavage reagents that are used to detect RNA-factor interactions. With these considerations in mind, we have used several methods to detect and characterize the interaction of factors with specific RNA regions (16).

Protection from oligonucleotide-directed RNase H cleavage. First, we have developed an RNase H-directed cleavage assay, which is diagrammed in Figure 1. Synthetic oligonucleotides complementary to specific RNA regions and RNase H are added directly to the splicing reaction at various times during in vitro processing, which enables an RNase H-directed site-specific cleavage reaction to occur (5, 17, 18). High concentrations of oligonucleotides allow cleavage of naked RNA even in regions of significant secondary structure. As diagrammed in Figure 1, if, in a particular RNA species, the region complementary to the oligonucleotide is free of associated factors, the RNA will hybridize to the oligonucleotide and be subsequently cleaved by RNase H. However, a factor bound to a specific RNA region can prevent hybridization of the oligonucleotide and/or RNase H cleavage. This is a particularly advantageous procedure for analyzing multiple, related RNA species. More recently this method has been used to study the interactions of specific factors with yeast pre-mRNAs (M. Rosbash, personal communication) and to identify a factor interacting with the polyadenylation site of SV40 RNA (M. Wickens, personal communication).

Using oligonucleotides complementary to various regions of the human β-globin IVS1 we have shown that very early during an in vitro splicing reaction (within 5 min) a factor specifically associates with the branch point; subsequently a factor becomes associated with the 5' splice site. In contrast, other intron regions are not detectably bound to factors and are therefore susceptible to RNase H-directed cleavage. Remarkably, the IVS1-containing RNA processing

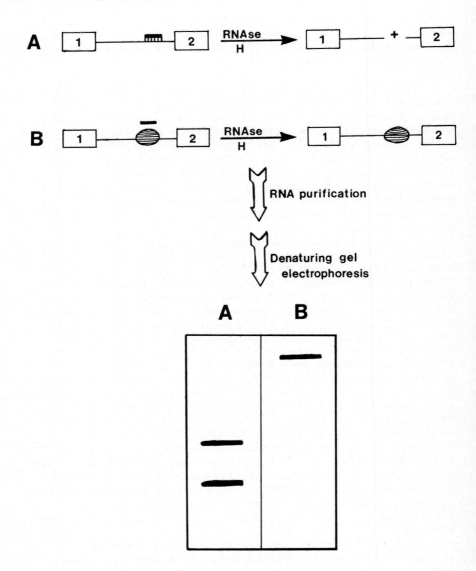

FIGURE 1. Schematic representation of the
oligonucleotide-directed RNase H cleavage assay.

products display a pattern of protection from RNase H-directed cleavage analogous to that found with the pre-mRNA. Thus, once established, these specific RNA-factor interactions are stable and maintained in the intron-containing RNA processing products (16).

   Protection from RNase A digestion. We have studied the interaction of factor(s) with the branch point in further detail using an RNase A protection assay (16). RNase A is added directly to an in vitro splicing reaction, incubation at 30°C continued, and the ^{32}P-RNA fragments that are resistant to RNase A digestion detected by denaturing polyacrylamide gel electrophoresis. The branch point-factor interaction is extremely stable; thus, at high concentrations of RNase A the primary RNase A-resistant product is the branch point region. This assay has been used to study the biochemical and sequence requirements for the association of a factor(s) with branch point. The specific, stable interaction of a factor(s) with the branch point is atypical of most nucleic acid-factor interactions in that there is a specific requirement for ATP hydrolysis (16). Significantly, cleavage at the 5' splice site which occurs during the first step of splicing, also has a specific requirement for ATP hydrolysis (10, 19). With regard to sequence requirements, branch point protection does not require the vast majority of exon sequence, the 5' splice site, or the authentic branch point sequence. However, branch point protection, like the first step in splicing (14, 20, 20a) is absolutely dependent upon the 3' splice site consensus sequence (16).

   Identifying and characterizing the biochemical components that are necessary and sufficient for selection of the branch point accompanied by the stable association of the branch point with factor(s) will be an important step towards understanding splice site selection. Micrococcal nuclease inactivation studies indicated that at least one of the factors involved in branch point protection contains an essential nucleic acid component (16). Immunoprecipitation experiments indicated that the U2 snRNP, an essential splicing factor (21, 22) becomes associated with the branch point region (22). However, our recent fractionation indicated that in addition to the U2 snRNP, other factors, probably including non-snRNP components, are necessary for the full pattern of branch point protection (B. Ruskin, P. Zamore, and M. Green, unpublished data).

2. Identification and Analysis of RNP Complexes Generated
During Pre-mRNA Splicing In Vitro

In the following section we will summarize how we identified RNP complexes of the human β-globin pre-mRNA and RNA
processing products, and describe and compare the results of
various experimental approaches to analyze the RNP complexes
that are generated during pre-mRNA splicing in vitro in a
HeLa cell nuclear extract. These experimental approaches
include: (i) immunoprecipitation with specific anti-snRNP
antibodies, and (ii) protection from RNase A digestion and
enzymatic debranching (23). Furthermore we will compare
these results to those obtained in the crude nuclear
extract.

    Identification of RNP complexes of pre-mRNA and RNA
processing products. To identify RNP complexes of pre-mRNA,
splicing intermediates and products, ^{32}P-labeled SP6/β-
globin pre-mRNA was spliced in vitro for 90 min in a HeLa
cell nuclear extract and the reaction mixture was fractionated by sucrose gradient centrifugation. The ^{32}P-labeled
RNA was purified from the gradient fractions and analyzed by
denaturing polyacrylamide gel electrophoresis (Fig. 2). The
sedimentation profile shows that under these conditions the
various RNP complexes are in the form of rapidly sedimenting
RNP complexes that can be partially fractionated. The pre-
mRNA RNP complexes distributed throughout the gradient, but
displayed peaks at 40S and 60S. The RNP complexes containing
the splicing intermediate RNAs (exon 1 and IVS1 lariat-exon
2) sedimented at 60S. The RNP complexes of the splicing products, i.e. of accurately spliced product RNA (exon 1-exon
2) and of excised lariat RNA, were distributed heterogenously across the gradient betweeen 10S and 60 S.
    The sedimentation properties of the various RNP com-
plexes were differentially affected by increasing the ionic
strength of the sucrose gradient (24). In particular, the
cosedimentation of the two splicing intermediate RNAs under
a variety of conditions (up to 0.9M KCl and with the addition of 10mM EDTA) strongly suggests that these two RNA
species are held together in a single RNP complex (24).

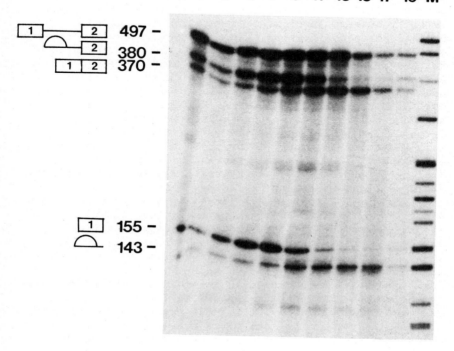

FIGURE 2. Sucrose gradient sedimentation analysis of RNP complexes generated during in vitro splicing of human β-globin pre-mRNA in HeLa cell nuclear extract. ^{32}P-labeled SP64-Hβ RNA was spliced in vitro for 60 min and analyzed by sucrose gradient sedimentation at 60 mM KCl as described (24). RNA prepared from a sample of the total splicing reaction (lane S) and from aliquots of the gradient fractions (numbered 1-20 from the bottom to the top of the gradient) was resolved by denaturing gel electrophoresis and detected by autoradiography. The sedimentation markers are indicated above the lanes. A schematic representation of the pre-mRNA, splicing intermediates, and products as well as the sizes of the various RNA species, is shown on the left. M, MspI-digested pBR322 DNA markers.

Analysis of 60S pre-mRNA RNP complex by protection from RNase A digestion. To determine whether factors bind to specific pre-mRNA regions during the formation of a functional 60S pre-mRNA RNP complex we used protection from RNase A digestion as an assay (16). 60S pre-mRNA RNP complexes formed after 20 min of in vitro splicing were isolated by sucrose gradient sedimentation and subjected to a partial RNase A digestion. The 60S RNP complex gave rise to multiple RNase A-resistant fragments, the most abundant of which were analyzed by RNase T1 digestion, followed by RNase A secondary digestion (5, 25). Based on these experiments we concluded that the branch point region of the pre-mRNA is most stably protected from RNase A digestion, presumably by an RNA binding factor(s). Preferential protection of sequences at the 3' end of the intron from nuclease digestion had been previously observed in the crude nuclear extract (16, 22, 26). The agreement between these two approaches suggests that structurally intact RNP complexes can be isolated by sucrose gradient sedimentation.

Analysis of splicing complexes by immunoprecipitation. A variety of studies have demonstrated that snRNPs are essential splicing factors whose mechanism of action involves binding to specific pre-mRNA regions (18, 21, 22, 26-28). Antibodies specific for either the Sm class of snRNPs (U1, U2, U4, U5, and U6 snRNPs; anti-Sm) or for U1 snRNP [anti-(U1)RNP] were used to probe for these components in the various RNP complexes isolated by sucrose gradient sedimentation. In Table 1 the immunoprecipitation efficiencies of the various RNP complexes are compared for both antibodies. Both 40S and 60S pre-mRNA RNP complexes were efficiently immunoprecipitated by anti-Sm, as well as the 60S lariat intermediate RNP complex. Equimolar recovery of both exon 1 and IVS1 lariat-exon 2 RNAs again argued that these two RNA species are held together in a single RNP complex. In contrast, the RNP complexes containing accurately spliced RNA were immunoprecipitated relatively inefficiently. RNP complexes containing the excised IVS1 displayed a characteristic pattern of anti-Sm immunoprecipitation. The larger forms (>30S) were efficiently immunoprecipitated, while the smaller forms (<30S) were not.

Using (U1)RNP specific antibodies we could show that both 40S and 60S pre-mRNA RNP complexes contain (U1)snRNP determinants. The 60S lariat intermediate RNP complex, however, was immunoprecipitated very inefficiently with the anti-(U1) antibody. Since we detected (U1)snRNP determinants in a distinct fraction of the excised IVS1 lariat RNP complexes (15S) these determinants are most likely conserved during the splicing pathway, but not accessible to the antibody in the 60S lariat intermediate RNP complex. Similarly as with the anti-Sm antibody, the RNP complexes of the accurately spliced RNA were relatively inefficiently immunoprecipitated by the anti-(U1)RNP antibody.

TABLE 1
IMMUNOPRECIPITATION[a] AND DEBRANCHING PROTECTION OF
PURIFIED RNP COMPLEXES

| RNP complex | Antiserum | | Debranching Protection |
	α-(U1)RNP	α-Sm	
40S pre-mRNA	++	++	NA
60S pre-mRNA	+	++	NA
60S lariat intermediate (exon 1/IVS1 lariat-exon 2)	–	++	resistant
20–60S spliced RNA (exon 1–exon 2)	–	–	NA
>30S excised lariat	–	+	resistant
<30S excised lariat	+ (15S)	–	sensitive

[a] ++, efficiently immunoprecipitated;
+, immunoprecipitated;
–, inefficiently or not immunoprecipitated;
NA, not applicable.

Analysis of splicing complexes by protection from enzymatic debranching. To determine whether specific protection of the branchpoint region in the pre-mRNA RNP complex, as determined by RNase A protection, is also retained in the lariat intermediate and in the excised lariat RNP complexes, we used resistance to enzymatic debranching as an assay (Fig. 3).

190   Bindereif, Ruskin, and Green

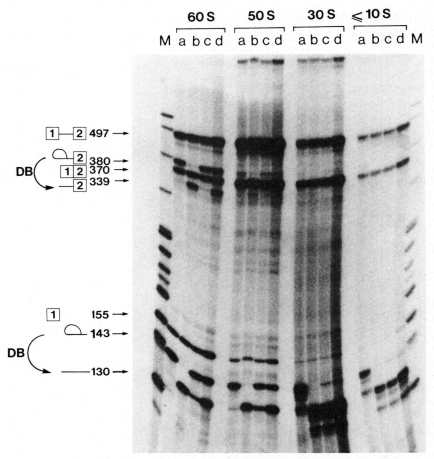

FIGURE 3.   Protection of lariat RNA species in RNP complexes from enzymatic debranching. Debranching (DB) protection was tested in gradient fractions from the 60S, 50S,
30S, and smaller-than-10S regions as described. For each
gradient fraction, the RNA composition is shown (lanes a),
as well as the RNA products after deproteinization and enzymatic debranching (lanes b) and after enzymatic debranching
without prior deproteinization (lanes c). An additional control are the RNA products after enzymatic debranching of a
1:1 mixture of deproteinized gradient fraction and untreated
gradient fraction (lanes d). M, MspI-digested pBR322 DNA
markers.

Enzymatic debranching resulted in conversion of the
deproteinized lariat RNAs to linear RNAs (Fig. 3). To obtain
partially purified RNP complexes containing lariat RNA
species, we fractionated 90-min splicing reaction mixtures
by sucrose gradient centrifugation and individually tested
fractions of 60S, 50S, 30S, and smaller than 10S. As controls, deproteinized RNA from the same gradient fraction and
a mixture of gradient fraction and deproteinized RNA were
assayed under identical conditions (Fig. 3). Remarkably, the
lariat intermediate RNA in the 60S RNP complex was completely resistant to enzymatic debranching. In contrast,
debranching resistant and sensitive forms of the excised
lariat RNP complex could be separated by gradient sedimentation. RNP complexes larger than 30S of the excised lariat
RNA were predominantly resistant to enzymatic debranching,
whereas RNP complexes smaller than 30S were sensitive.
Increasing the salt concentration to 0.5M KCl converted most
of the excised lariat RNP complexes to a small (15S),
debranching-sensitive form (24). Remarkably, the RNP complex
containing the lariat intermediate was completely stable to
enzymatic debranching even when isolated from 0.9M KCl (24).

DISCUSSION

The identification and partial characterization of RNP
complexes containing pre-mRNA, splicing intermediates and
products allowed us to draw several conclusions about the
splicing pathway. Specific, stable RNA-factor interactions
are established early and are retained throughout the splicing reaction.
As previously discussed (24), the 60S pre-mRNA RNP complex most likely represents the functional splicing substrate. 40S and smaller pre-mRNA RNP complexes may be intermediates of a pre-mRNA RNP assembly pathway. The product of
the first step in the pre-mRNA splicing reaction, i.e. 5'
splice site cleavage and lariat formation, is a 60S RNP complex containing both splicing intermediate RNA species (exon
1 and IVS1 lariat-exon 2). Both 60S pre-mRNA and lariat
intermediate RNP complexes are associated with factors at
the branch point region. Therefore structural features are
probably conserved during the first step of the reaction.
The second step, i.e. 3' splice site cleavage and exon 1-
exon 2 ligation, results in the excision of the IVS1 lariat.
Most of the Sm determinants present in the 60S lariat intermediate RNP complex seem to segregate with the IVS1 lariat

RNP complexes rather than with the spliced RNA RNP complexes. Excised IVS1 lariat can be found in RNP complexes sedimenting heterogenously between 10S and 60S which have distinct properties based on immunoprecipitation and debranching protection assays (see Table 1). The size distribution of these excised IVS1 lariat complexes probably reflects the ordered dissociation of factors from the branchpoint of the excised lariat complex leading to unmasking of (U1)snRNP determinants and to smaller lariat RNP complexes that can be debranched and eventually further degraded. This may be related to the in vivo intron degradation pathway.

Goals for the future are to identify the protein and snRNA components of the functional pre-mRNA RNP complex, to determine how these components are assembled, disassembled, and how they distribute among the various RNP complexes generated during the splicing reaction, and ultimately to reconstitute functional splicing complexes from pre-mRNA and purified components.

REFERENCES

1. Flint J (1983). RNA splicing in vitro. In Jacob ST (ed): 'Enzymes in Nucleic Acid Synthesis and Modification, 2', Boca Raton, Florida, CRC Press, p 76-110.
2. Ziff EB (1980). Transcription and RNA processing by the DNA tumour viruses. Nature 287: 491-499.
3. Tooze J (1981). 'DNA Tumor Viruses'. Cold Spring Harbor Press.
4. Mount SM (1982). A catalogue of splice site junctions. Nucl. Acids Res. 10: 459-472.
5. Ruskin B, Krainer AR, Maniatis T, Green MR (1984). Excision of an intact intron as a novel lariat structure during pre-mRNA splicing in vitro. Cell 38: 317-331.
6. Gruss P, Khoury, G (1980). Rescue of a splicing defective mutant by insertion of an heterologous intron. Nature 286: 634-637.
7. Kuhne T, Wieringa B, Reiser J, Weissmann C (1983). Evidence against a scanning model of RNA splicing. EMBO J 2: 727-733.
8. Pederson T (1983). Nuclear RNA-protein interactions and messenger RNA processing. J Cell Biol. 97: 1321-1326.
9. Hardy SF, Grabowski PJ, Padgett RA, Sharp PA (1984). Cofactor requirements of splicing of purified messenger RNA precursors. Nature (London) 308: 375-377.

10. Krainer AR, Maniatis T, Ruskin B, Green MR (1984). Normal and mutant human β-globin pre-mRNAs are faithfully and efficiently spliced in vitro. Cell 36: 993-1005.
11. Hernandez N, Keller W (1983). Splicing of in vitro synthesized messenger RNA precursors in HeLa cell extracts. Cell 35: 89-99.
12. Padgett RA, Hardy SF, Sharp PA (1983). Splicing of adenovirus RNA in a cell free transcription system. Proc. Natl. Acad. Sci. USA 80: 5230-5234.
13. Brody E, Abelson J. (1985). The 'spliceosome': yeast pre-messenger RNA associates with a 40S complex in a splicing-dependent reaction. Science 228: 1344-1349.
14. Frendewey D, Keller W (1985). Stepwise assembly of a pre-mRNA splicing complex requires U-snRNPs and specific intron sequences. Cell 42: 355-367.
15. Grabowski PJ, Seiler SR, Sharp PA (1985). A multicomponent complex is involved in the splicing of messenger RNA precursors. Cell 42: 345-353.
16. Ruskin B, Green MR (1985). Specific and stable intron-factor interactions are established early during in vitro pre-mRNA splicing. Cell 43: 131-142.
17. Donis-Keller (1979). Site specific cleavage of RNA. Nucl. Acids Res. 7: 179-192.
18. Kraemer A, Keller W, Appel B, Luehrmann R (1984). The 5' terminus of the RNA moiety of U1 small nuclear ribonucleoprotein particles is required for the splicing of messenger RNA precursors. Cell 38: 299-307.
19. Grabowski PJ, Padgett RA, Sharp PA (1984). Messenger RNA splicing in vitro: an excised intervening sequence and a potential intermediate. Cell 37: 415-427.
20. Ruskin B, Green MR (1985). Role of the 3' splice site consensus sequence in mammalian pre-mRNA splicing. Nature (London) 317: 732-734.
20a.Reed R, Maniatis T (1985). Intron sequences involved in lariat formation during pre-mRNA splicing. Cell 41: 95-105.
21. Krainer AR, Maniatis T (1985). Multiple factors including the small ribonucleoproteins U1 and U2 are necessary for pre-mRNA splicing in vitro. Cell 42: 725-736.
22. Black DL, Chabot B, Steitz JA (1985). U2 as well as U1 small ribonucleoproteins are involved in pre-messenger RNA splicing. Cell 42: 737-750.
23. Ruskin B, Green MR (1985). An RNA processing activity that debranches RNA lariats. Science 229: 135-140.
24. Bindereif A, Green MR (1986). Ribonucleoprotein complex formation during pre-mRNA splicing in vitro.

Mol. Cell. Biol., in press.
25. Volckaert G, Fiers V (1977), Microthin-layer techniques
for rapid sequence analysis of ^{32}P-labeled RNA: double
digestion and pancreatic ribonuclease analyses. Anal.
Biochem. 83: 228-239.
26. Chabot B, Black DL, LeMaster DM, Steitz JA (1985). The
3' splice site of pre-messenger RNA is recognized by a small
ribonucleoprotein. Science 230: 1344-1349.
27. Mount SM, Petterson I, Hinterberger M, Karmas A, Steitz
JA (1983). The U1 small nuclear RNA-protein complex selectively binds a 5' splice site in vitro. Cell 33: 509-518.
28. Padgett RA, Mount SM, Steitz JA, Sharp PA (1983). Splicing of messenger RNA precursors is inhibited by antisera to
small ribonucleoprotein. Cell 35: 101-107.

Transcriptional Control Mechanisms, pages 195–208
© 1987 Alan R. Liss, Inc.

TRANSLATION OF POLYCISTRONIC
mRNAs IN MAMMALIAN CELLS

Randal J. Kaufman, Patricia Murtha,
and Monique V. Davies

Genetics Institute
Cambridge, MA 02140

ABSTRACT

The efficiency of translation of polycistronic mRNAs in mammalian cells was studied.
Transcription units, constructed to contain
one, two, or three open reading frames, were
introduced transiently into COS monkey cells
and stably into Chinese hamster ovary cells.
Analysis of the levels of mRNA and protein
synthesis demonstrated that the mRNAs generated
encode multiple proteins. The efficiency of
translation is reduced approximately 10-fold
by the insertion of an upstream open reading
frame. The results support a modified 'scanning'
model for translation initiation which allows for
reinitiation of translation at internal AUGs.
The utility of vectors containing polycistronic
transcription units for the expression of
heterologous genes in mammalian cells is discussed.

INTRODUCTION

The study of the mechanism by which eukaryotic
ribosomes recognize particular sites in messenger RNA
has generated a substantial amount of evidence that
supports a 'scanning' model for translation
initiation (1).   This model suggests that a 40S
ribosomal subunit binds the 5' end of the mRNA and
migrates in the 3' direction until it encounters the

first AUG triplet which, if present in an appropriate
context, can efficiently serve as the initiator
codon. The hypothesis that 40S ribosomal subunits
bind at the capped 5' end of the mRNA and migrate on
mRNA prior to assembly of the 80S ribosome is
supported by the stimulatory effect of the m^7G cap
(2), the inability of ribosomes to bind circular
mRNAs (3,4), and the negative influence of
significant secondary structure and of additional
AUGs inserted between the 5' end of the mRNA and the
initiation codon (5,6,7,8). The fact that eukaryotic
ribosomes usually translate only the 5'-proximal open
reading frame (ORF) in polycistronic mRNAs is also
suggested by a scanning model. Several studies have
demonstrated translation initiation from internal AUG
codons in experimental constructs. These reports
have analyzed the effect of insertion or deletion of
initiator or terminator codons upstream from the
translation start site of a particular (ORF)
(5,6,9,10). These studies demonstrate that the
insertion of an upstream AUG, which is out of frame
with the downstream ORF, can severely suppress
translation of the downstream ORF. Insertion of a
termination codon can reverse the suppression. Thus,
these studies suggest a modified scanning model with
the potential for reinitiation at internal,
downstream AUGs. However, to date there have been
few reports that polycistronic mRNAs can actually
translate two or more proteins from non-overlapping
ORFs in the same transcription unit in mammalian
cells (11,12).

    We have examined the potential for mammalian
cells to translate polycistronic mRNAs. Plasmids
harboring transcription units containing one, two, or
three ORFs encoding protein products that can be
easily monitored, have been constructed and
introduced into Chinese hamster ovary cells by stable
transformation or into COS monkey cells by transient
transfection. The results demonstrate that both cell
types are capable of translating multiple ORFs on a
single polycistronic transcript. However, it was
found that the amount of translation of an ORF was
dramatically reduced by the presence of an upstream
ORF.

METHODS

Cell Culture

DHFR deficient CHO cells were transfected with the ADA-DHFR expression plasmid and selected initially for ADA and then amplified in increasing concentrations of 2'deoxycoformycin as described (16). The two cell lines (1-0.1 and 2-0.1) were designated 1-AU.1 and 2-AU.1 previously (16). COS cells were transfected as described (13).

Cells were labeled with ^{35}S-methionine and immunoprecipitations carried out as described (13,16). Total RNA was prepared by guanidine extraction and was analyzed by blot hybridization following electrophoresis through agarose gels (17).

Vector Construction

The construction of the ADA-DHFR expression vector used for the CHO cell experiments has been described (14). This plasmid (designated p9ADA5-29 in ref. 14) contains the mouse ADA cDNA sequence. The basic plasmid (pMT2) used for the COS cell experiments were derived from p91023(B) (15) by replacing the tetracycline resistance marker with an ampicillin resistance marker and is designated pMT2 (See Fig 3). The COS cell experiments were conducted with a human adenosine deaminase clone that was isolated from pSV2(HdIII) (18) by HindIII digestion and inserting the ADA coding segment into the unique EcoRI site in pMT2. The dicistronic transcription unit in p9ADA5-29 is identical to the ADA-DHFR dicistronic transcription unit in pMT2-ADA except p9ADA5-29 encodes the murine ADA and pMT2-ADA encodes the human ADA. The CSF coding region was derived from pCSF-1 (15) by EcoRI digestion and insertion of the 800 bp fragment into the EcoRI site between the ADA and DHFR sequences.

RESULTS

Translation of a dicistronic mRNA in stable transformed CHO cells

In order to determine whether a dicistronic mRNA
can generate two proteins, the following approach was
taken. Plasmid p9ADA5-29 (see Methods and ref.14),
which encodes a dicistronic ADA-DHFR transcription
unit, was introduced into DHFR deficient Chinese
hamster ovary cells and transformants isolated after
growth in media which selects for expression of ADA
(16). Two transformants were randomly chosen and
propagated in increasing concentrations of
2'-deoxycoformycin, a tight-binding inhibitor of
ADA. This resulted in cells which were resistant to
100 uM 2'-deoxycoformycin, and which contained
approximately 500 copies of the plasmid DNA
integrated into the CHO chromosome (16). In order to
monitor DHFR expression from the dicistronic
transcription unit, the cells were grown under
conditions that require functional DHFR, i.e. growth
in the absence of added nucleosides, in the presence
and absence of increasing concentrations of
methotrexate (MTX).   This provides a very sensitive
assay for functional DHFR.   Table 1 shows the
results of a plating efficiency experiment where the
original ADA transformant and the highly amplified
cells were propagated in increasing concentrations of
MTX. The result demonstrates that 8% of the original
transformants grew in the DHFR selective media.
After amplification, this value increased to 125%.
The increase above 100% was a result of the greater
toxicity of the ADA selective media. As the MTX
concentration was increased, the plating efficiency
decreased, indicating that growth was actually
resultant from functional DHFR. Low levels of MTX
had little effect on the plating efficiency whereas
0.2 uM reduced the plating efficincy to less than
1%. This result indicates a significant level of
DHFR expression from the ADA-DHFR expression vector,
in the absence of any direct selection for DHFR
expression. Thus, cells which have amplified the
ADA-DHFR dicistronic transcription unit express both
ADA and DHFR proteins that confer resistance to the
appropriate selective agents.

TABLE 1
PLATING EFFICIENCY VS. METHOTREXATE CONCENTRATION OF THE ADA TRANSFORMANT OF BEFORE AND AFTER SELECTION FOR ADA AMPLIFICATION

[Methotrexate]

1.0uM	0.0 uM*	0.0 uM**	0.02 uM	0.02 uM
Cell line				
1-0.1 <.05	100	8.6	<0.05	<0.05
1-100 <.05	100	125	79	0.8

The plating efficiency for the original ADA transformant (1-0.1) grown in 0.1mM 2' deoxycoformycin and the amplified transformant (1-100) grown in 100 mM 2' deoxycoformycin were determined by plating 200 and 2000 cells per 10 cm dish in increasing concentrations of methotexate. Colonies were stained and counted 12 days after plating. Values are expressed as a percent of the control which was 23% for the clone 1-0.1 and 22% for the clone 1-100. *Indicates plating efficiency in 1.1 mM adenosine, 1mm uridine, 10 ug/ml deoxyadenosine, and 10 ug/ml thymidine with either 0.1 or 100 uM 2' deoxycoformycin for 1-0.1 and 1-100, respectively (16). ** Indicates plating efficiency in the efficiency in the DHFR selection media (nucleoside free alpha media). The 125% value for the 2' deoxycoformycin resistant line in alpha media reflects the ability of these cells to grow better in the absence of high concentrations of adenosine.

DHFR expression has been directly demonstrated by ^{35}S-methionine radiolabeling and immunoprecipitation of the cell extracts with a rabbit anti-murine DHFR antibody (Fig 1). A 20 Kd band was found to comigrate with authentic murine DHFR, and its level increased as cells were selected for increased copy number of the ADA-DHFR dicistronic transcription unit. The level of DHFR synthesis was comparable to the level in CHO cells selected for resistance to 0.1 uM MTX (E6-0.1 in Fig 1), which is consistent with the level of MTX resistance in the cells expressing the ADA-DHFR dicistronic mRNA.

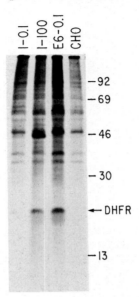

Figure 1

      Analysis of DHFR Synthesis in ADA Amplified and Control CHO Cells

      Cells were labeled 45 min with ^{35}S-methionine and cell extracts prepared for immunoprecipitation as described (17) with a rabbit anti-DHFR antibody.   The cell lines indicated are clone 1 in 0.1 uM (1-0.1) and 100 uM (1-100) 2'deoxycoformycin, another CHO cell line resistant to 0.1 uM methotrexate after transfection with pMT2 (E6-0.1), and the original CHO cells (CHO).

      In order to confirm that the DHFR expression observed was derived from the ADA-DHFR dicistronic mRNA, the cells were examined by DNA and RNA blot hybridization analysis. First, Southern blot analysis indicated a single DHFR sequence had integrated and that no rearrangement of sequences surrounding the integrated DHFR DNA had occurred (data not shown).   Second, RNA blot hybridization demonstrated the appearance of a single mRNA encoding ADA which has the expected size if derived from the dicistronic transcription unit.   Upon hybridization

to a DHFR probe, the same species is detected with no
detectable species of lower molecular weight (Fig
2).   Comparison to a cell line resistant to 0.1 uM
MTX (E6-0.1 in Fig 2), which expresses DHFR from a
separate monocistronic transcription unit, similar to
the one in pMT2, demonstrates that DHFR expression at
this level would generate a detectable transcript by
RNA blot analysis.   Comparison of the abundance of
the DHFR mRNA species in these two cell lines and of
the relative amount of DHFR synthesis (Figs 1 and 2)
indicate that the monocistronic mRNA translates DHFR
approximately 10-fold more efficiently than the
dicistronic mRNA derived from p9ADA5-29.

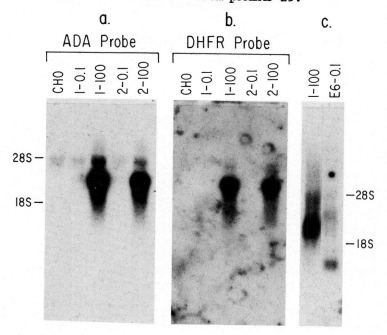

Figure 2
RNA blot hybridization analysis of original and ADA
amplified cell lines.
    Total RNA was isolated and electrophoresed on
formaldehyde-formamide denaturing agarose gels and
transfered to nitrocellulose paper.   Hybridization
was to a nick-translated probe from a fragment of the
ADA coding region (a) or to the DHFR coding region

(b,c). The filter represented in a was washed and rehybridized with the DHFR probe (b). The cell lines indicated are as described in Methods. Indicated are the 18 and 28 S ribosomal marker bands.

Translation of dicistronic and tricistronic mRNAs in transiently transfected COS monkey cells.

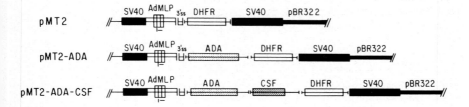

Figure 3.   Diagramatic Representation of Expression Vectors

The construction of these plasmids is decribed under Methods. pMT2 is a derivative of p91023(B) (15) which utilizes the adenovirus 2 major late promoter for transcription initiation and contains the majority of the adenovirus tripartite leader sequence, an introduced 3' splice site, and the origin of replication, transcriptional enhancer, and the early polyadenylation site from SV40. pMT2 also contains the murine dihydrofolate reductase (DHFR) coding region. pMT2-ADA and pMT2-ADA-CSF are identical to pMT2 except for the insertion of the human adenosine deaminase (ADA) coding region (18) with or without the coding region for human granulocyte-macrophage colony stimulating factor (CSF) (15). All coding regions are present in the sense orientation with respect to the direction of transcription.

The results from CHO cells indicate that polycistronic mRNAs can be translated but the efficiency of the downstream ORF is dramatically reduced. In order to determine if the introduction of another ORF, upstream of DHFR, has further suppressive effects, pMT2-ADA-CSF was constructed (Fig 3). In order to more conveniently assay for translation of the polycistronic mRNAs, these DNAs were transiently introduced into COS monkey cells. 72 hr post-transfection, the cells were labeled 1 hr with ^{35}S-methionine and cell extracts prepared for immunoprecipitation and SDS-polyacrylamide gel electrophoresis (Fig 4). It was found that DHFR and ADA represent a major proportion (approximately 5%) of the total protein synthesis in cells tranfected with pMT2 and pMT2-ADA, respectively (Fig 4, panel a). DHFR synthesis could not be detected in the proteins synthesized in cells transfected with pMT2-ADA and pMT2-ADA-CSF by this analysis. These results demonstrate the severe suppressive effect of upstream ORFs on DHFR translation. Analysis by immunoprecipitation with anti-DHFR and anti-ADA antibodies is shown in Fig 4 (panels b and c). The introduction of the CSF ORF results in a 3-4 fold further reduction in DHFR synthesis (panel c) when compared to the synthesis of ADA (panel b). RNA blot analysis has demonstrated that these plasmids direct the synthesis of equivalent amounts of single mRNAs of the expected size and that contain the expected ORFs. These results demonstrate a 10-fold reduction in DHFR translation by insertion of an upstream ADA ORF and an additional 3-4 fold reduction by the additional insertion of a CSF ORF.

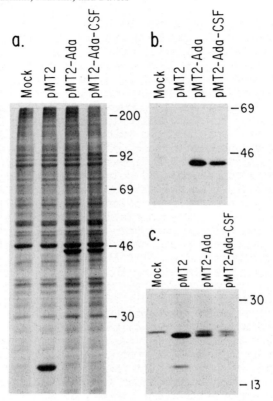

Figure 4
Analysis of total and immunoprecipitated ADA and DHFR
proteins synthesized in transfected COS cells
    COS cells were transfected with the plasmids
indicated as described in Methods. 72 hr post
transfection, the cells were labeled with
^{35}S-methionine for 1 hr and cell extracts taken for
analysis by SDS-reducing polyacrylamide gel
electrophoresis before and after immunoprecipitation
with either an anti-ADA antibody (panel b) or an
anti-DHFR antibody (panel c). Panel a represents the
total protein synthesis. The bands migrating at 20
Kd and 44 Kd are DHFR and ADA respectively. The
lanes are labeled with the DNA which was transfected
into the cells. Mock refers to cells which did not
receive DNA. The precipitation of DHFR in pMT2 was
not quantitative. The band migrating slightly above
the murine DHFR in panel c is the monkey DHFR.

DISCUSSION

Hybrid transcription units have been constructed which contain multiple ORFs encoding DHFR, ADA, and granulocyte-macrophage CSF. Expression of these proteins from a single mRNA has been studied in stable transformed Chinese hamster ovary cells and in transiently transfected COS monkey cells. The amount of translation is dramatically reduced by insertion of upstream ORFs. The efficiency of DHFR translation decreased approximately 10-fold in COS and CHO cells by the insertion of an upstream ADA coding region. Insertion of another upstream ORF, that for CSF, resulted in a 3-4-fold further decrease in DHFR translation. These results are consistent with previous work, which has shown that AUG codons, inserted upstream of the 'correct' AUG initiation codon, reduce translation at the 'correct' site (5,6,9,10). Although the potential for independent internal initiation cannot be ruled out at this time, these observations support a modified 'scanning' model of translation initiation, which allows for translation reinitiation (16), rather than independent internal initiation. This hypothesis suggests that the 40S ribosome remains associated with the mRNA after translation of the proximal ORF and resumes scanning until it reaches the next AUG codon present. The frequency for reinitiation at DHFR as shown here for the ADA-DHFR dicistronic transcript was 10%. The sequence context of the internal AUG can also affect reinitiation, however, the optimal contexts for primary initiation and reinitiation appear to be the same (16). These findings bring up the interesting potential of many mRNAs having the ability to encode multiple proteins. This possibility requires the examination of mRNAs for the occurrence of multiple non-overlapping ORFs.

DHFR has been useful as an amplifiable marker and has been exploited to obtain high level expression of heterologous genes in mammalian cells (17). However, cotransfection and coamplification of a heterolgous gene with a DHFR gene frequently

results in deletion of the desired heterologous gene
(17). The use of polycistronic expression vectors,
that encode transcripts containing the desired coding
region upstream of the DHFR coding region, may
provide an approach to solve this problem; the
heterologous gene and the DHFR gene would be linked
on the same transcription unit, and as a result,
deletion should be minimized.

The DHFR coamplification approach to
heterologous gene expression has generally been
limited to Chinese hamster ovary cells which are
deficient in DHFR. The dicistronic ADA-DHFR vector
described here, provides a unique opportunity to
introduce and amplify to very high copy number
foreign genes in a variety of mammalian cells.
First, since ADA can function as a dominant
selectable and amplifiable marker for gene transfer
in a variety of cells (16), initial transformants can
be selected for ADA expression and the DNA amplified
by growth in increasing concentrations of
2'-deoxycoformycin. Cells selected to contain a high
degree of amplification of the ADA-DHFR polycistronic
transcription unit should also produce a significant
amount of DHFR. Then, methotrexate resistance
selection could be applied to further amplify the
gene copy number. The potential usefulness of this
doubly amplifiable vector is currently being
evaluated.

REFERENCES

1.   Kozak, M. (1980) Evaluation of the 'scanning
      model' for initiation of protein synthesis in
      eukaryotes. Cell 22, 7-8.
2.   Shatkin, A. J. (1976) Capping of eukaryotic
      mRNAs. Cell 9, 645-653.
3.   Kozak, M. (1979) Inability of circular mRNA to
      attach to eukaryotic ribosomes. Nature 280:
      82-85.
4.   Konarska, M., Filipowicz, W., Domday, H., and
      Gross, H. J. (1981) Binding of ribosomes to
      linear and circular forms of the 5' terminal
      leader fragment of tabacco mosaic virus RNA
      Eur. J. Biochem. 114, 221-227.

5. Kozak, M. (1984) Selection of initiation sites by eukaryotic ribosomes: effect of inserting AUG triplets upstream from the coding sequence for preproinsulin. Nucl. Acids, Res. 12, 3873-3893.

6. Liu, C-C., Simonsen, C. C., and Levinson, A. D. (1984) Initiation of translation at internal AUG codons in mammalian cells. Nature 309, 82-85.

7. Pelletier, J., and Sonenberg, N. (1985) Insertion mutagenesis to increase secondary structure within the 5' noncoding region of a eukaryotic mRNA reduces translational efficiency Cell. 40: 515-526.

8. Kozak, M. (1986) Influences of mRNA secondary structure on initiation by eukaryotic ribosomes Proc. Natl. Acad. Sci. USA 83: 2850-2854.

9. Dixon, L., and Hohn, T. (1984) Initiation of translation of the cauliflower mosaic virus genome from a polycistronic mRNA: evidence from deletion mutagenesis. EMBO J. 3, 2731-2736.

10. Hughes, S., Mellstrom, K., Koslik, E., Tamanoi, R., and Brugge, J. (1984) Mutation of a termination codon affects src initiation. Mol. Cell. Biol. 4, 1738-1746.

11. Mertz, J.E., Murphy, A. and Barkan, A. (1983) Mutants deleted in the agnogene of simian virus 40 define a new complementation group. J. Virol. 45: 36-46.

12. Mertens, P.P.C. and Dobos, P. (1982) Messenger RNA of infectious pancreatic necrosis virus is polycistronic. Nature 297: 243-246.

13. Kaufman, R. J. (1985) Identification of the components necessary for adenovirus translational control and their utilization in cDNA expression vectors. Proc. Natl. Acad. Sci. 82: 689-693.

14. Yeung, C-Y., Ignolia, D.E., Roth, D.B. Shoemaker, C., Al-Ubaidi, M.R., Yen, J-Y., Ching, C., Bobonis, C., Kaufman, R.J., and Kellems, R.E. (1985) Identification of functional murine adenosine deaminase cDNA clones by complementation in Escherichia coli. J. Biol. Chem. 260: 10299-10307.

15.   Wong, G.G., Witek, J.S., Temple, P.A., Wilkens, K.M., Leary A.C., Luxenberg, D.P., Jones, S.S., Brown, E.L., Kay, R.M., Orr, E.C., Shoemaker, C.S., Golde, D.W., Kaufman, R.J., Hewick, R.M., Wang, E.A., and Clark, S.C. (1985) Human GM-CSF: Molecular cloning of the complementary DNA and purification of the natural and recombinant proteins. Science 228: 810-815.

16.   Kaufman, R.J., Murtha, R., Ingolia, D.E., Yeung, C-Y., and Kellems, R.E. (1986) Selection and amplification of heterologous genes encoding adenosine deaminase in mammalian cells. Proc. Natl. Acad. Sci. 83:   .

17.   Kaufman, R.J., Wasley, L.C., Spiliotes, A.J., Gossels, S.D., Latt, S.A., Larsen, G.R., and Kay, R.M. (1985) Coamplification and coexpression of human tissue-type plasminogen activator and murine dihydrofolate reductase in Chinese hamster ovary cells. Mol. Cell. Biol. 5: 1750-1759.

18.   Orkin, S.H., Goff, S.C., Kelley, W.N., and Daddona, P.E. (1985) Transient expression of human adenosine deaminase cDNAs: Identification of a nonfunctional clone resulting from a single amino acid substitution. Mol. Cell. Biol. 5: 762-767.

Transcriptional Control Mechanisms, pages 209–220
© 1987 Alan R. Liss, Inc.

ELECTRON MICROSCOPIC AND BIOCHEMICAL EVIDENCES FOR THE PRESENCE OF DISTINCT SUBPOPULATIONS OF SV40 MINICHROMOSOMES

E. Regnier, E. Weiss, P. Schultz, P. Colin, J.C. Homo, C. Ruhlmann, P. Oudet

LGME du CNRS, U.184 de l'INSERM - Faculté de Médecine 11, rue Humann - 67085 STRASBOURG Cédex - FRANCE

ABSTRACT SV40 minichromosomes, prepared in physiological or low ionic strength buffers in the presence of low concentrations of divalent cations, contain significantly less SV40 late proteins than the same material prepared in the presence of EDTA. The difference in late proteins does not affect the gap structure present on 25% of the molecules, nor the presence of nucleosomes precisely localised on both side of the gap. The extracted transcriptional activity per microgram of SV40 DNA is unchanged, but reduced in proportion to the amount of extracted material. The presence of a low concentration of divalent cations during the preparation of the SV40 minichromosomes allows the extraction of a more homogeneous subpopulation with regard to the protein content, while retaining similar structural and biological features when compared to the material extracted in physiological ionic strength buffers, in the presence of EDTA.

INTRODUCTION

SV40 minichromosomes are considered a model system for the study of chromatin structure and its modulations during transcription or replication. The minichromosomes are classically obtained from CV1 or BSC1 cells 35-44 hours after infection (1). The SV40 DNA contains an origin of replication, genes expressed early and late after infection as well as regulatory elements of transcription and replication (for a review see 2). This DNA is associated with the four histones H2a, H2b, H3, and H4 and results in the packaging of the SV40 genome in 20-24 nucleosomes. The histone H1 is present on minichromosomes isolated at low or physiological ionic strength at a ratio corresponding roughly to one H1 molecule per nucleosome. The late viral proteins VP1, VP2, VP3 are present in significant amounts and could play a role, although as yet undefined, in the structure of the nucleo-protein complex. Other proteins such as the T-antigen are also present but in a minute amount and do not seem to be responsible for the overall structure of the bulk of the minichromosomes.

The SV40 minichromosome population is not homogeneous. A transient subset representing 25% of the extracted mate-rial contains a nucleosome-free region or gap covering the origin of replication and the transcriptional early and late promoter elements (the ORI region). These gapped molecules were shown to be preferentially sensitive to DNases in the ORI region (3) and are sensitive to restriction enzymes over the whole genome (4). It has been shown, using different approaches, that the SV40 transcriptional complexes present a gap covering the ORI region (5, 6).

Previous investigations have suggested that the nucleo-somal chromatin structure is a feature of transcribing or post-replicating minichromosomes in 130 mM NaCl, i.e. cha-racteristic of active molecules (1). Recent observations have suggested that the subpopulation of torsionally strai-ned molecules, which can be relaxed by an excess of topoiso-merase, contain the transcribing minichromosomes (7).

Divalent cations have been shown to play a crucial role in the compaction of chromatin structure and in the protein structure of DNA binding proteins (8). The presence of diva-lent cations is also required for transcription and repli-cation.

We have investigated the possible role of divalent cations in the extraction of minichromosomes and the charac-teristics of the different subpopulations previously descri-bed. Despite a significant change in the protein content of material prepared in the presence of 0.1 mM $MgCl_2$, the gap-ped minichromosomes present the same accessibility to restriction enzymes and micrococcal nuclease as those prepa-red in the presence of EDTA. The transcriptional complexes present the same characteristics under both conditions of extraction.

METHODS

Labelled SV40 minichromosomes were prepared and purified according to the protocol developed by Varshavsky et al. in which the EDTA was replaced by 0.1 mM $MgCl_2$ (9). DNA, protein analysis and nucleases digestions used published procedures (4). Electron microscopic characterisation and mass determination were performed as described by Schultz et al. (10). The characterization of the transcriptional activity in non-denaturing or denaturing conditions was as described by Weiss et al. (6 and ref. 17 therein).

RESULTS

SV40 minichromosomes were prepared from CV1 infected cells 40 hours after infection. They were extracted by diffusion from purified nuclei in physiological ionic strength (9). Either 10 mM EDTA or 0.1 mM $MgCl_2$ was used during the different steps. In either case, after sedimentation on a 5-30% sucrose gradient, the SV40 DNA peak is detected at 75 S (Fig. 1) indicating that the overall structure is preserved in the presence or absence of divalent cations. The sedimentation coefficient appears unchanged, but the amount of material extracted from infected nuclei differs significantly. In six independent experiments, 3 to 6 times less material was extracted in the presence of $MgCl_2$ when compared to the extraction in the presence of EDTA.

The Mg-minichromosomes have a different protein pattern. Aliquots of the peaks shown on Fig. I were concentrated and applied to a SDS gel (Fig. 2). Since the material applied corresponds to the same amount of DNA, we can compare directly the protein pattern obtained. As a first approximation, the five histones H1, H2a, H2b, H3 and H4 are present in identical amounts suggesting that the basic chromatin structure is present in both type of preparations. The presence of the histone H1 shows clearly that we are not dealing with the mature viral form, since this histone is absent in the purified viruses shown as a control in Fig. 2a. The presence of late proteins VP1, VP2, VP3 on purified minichromosomes has been described previously (11). The comparison in Fig. 2 of lanes b and c shows that the late proteins are present in a significantly lower amount when the material is prepared in presence of 0.1 mM $MgCl_2$ (lane c) than on minichromosomes prepared in the absence of divalent cations (lane b).

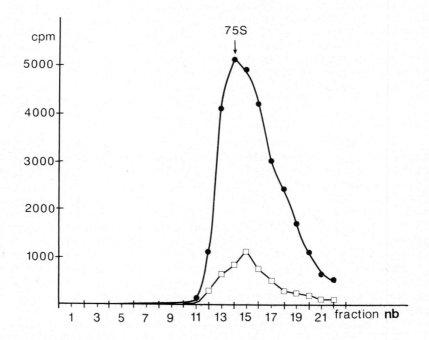

FIGURE 1. In vivo [H -labeled SV40 minichromosomes were prepared in the presence of EDTA or MgCl . The nuclear extracts were sedimented on 5-30% sucrose in the presence of EDTA or MgCl respectively. The black circles indicate the profile of the DNA counts of the material prepared in EDTA. The empty squares correspond to the counts of the minichromosomes prepared in presence of 0.1 mM MgCl .

We can quantify the amount of proteins by densitometry of SDS gels stained with Coomassie blue. The ratio of the intensities of the VP1 to the four histone bands combined was 0.26 for the Mg-minichromosomes and 0.68 for the EDTA-minichromosomes. If we assume a linearity in the sensitivity of these measurements, this would correspond to 1.2 VP1 molecules per nucleosome (eight histones) on EDTA material and 0.6 molecules when the minichromosomes are prepared in presence of MgCl$_2$.

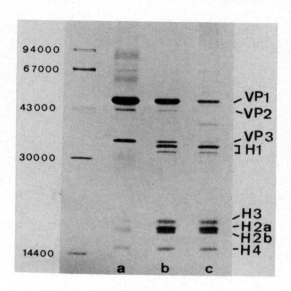

FIGURE 2. SDS gel electrophoresis of proteins from (a) purified viruses (b) EDTA minichromosome peak and (c) MgCl$_2$ minichromosome peak.

The Mg-minichromosomes present a more homogeneous mass histogram. When adsorbed on EM grids in the presence of 130 mM NaCl, the stained minichromosomes appear as a globular structure where individual nucleosomes are difficult to resolve. As shown in Fig. 3 this appearance is unchanged when the minichromosomes are extracted and purified in the presence of 0.1 mM MgCl$_2$.

Unstained specimens, in the presence of TMV as an internal standard, were analysed using the mass determination technique developed on a STEM equiped with a cold stage (10). In the case of EDTA minichromosomes (Fig. 4a) we have distinguished two classes of structures the first open and the second more dense (10). The histogram of the Mg-minichromosomes (Fig. 4b) appears more homogeneous (between 6-8 M.daltons) and corresponds to the lighter peaks of the EDTA-minichromosome distribution (Fig. 4a).

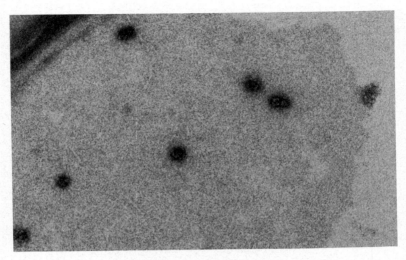

FIGURE 3. Stained preparation of Mg-minichromosomes adsorbed on grid in presence of 130 mM NaCl. The diameter of the particles ranges between 30-35 nm.

As seen on the protein gel (Fig. 2) the Mg-material contains less proteins per DNA molecule. As VP1 is the major protein this mass difference is essentially due to a lower VP1 content (2-3 times less). This possibility is further supported by the fact that when VP1 is detached from the minichromosomes in presence of 0.4-0.5 M NaCl, the mass histogram is displaced to 5 M.daltons.

Twenty five percent of the Mg-minichromosomes contain a nucleosomal free region. When opened in a low ionic strength buffer with EDTA, the Mg-minichromosomes as for the EDTA-minichromosomes, have the classical beads on a string EM appearance. In both cases there are 20 to 24 nucleosomes and a nucleosome-free region or gap is observed on 25% of the molecules (Fig. 5). This gap region is precisely located as previously described over the ORI region (4).

The gapped minichromosomes are preferentially sensitive to restriction enzymes (4). After digestion with BamHI, we thus obtain fractions enriched in gapped molecules on a second sucrose gradient. These fractions were digested with micrococcal nuclease, deproteinized, redigested to completion with the same restriction enzyme, and end labelled by hybridization to a small radioactive probe. Using this procedure, we are able to show that gapped fractions (Fig. 6a)

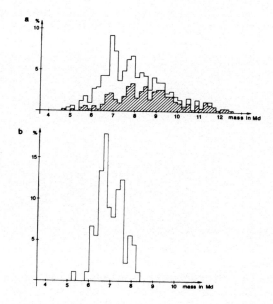

FIGURE 4. Mass histogramm of EDTA prepared minichromosomes (a) and MgCl$_2$ prepared minichromosomes (b). TMV particles were used as an internal standard. The dashed histogramm in (a) corresponds to identified more compact structures (10).

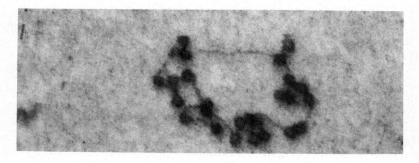

FIGURE 5. Stained gapped Mg-minichromosome. 21 nucleosomes of 13.5 nm in diameter are visible. The extended
region corresponds to the gap.

are preferentially sensitive in the naked region (ORI). This hypersensitivity is present, but to a lower extent, in fractions containing only 15% of gapped molecules.

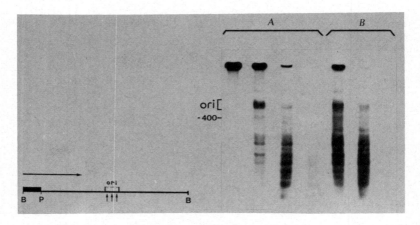

FIGURE 6. Micrococcal nuclease kinetics of fractionated minichromosomes after BamHI digestion. The cutting sites were mapped using the radioactive BamHI-PstI fragment (B, P; coordinates 2533-3204) denoted by a solid box on the linearised SV40 DNA map. The ORI region and -400 base pairs from the end of the open free region are indicated. The segment between the -400 position and the ORI region is protected in A (fractions containing 60% of gapped molecules) and is partially sensitive in B (fractions containing only 15% of gapped molecules).

The precise localisation of the nucleosome-free region (4) implies that the nucleosomes present on both sides of the gap cover specific sequences. This was confirmed by the experiment of Fig. 6a where 400 base pairs (two nucleosomes) are specifically protected both on the early and on the late side of the gap. The protected and sensitive regions do not arise from the sequence specificity of the micrococcal nuclease enzyme since they are not observed in kinetics realized on deproteinized SV40 DNA (not shown) and they are more confused on minichromosome fractions containing only 15% of gapped molecules. The fact that around 400 base pairs are specifically protected supports the suggestion that the first two nucleosomes, on both sides, are precisely localized. The same experiment realised on EDTA prepared minichromosomes gave essentially the same observation (not shown).

Characterisation of the transcriptional complexes extracted at physiological ionic strength in presence or in absence of 0.1 mM $MgCl_2$. In the population of extracted minichromosomes, those associated with transcriptional complexes correspond to a low percentage (around 1%). This subpopulation can be assayed either while keeping the chromatin structure intact or in the presence of sarkosyl and ammonium sulfate that dissociate the histone and most of the non-histones from the stable RNA polymerase-DNA complex.

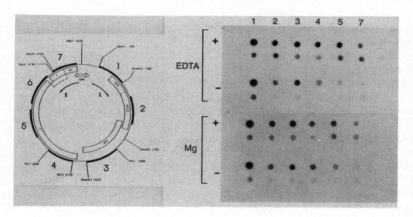

FIGURE 7. Radioactive RNA elongated on (EDTA) or (Mg) prepared minichromosomes was hybridised to the single stranded probes 1, 2, 3, 4, 5, 7 immobilized on nitrocellulose filters. The experiments were done in the presence (+) or absence (-) of sarkosyl and ammonium sulfate. The upper lane, in each case, corresponds to the late coding strands and the lower lane to the early coding strands probes. The localisations of the probes along the SV40 genome are represented as thick segments on the map.

The total transcriptional activity of Mg-minichromosomes represents around 15% of the activity present on EDTA-minichromosomes. This percentage corresponds precisely to the extractability of the SV40 minichromosomes in the presence or absence of divalent cations. The proportion of transcriptional complexes is very similar in both cases.

To further characterize these complexes, the active RNA-polymerase molecules were localised after a 15 min elongation at 32° in presence (+) or in absence (-) of sarkosyl

and ammonium sulfate. The labelled RNA was hybridised to
immobilised single stranded DNA probes (1-7) as indicated on
fig. 7, the late coding strand was present on the upper lane
and the early coding strand on the lower one. RNA corresponding to the same amount of DNA was used.

The late strand is preferentially transcribed in all
cases : in denaturing and non denaturing conditions, for
EDTA- or Mg-minichromosomes. The active RNA polymerases are
present on the late genes (probes 1, 2, 3) but also on the
early genes (non coding strand) showing that the enzyme
continues to transcribe after the end of the late genes.
This transcription drops abruptly in the region of probe 7,
suggesting that most of the RNA polymerases are released
from the template before reaching this region.

In non denaturing conditions (fig. 7 (-)), the transcription of the late strand over the early genes (probes
4-7) is reduced when compared to the elongation occuring in
absence of a chromatin structure (+). If the RNA polymerases
are not released in these incubation conditions, this observation suggests that the enzyme does not transcribe the
early region as efficiently as the late region.

From all these observations, it can be proposed that the
characteristics of the transcriptional complexes are identical whether the minichromosomes are prepared in the presence
or in the absence of divalent cations. The excess of late
proteins (1.25 VP1 molecules per nucleosome, see above.), if
they are present on the transcribing minichromosomes, does
not seem to modify the potential for transcription of the
isolated nucleoproteic structures. In addition, the presence
of a chromatin structure packing the SV40 genome, does not
inhibit significantly the progression of the RNA polymerase
on the late genes that are preferentially transcribed.

DISCUSSION AND CONCLUSION

The presence of 0.1 mM $MgCl_2$ during the extraction and
the purification of the SV40 minichromosomes results in a
modification of their protein content : they contain about
0.6 VP1 molecule per nucleosome. This could result from
either the specific extraction of a subpopulation of minichromosomes containing such a low VP1 to histones ratio or
the possibility that the late proteins are less exchanged or
stabilised in the presence of divalent cations. The fact
that the overall structure and transcriptional activity
were not modified in presence of 2-3 times less VP1 could be

due to the weak interaction of VP1 with the minichromosome structure as demonstrated by the fact that some VP1 proteins are spontaneously released even at physiological ionic strength by purification on sucrose gradients or exclusion chromatography.

The concentration of 0.1 mM $MgCl_2$ was chosen so as to obtain intact minichromosomes with more than 75% of super-helical DNA. At higher $MgCl_2$ concentrations the DNase activity introduces nicks on the SV40 DNA.

The action of divalent cations is not restricted to $MgCl_2$, as we also used different concentrations of $CuCl_2$, $MnCl_2$ and $ZnCl_2$. The minichromosomes obtained in these conditions have a very similar protein content as described for the Mg-minichromosomes.

This effect is not simply a salt or neutralisation effect since the experiments were realised in the presence of 130 mM NaCl and only 0.1 mM $MgCl_2$. The implication of specific interactions is further supported by the fact that an identical effect was obtained with only 0.004 mM $ZnCl_2$. It could thus be proposed that divalent cations are necessary to maintain specific interactions or structures and have to be present during the extraction and purification of nucleoprotein structures.

ACKNOWLEDGEMENTS

We thank Prof. Pierre Chambon for constant interest and helpful discussions, Geoffrey Richards for critical reading of the manuscript. We are grateful to Claudine Kister, Bernard Boulay for help in preparing the manuscript. This work was supported by the CNRS (ATP 6984), the INSERM, the Fondation pour la Recherche Médicale, the Ministère de la Recherche et de la Technologie and the Association pour le Développement de la Recherche sur le Cancer.

REFERENCES

1.   Tooze J (1982). "DNA Tumor Viruses". Cold Spring Harbor Laboratory, New York.
2.   Serfling E, Jasin M and Schaffner W (1985). Enhancers and eukaryotic gene transcription. TIG, 224.
3.   Jongstra J, Reudelhuber TL, Oudet P, Benoist C, Chae C, Jeltsch JM, Mathis DJ and Chambon P (1984). Induction of

altered chromatin structure by SV40 enhancers and promoter elements. Nature 307:708.

4. Weiss E, Ghose D, Schultz P and Oudet P (1985).T antigen
is the only detectable protein on the nucleosome- free
region of isolated   simian virus 40 minichromosomes.
Chromosoma 92:391.

5. Choder M, Bratosin S and Aloni Y (1984). A direct analysis of transcribed minichromosomes : all transcribed
SV40 minichromosomes have a nuclease hypersensitive
region within a nucleosome free domain. EMBO J 3:2929.

6. Weiss E, Ruhlmann C and Oudet P (1986). Transcriptionally active SV40 minichromosomes are restriction enzyme
sensitive and contain a nucleosome-free region. Nucleic
Acids Res., in press.

7. Luchnik AN, Bakayev VV, Zbarsky HB and Georgiev GP
(1982). Elastic torsional strain in DNA within a fraction of SV40 minichromosomes : relation to transcriptionally active chromatin. EMBO J 1:1353.

8. Hartshorne TA, Blumberg H and Young ET (1986). Sequence
homology of the yeast regulatory proteins ADR1 with
Xenopus transcription factor TFIIIA. Nature 320:283.

9. Varshavsky AJ, Nedospasov SA, Schmatchenko VV, Bakayev
VV, Chumakov PM and Georgiev GP (1977). Compact form of
SV40 viral minichromosome is resistant to nuclease :
possible implications for chromatin structure. Nucleic
Acids Res. 7:705.

10. Schultz P, Weiss E, Colin P, Regnier E and Oudet P
(1986). Caracterisation of SV40 chromatin by mass determination on STEM. Submitted to Chromosoma.

11. Keller W, Muller U, Eicken I, Wendel I and Zentgraf H
(1977). Biochemical and ultrastructural analysis of SV40
chromatin. XLII Cold Spring Harbor Symposium of Quant.
Biol., p.226.

Transcriptional Control Mechanisms, pages 221–234
© 1987 Alan R. Liss, Inc.

THE ROLE OF PHOSPHORYLATION
IN THE
ACTIVATION FOR CHROMATIN ASSEMBLY OF
XENOPUS NUCLEOPLASMIN[1]

Roger Chalkley, Matt Cotten and Linda Sealy

Department of Molecular Physiology and Biophysics
Vanderbilt University
Nashville. TN.

ABSTRACT Nucleoplasmin isolated from the unfertilized eggs of _Xenopus_ _laevis_ is capable of highly efficient _in vitro_ assembly of chromatin. In contrast, nucleoplasmin isolated from oocytes is vastly inferior. We have found that during the final maturation process immediately prior to laying, nucleoplasmin becomes massively phosphorylated (approximately 95 phosphate groups per molecule). This modification appears to be playing a major role in enabling nucleoplasmin to become competent for chromatin assembly

INTRODUCTION

A great deal of effort is currently being expended to develop systems capable of correctly assembling chromatin. Success is being achieved using two major approaches. On the one hand crude extracts from cells capable of assembling chromatin has led to efficient assembly (Nelson et al.

[1]This work has been supported by grants from the National Institutes of Health to R.C. (GM 34066, GM 28817) and to the Diabetes and Endocrinology Research Center (AM 25295) and from the American Cancer Society to L.S. (NP 518). M.C. was supported by grant GM 07728 from the Cell and Molecular Biology Training Program of the U.S. Public Health Service. L.S. is a scholar of the Leukemia Society of America.

1979, Laskey et al. 1977), albeit thus far at the
expense of some uncertainty as to the details of the assembly process. The other major approach has entailed the
isolation of factors which are presumed to play a critical
role in assembly in vivo (Laskey and Earnshaw, 1980) and to
attempt to use these factors in in vitro assembly assays.
This approach has the advantage of using defined components
so that the details of assembly can be studied with some
precision. However, one has to constantly exercise caution
in the analysis of the results, since one is highly dependent upon the initial decision of what factors to purify
and whether critical, perhaps unknown, components have been
inadvertently omitted. We have chosen to use the latter
approach because of its overall simplicity, with the inten-
tion of returning to the extracts for additional factors if
we found that the assembled product was less than optimal
in its behavior.

    We have based our studies of in vitro assembly on the
use of the putative assembly agent, nucleoplasmin (Np).
This acidic protein was first identified by Laskey et al.
1978, who showed that it could be used to assemble histone
onto DNA to produce a material which had many of the characteristics of native chromatin. However,in one regard this
material was somewhat unusual in as much as it required a
several-fold excess of histone over the amount of DNA for
adequate assembly. This is clearly unphysiological since
there is plenty of evidence that both histone and DNA are
made in a balanced 1:1 ratio in vivo. One of the first
points we were to address was the matter of this imbalance
in input histone and DNA.

                        RESULTS

Purification of Nucleoplasmin.

    The most active preparations of nucleoplasmin are
obtained from unfertilized eggs of Xenopus laevis. The
standard procedure then involves the preparation of a clar-
ified extract, removal of lipid and a heat treatment (80°),
after which Np is one of the relatively few proteins which
is still soluble. The soluble material is then fractionated
on a DEAE-cellulose column and the Np identified in the
first place by its electrophoretic mobility on SDS polyacrylamide gels. That material which is essentially free of
Coomassie staining contaminants is collected. At this stage
of the purification, the Np is capable of promoting an

extensive assembly of histones onto DNA. However we have
observed, as did Laskey and his colleagues that a large
excess of histone is necessary to achieve full assembly. In
addition we noted that the Np seemed to vary in this regard
from preparation to preparation. Accordingly, we suspected
that there was an impurity present which was interfering
with assembly. Additional experiments indicated that this
was indeed correct, and that the variable impurity was able
to inhibit topoisomerase activity, (which is necessary to
measure the extent of assembly). In addition this agent can
also bind histones.

Further purification of Np through phenyl sepharose led
to a resolution and separation of assembly and topo-
inhibiting activity. The Np so prepared is capable of
directing assembly at a DNA:histone ratio of unity, the
physiological ratio. Presumably the extra histone needed
for assembly using the less pure Np was needed to titrate
the topo-inhibiting substance.

Acetylated Histones are a good Substrate for in vitro Assembly.

Newly deposited histones H3/H4 are acetylated at the
time of deposition and the modification is removed relatively promptly thereafter (Jackson et al. 1975). Since it
is relatively easy to obtain histone which have a wide
range of such modifications we have assayed whether acetylation status could affect in vitro assembly. As shown in
figure 1, rather considerable differences are found in the
ability of variably acetylated core histones to assemble
onto DNA. Highly acetylated histones appear to be superior
inasmuch as the same degree of assembly is obtained with
approximately twofold less histone. Indeed, only with the
more highly acetylated histone does assembly occur at the
ratio of histone to DNA seen in the cell. It is not at all
clear why the acetylated histone should be so much more
superior in its assembly behavior. Certainly Np is able to
bind both forms of histone with facility, though we cannot
say for certain that the affinity of the two forms for Np
is identical. Conceivably a lower affinity of histone for
Np might lead to a greater ease of transfer to the DNA. A
second possibility is that the acetylation plays a role in
organizing the histone octamer in its correct conformation
on the Np surface. Certainly assembly without organization
leads to extensive precipitation and indeed the primary
function of an assembly system is almost certainly that of
organizing the histone prior to deposition. This last idea

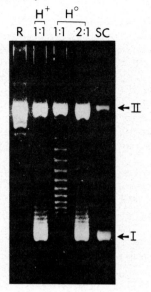

FIGURE 1. Effect of histone acetylation status on assembly using nucleoplasmin. Chromatin was assembled from pBR322 DNA (400 ng), nucleoplasmin (540 ng) and histone (360 ng) using our standard conditions in the presence of topoisomerase I. After 2 hr the reaction was terminated in SDS and analyzed on 2% agarose. The extent of assembly under these conditions for hyperacetylated (H^+) and control (H^o) histone are shown. The assembly in the presence of additional control histone (720 ng histone and 1080 ng nucleoplasmin) is also shown (2:1).

is also subject to analytical testing using crosslink-ing procedures.   Thus in all of our experiments we have employed highly acetylated histones.

Nucleoplasmin from Oocytes is a Poor Assembly Agent.

Our initial experiments utilized Np obtained from unfertilized Xenopus eggs. However collection of eggs is tedious and time consuming. Accordingly we decided that it would be advantageous to isolate Np directly from oocytes. The preparative procedures apply equally well to this mate-rial and good yields of Np can be obtained. Oocyte Np was assayed in the standard manner using hyperacetylated acety-lated histone. The results were surprising and are shown in figure 2. Oocyte Np is much inferior to egg Np in its ability to promote nucleosome formation. Not only is a much larger amount of the Np-histone complex necessary to obtain what is at best only partial assembly, but the oocyte material is much less able to prevent precipitation of inappropriately assembled material. Sucrose density gradient analysis of the ability of the two forms of Np to bind histone

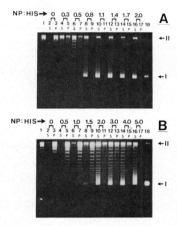

FIGURE 1. In Vitro Chromatin Assembly Capabilities of Egg or Oocyte Nucleoplasmin. Topoisomerase I treated, relaxed pBR322 DNA (A and B, lane 1) was incubated with hyperacetylated core histones previously mixed with increasing amounts of egg Np A, lanes 2-17) or oocyte Np (B, lanes 2-17). Mass ratios of Np to histone are shown above the lanes. After 2 hr at room temperature, assembly reactions were centrifuged at 12000g for 5 min, and both supernatant (S) and pellet (P) material were analyzed on 2% agarose gels shown in panels A and B. Supercoiled form I pBR322 DNA is shown in panels A and B, lane 18. In this gel system, circular pBR322 DNA containing one or two positive or negative supercoils is not resolved from nicked, circular (form II) DNA.

indicates that they both generate complexes of essentially the same size, in addition we find that the increased S value of the complexes is maintained throughout the centrifugation indicating that the affinity of the histones for the two Nps is very high (unpublished observations). Thus we conclude that the different Nps are both capable of binding histones and that the pronounced difference in assembly reflects either a difference in ability to

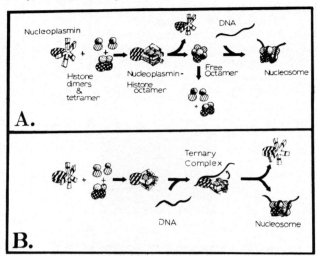

FIGURE 3. Two Schemes for the Role of Nucleoplasmin in
the Deposition of Histones onto DNA. (A). Nucleoplasmin
serves as a means for organizing histones into an octamer.
The octamer dissociates and binds to DNA before significant
dissociation of the octamer to dimers and tetramers occurs.
(B). Nucleoplasmin serves as a means for organizing histones into octamers, but then delivers the histones to the
DNA via a ternary complex, thus obviating breakdown of the
octamer into its component parts during the assembly process.

form the correct octameric structure on the surface of
the Np, or a difference in the relative ability top transfer such a histone complex to the DNA.
     We can envisage two general schemes for the transfer of
assembled histone octamers onto DNA. These are outlined in
figure 3. The main difference is in the nature of the
intermediates involved in the deposition process. The
first mechanism shows Np organizing the histone octamer
which then dissociates from the Np-histone complex to form
a transient free octamer which binds DNA before it dissociates to give the forms of histone (H2A/H2B and (H3/H4)$_2$)
which are the stable forms of histone complexes at the
ionic strenghts used in our assembly reactions. The second
mode of assembly involves the formation of a ternary complex between DNA, Np and histones. As will be discussed
later the structure of the Np molecule appears to be particularly suited to deliver histones to DNA. We anticipate
that the histones are then shifted from the Np directly

onto the DNA, without an independent existence. Thus the most likely point for the molecular difference between the two Nps lies in the ability to organize the histones into correct octamers.

The Two Forms of Nucleoplasmin Differ in Electrophoretic Mobility.

Analysis of egg and oocyte Nps on SDS gels revealed that the egg form migrates at a slightly slower rate and that it exhibits a considerably greater heterogeneity. This immediately suggested to us that the difference between the two forms of Np lies in some form of post-translational modification. Certainly there is insuffient time durung the laying process to completely resynthesize a new complement of Np which is one of the most abundant proteins in the oocyte germinal vesicle. Analysis on 2-D gels indicated that the isoelectric point of egg Np is substantially more acidic than that seen for the oocyte form. The more common forms of post-translational modification which would lead to a more acidic protein include glycosylation, ADP ribosylation, acetylation and phosphorylation. The unusually low A_{260} precluded significant ADP-ribosylation. Further, only non-specific binding to a con-A column was observed, arguing against the presence of mannose-containing carbohydrates. There are however a number of reports of a change in mobility on SDS gels when proteins are quite extensively phosphorylated (Shih et al., 1979; Stadel et al 1983; Wegener and Jones, 1984; Zoller et al, 1979). Also earlier reports had indicated that Np takes up radiolabelled phosphate during in vitro incubations. In an initial series of experiments we were able to confirm this observation, though the phosphorylated Np did not show any change in mobility compared to the precursor material on our SDS gels.

Incubation of the egg and the oocyte forms of Np with phosphatase leads to a dramatic change in mobility (see figure 4) so that the egg form of Np now approaches that of the oocyte form which has also shown a small increase in mobility after such treatment. We have found that significant mobility shifts are dependent upon using stoichiometric amounts of phosphatase perhaps reflecting a strong binding between Np and the phosphatase, though this has not been investigated in great detail. Nonetheless, we were concerned that the mobility change might be a reflection of minor proteolytic contaminants in the vast excess of

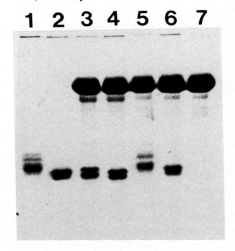

FIGURE 4. Phosphatase Treatment of Egg and Oocyte
Nucleoplasmin. Lane 1, egg nucleoplasmin; lane 2, oocyte
nucleoplasmin; lane 3, egg nucleoplasmin treated with calf
intestinal phosphatase; lane 4, oocyte nucleoplasmin
treated with calf intestinal phosphatase; lanes 5 and 6,
egg or oocyte nucleoplasmin treated with calf intestinal
phosphatase in the presence of 15 mM sodium molybdate; lane
7, calf intestinal phosphatase. Samples were resolved by
SDS-PAGE followed by staining with coomasie blue.

enzyme. This was excluded by experiments in which we
performed the digestion in the presence of large amounts of
enzyme together with potent inhibitors of phosphatase
action. In this case no mobility shift is observed. Since
the mobilities of the two forms of Np are still not identical even after extensive digestion with phosphatase, we
wondered if this reflected a second type of modification or
if there were steric inhibitions to complete removal of
phosphate groups. This point was addressed by first treat-
ing Np samples with phosphatase as described above. The
samples were then denatured by exposure to SDS and then
re-exposed to phosphatase after removal of the detergent.
We find that now both egg and oocyte Np migrate with the
same mobility, though a small measure of heterogeneity
remains (Cotten et al., Biochemistry, in press) When the
results of this type of experiment are displayed on acid-
urea gels we observe an increase in electrophoretic mobility of 21%. At the pH (2.2) of this system the charge on an
unmodified nucleoplasmin monomer of molecular weight 33000

is +45. Thus the full modification of the egg Np represents a decrease equivalent to -9.5 charges. Finally, since the charge on a phosphate group in this system is -0.5, the total number of phosphate groups on the egg Np monomer is 19 compared to a value of 7 for oocyte Np. Thus, for the intact egg Np pentamer there are on the order of 95 phosphate groups associated with the active form of this molecule.

Egg Nucleoplasmin is Modified with Phosphate in that Zone Which Binds Histones.

Nucleoplasmin can be cleaved with pepsin to yield an amino-terminal fragment of M.W. 16000 which contains the site for the binding of histones. In addition a somewhat smaller carboxy-terminal fragment is first produced, which contains the signal for nuclear localization (Dingwall et al., 1982) and this peptide is then cleaved to yield two fragments of M,W. 12000 and 4000. These various fragments are cleanly separated on an acid-urea system. Accordingly we have treated the various Np samples with pepsin either with or without a prior phosphatase digestion, and displayed the results on an acid-urea gel as shown in figure 5. It is clear that the bulk of the heterogeneity is found in the larger, more acidic amino-terminal fragment, and that this heterogeneity is largely abolished after phosphatase treatment. In contrast the carboxyterminal fragments move extremely rapidly, indicating an unusual degree of positive charge; also, we find only a small amount of phosphatase-sensitive microheterogeneity in this part of the molecule.

The picture that emerges then is of a fairly negatively charged molecule which has an amazing internal distribution of charge so that the amino-terminal half of the molecule contains almost all of the negative charge (including up to approximately 95 phosphate groups per pentameric molecule). It is this region of the molecule which is involved in binding and organizing histones prior to deposition onto DNA. In contrast the other end of the molecule is highly positively charged and has been shown to contain the nuclear localization sequences. We suspect that this half of the molecule may also contain information permitting the histone-Np complex to approach the DNA molecule correctly.

Phosphorylation of Nucleoplasmin Improves Assembly
Capacity.

As indicated above, the oocyte form of Np is a much
less efficient agent for assembly of histones than that
obtained from eggs. Since the egg form of Np is much more
highly phosphorylated, especially in that part of the molecule involved in binding and organizing histones, it seemed
reasonable to ask if the superior assembly capacity of egg
Np was a direct consequence of its more extensive phosphorylation. Initially we were not able to generate a mobility
shift in oocyte Np during an in vitro incubation of oocyte
Np in stimulated oocyte extracts (though this has now been

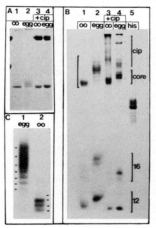

FIGURE 5. Panel A. Trypsin digestion of normal and CIP-
treated nucleoplasmins. Lane 1, oocyte nucleoplasmin
digested with trypsin as described in the text. Lane 2, egg
nucleoplasmin digested with trypsin. Lane 3, CIP-treated
oocyte nucleoplasmin digested with trypsin. Lane 4, CIP-
treated egg nucleoplasmin digested with trypsin. Trypsin
digestion and CIP treatment were performed as described in
the methods section. Samples were resolved by NaDodSO₄PAGE.
   Panel B. Pepsin digestions of nucleoplasmin resolved by
acid-urea gel electrophoresis. Lane 1, oocyte nucleoplasmin digested with pepsin. Lane 2, egg nucleoplasmin
digested with pepsin. Lane 3, CIP treatment of oocyte
nucleoplasmin pepsin digestion products. Lane 4 CIP treatment of egg nucleoplasmin pepsin digestion products. Lane
5, hyperacetylated core histones. The brackets indicate
peptides derived from the pepsin activity upon the phosphatase (CIP) derived from the amino-terminal core of nucleo-

plasmin (core) and derived from the 16 Kd and 12Kd carboxy-
terminal tail of nucleoplasmin (16 and 12) as verified by a
second dimensional electrophoresis in an NaDodSO$_4$PAGE.

achieved using partially purified maturation promoting
factor). Accordingly, we chose to assay the assembly capacity of egg Np after removal of some of its attendent phosphate with phosphatase. This experiment was undertaken with
the clear recognition that this approach does not remove
all of the phosphate without an additional denaturation
step, as discussed above. Nonetheless we realised that if a

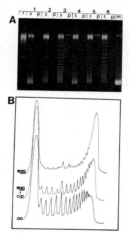

FIGURE 6. Chromatin assembly activity of dephosphory-
lated egg nucleoplasmin. Egg nucleoplasmin was treated
with calf intestinal phosphatase repurified by chromatography on DEAE-cellulose as described in the methods section.
Chromatin assembly was assayed by the induction of supercoils in a relaxed circular DNA (pBR322) using either hyperacetylated core histones at a histone to DNA ratio of 1;
or control core histones at a histone to DNA ratio of 2 as
previously described (7) with a nucleoplasmin to histone
ratio of 1.4 for both types of histones. Following the
assembly period (2 hours, room temperature, in the presence
of topoisomerase I) , the reactions were centrifuged at
15,000 x G for 5 minutes to identify precipitated products
and the material in the pellet and supernatant was adjusted
to 0.3 M sodium acetate, 0.2 mg/ml proteinase K, 0.5 %
NaDodSO$_4$, incubated for 40 minutes at 37°C and the DNA was
recovered by ethanol precipitation. The resulting DNA
samples were resolved by electrophoresis on a 2% agarose

gel in 36 mM Tris, 30 mM sodium phosphate, 1 mM EDTA, pH
7.7 and the DNA pattern was located by staining with ethidium bromide.

Panel A: lane r, relaxed, form I DNA (the starting
material for the assembly reaction); lanes p and s; the
pellet and supernatant fraction from each reaction; lane 1,
egg nucleoplasmin assembly with hyperacetylated histone;
lane 2, phosphatase-treated egg nucleoplasmin assembly with
hyperacetylated histones; lane 3, oocyte nucleoplasmin
assembly with hyperacetylated histones; lane 4, egg nucleoplasmin assembly with control histones; lane 5, phosphatase-treated egg nucleoplasmin assembly with control histones; lanes 6, oocyte nucleoplasmin assembly with control
histones.

Panel B: The Polaroid type 55 negative of the assembly
reaction products (supernatant fractions) from control histone assembly was scanned densitometrically. Egg; the DNA
species produced by assembly with egg nculeoplasmin; Egg +
CIP, the DNA species produced by assembly with egg nucleoplasmin treated with calf intestinal phosphatase; Oo, the
DNA species produced by assembly with oocyte nucleoplasmin.

significant decrease in assembly ability were to be
obtained upon removing a substantial amount of the modifying phosphate groups, that this would provide strong presumptive evidence that phosphorylation was playing an
important role in the activation of Np during oocyte maturation.

The results of such an experiment are shown in figure
6. We show once again that the two forms of Np clearly differ in their capacity for assembly. However, after phosphatase treatment the assembly ability of the egg Np is now
severely impaired and more closely resembles that of the
oocyte form, thus providing evidence that phosphorylation
is indeed playing a critical role in the activation of oocyte Np.

CONCLUSION

Oocyte nucleoplasmin from Xenopus laevis is a protein
of molecular weight 160000. It consists of five similar, or
identical, subunits. Each subunit is composed of a highly
negatively charged amino-terminal half, which is modified
by the presence of about 7 phosphate groups largely occuring on serine residues. As such, oocyte nucleoplasmin is a

very poor agent for the assembly of chromatin _in vitro_.
Before such an oocyte can be fertilized it must undergo a
progesterone-dependent germinal vesicle breakdown, followed
by meiotic reduction as part of the overall process of
final oocyte maturation. As a part of this maturation process the nucleoplasmin molecules are massively phosphorylated until about 20 phosphate groups per monomer are present. Almost all of these phosphate groups are located in
the amino-terminal half of the molecule, the same zone to
which histone binding has been localized. We suspect that
the role of the extra phosphate groups is to organize histone into a correct octameric conformation prior to deposition. We have also found that most efficient deposition of
histones onto DNA occurs if the histones are in the acetylated form.

Finally, we note that the carboxy-terminal half of the
nucleoplasmin molecule is more highly positively charged
than most histones, and it seems highly likely that the
function of this portion of the molecule is to shuttle the
nucleoplasmin-histone complex onto the DNA. One curious
point emerges from these considerations, given the acutely
bipolar charge distribution of nucleoplasmin. Why don't the
nucleoplasmin molecules form head to tail aggregates? We
suggest that in the uncomplexed state of nucleoplasmin the
opposite charges are able to internally neutralize themselves. When histones approach the negatively charged part
of the molecule, the positively charged portion is shifted
into a new position capable of interacting with DNA.

REFERENCES

Dingwall, C., Sharnick, S.V. and Laskey, R.A.   (1982)
    Cell 30, 449-458.
Cotten, M., Sealy, L. and Chalkley, R. (1986) Biochemistry
    in press.
Earnshaw, W.C., Honda, B.M., Laskey, R.A. and Thomas,
    J.O. (1980) Cell 21, 373-383.
Laskey, R.A., Mills, A.D. and Morris, N.R. (1977) Cell
    10, 237-243.
Laskey, R.A., Honda, B.A., Mills, A.D. & Finch, J.T.
    (1978) Nature 275, 416-420.
Jackson, V., Shires, A., Tanphaichitr, N., and Chalkley,
    R. (1976) J. Mol. Biol., 104, 471-483.
Shih, T.Y., Weeks, M.O., Young, H.A. and Scolnick, E.M.
    (1979) Virology 96 64-79.
Stadel, J.M., Nambi, P., Shorr, R.G., Sawyer, D.R.,

234 Chalkley, Cotten, and Sealy

Caron, M. and Lefkowitz, R.J. (1983) Proc. Nat. Acad. Sci. <u>80</u> 3173-3177.
Wegener, A.D. and Jones, L.R. (1984) J. Biol. Chem. <u>259</u>, 1834-1841.
Zoller, M.J., Kerlavage, A.R. and Taylor, S.S. (1979) J. Biol. Chem. <u>254</u>, 2408-2412.

Transcriptional Control Mechanisms, pages 235–245
© **1987 Alan R. Liss, Inc.**

GENE ACTIVATION AND CHROMOSOME STRUCTURE AT HEAT SHOCK LOCI OF DROSOPHILA MELANOGASTER[1]

J. T. Lis

Section of Biochemistry, Molecular and Cell Biology,
Cornell University, Ithaca, NY 14853.

ABSTRACT   The expression of germline-integrated heat shock genes is strongly induced by heat shock. In addition, in response to heat shock, large puffs rapidly form on polytene chromosomes at the sites of integration. Analyses of in vitro modified heat shock genes reveal that only a short segment of the promoter-regulatory region is required to generate large puffs, provided that the promoter is connected to a large transcription unit. Using a new photo-crosslinking method, we have investigated changes in the in vivo interactions of specific proteins and DNA at heat shock loci upon induction. Both RNA polymerase II and topoisomerase I are recruited to heat shock genes after heat shock. Both are at high density on the mRNA coding regions (but not flanking regions) of these genes in induced cells. In uninduced cells, the low level of RNA polymerase II associated with the hsp70 gene is concentrated at the trancription start site. It appears that in uninduced cells, RNA polymerase II has access to the hsp70 promoter, but is incapable of transcribing the main body of the gene.

[1]This effort was supported by a grant from the National Institutes of Health, GM25232.

INTRODUCTION

Transcriptional activation of a eucaryotic gene can be divided into two broad and, at present, poorly understood processes. The first is initiation – the mechanism by which cis-acting regulatory sequences and diffusable activator proteins interact to permit proper engagement of RNA polymerase. The second is transcript elongation, which also is a mechanistically formidable process. DNA in the interphase nucleus is wrapped around histone cores to form nucleosomes which are in turn tightly packed in a helical arrangement. In response to gene activation, RNA polymerase II, which is itself larger than a nucleosome, must penetrate this structure and presumably lift the DNA off the nucleosome surface (at least locally) as it transcribes. From purely geometric considerations, this compact packing of chromatin must be altered at least transiently.

Indeed, dramatic changes in chromatin structure accompany gene activation (1,2). At the biochemical level, activated genes in intact nuclei become sensitive to digestion by nucleases. In the case of highly active genes the structure of the nucleosome itself appears to be dramatically altered and in some cases histones may even be absent from the DNA. On the more macroscopic scale of the polytene chromosome, puffs form at sites that are highly active trancriptionally. Moreover, indirect immunofluoresence shows that a variety of proteins are rapidly recruited to newly formed puffs. These proteins include RNA polymerase II and topoisomerase I, but also a variety of nonhistone chromosomal proteins about which little is understood biochemically. Taken together, these results indicate that the activation of genes can cause not only a striking cytological change (puffing), but also localized changes in molecular composition and chromatin structure.

Over the past several years my colleagues and I have used the heat shock genes of *Drosophila melanogaster* as a model for studying gene activation and the accompanying changes in chromatin structure. Much of our recent efforts have focused on two in vivo assays. The first is P element-mediated germline transformation of Rubin and Spradling (3) which we have employed to examine cis-acting elements that participate in the transcriptional activation of heat shock genes and in chromosome puffing. The second is a photo-crosslinking technique (4-7) that we developed to examine specific protein-DNA interactions during the reorganization of the chromosome fiber that underlies transcription and its activation. In this paper, I will summarize these findings and discuss these observations with regard to general mechanisms of gene activation and transcription.

RESULTS

Heat Shock-induced Transcription and Chromosome Puffing

Germline transformation (3) provides a means of examining not only the expression of an in vitro modified gene but also the affect of the introduced DNA sequences on the structure of polytene chromosomes. In response to heat shock, a large puff forms at the site of insertion of either hsp70-lacZ or hsp26-lacZ hybrid genes (8,9). The DNA sequences required for the puff formation are of two types. The first is a specific DNA sequence requirement, that is, a short DNA segment that contains a functional heat shock promoter region. Deletions that cripple the hsp70 promoter also disable the puff response (9). For example, consider transformants containing the following pair of 5' deletion mutants. Transformants of the −89 deletion possess two copies of the heat shock consensus sequence (10), both of which have been shown to be required for the normal heat shock response in Drosophila (9,11) and bind heat shock transcription factor cooperatively (12). In response to heat shock, these transformants produce high levels of ß-galactosidase and large puffs at the sites where this DNA has inserted. In contrast, transformants of a 5' deletion that extends an additional 16 bp (−73) produces neither heat induced ß-galactosidase nor new puffs. A similar analysis of 3' deletions show that no specific sequences downstream of site +65 on the hsp70 gene are required for expression or puffing (9). Thus, the

region of specific sequences required for heat shock
puffing appears to be within a 154 bp interval
containing the heat shock promoter-regulatory
region.
    The second DNA requirement is sequence general
in that the size of the puff is related to the
length, not sequence, of the heat shock transcript.
Puffs produced at sites of hybrid hsp70-lacZ genes
are at least as large as any of the naturally
occuring heat shock or developmental puffs (9,13).
The heat induced transcript of these hybrid genes is
at least 9 kb. In contrast, heat shock puffs are
barely detectable at sites containing a shortened
hsp70 gene which encodes a heat-induced trancript of
0.8 kb (9). The sites of insertion of other hybrid
hsp70 genes produce transcripts and puffs of sizes
intermediate between the above extremes (11,14).
    Although promoter strength and transcript
length are prime determinants of the chromosome puff
(9), the changes in chromosome structure induced by
gene activation are not limited to the promoter and
transcription unit. Nuclease sensitive regions are
known to extend well beyond the transcription unit
(2). Also, in situ hybridizations with a biotin-
peroxidase coupled detection system show that
sequences upstream of a heat shock promoter
decondense in response to heat shock (13).

Photo-crosslinking Reveals the In Vivo Distribution
of RNA Polymerase II on DNA Segments within Puffs.

    The cytological changes that occur during the
transition of a chromosomal band to a puff are, at
least in part, a manifestation of the changes in
chromatin organization and specific protein-DNA
interactions. To examine such interactions, we have
been experimenting with a photo-crosslinking
methodology applied to intact cells. The general
strategy is to generate crosslinks between proteins
and DNA by irradiating cells with intense UV light.
Specific protein-DNA complexes are then isolated by
immunoprecipitation with antiserum to the protein of
interest, and the coprecipitated DNA is assayed by
hybridization.
    To test this strategy, we examined the
interaction of bacterial RNA polymerase with
specific regulated genes. The choice of UV-light

and bacterial polymerase as the model protein was in part due to the success of Wu and his colleagues with in vitro crosslinking (15). Our in vivo studies demonstrated that DNA from transcriptionally-active genes are precipitated with RNA polymerase antibody much more efficiently than DNA from inactive genes (4). For example, the level of *lac* operon DNA recovered by this procedure is increased 30-60 fold by the addition of *lac* operon inducer. Controls behaved as expected: the recovery of the expressed genes is dependent both on the irradiation of cells and on the addition of the polymerase antiserum.

As a eucaryotic model, we examined the interaction of RNA polymerase II with a variety of Drosophila genes in Drosophila cell cultures (5). After irradiating Drosophila cell cultures with UV light, nuclei are prepared and lysed in 2% sarkosyl. The DNA is sheared to approximately 30 kb in length and banded in CsCl gradients containing sarkosyl to separate the DNA and crosslinked protein-DNA complexes from the noncovalently bound protein. The DNA is further cleaved either by sonication or by restriction endonucleases prior to incubation with antibody to the protein of interest, in this case RNA polymerase II antibody, which was generously provided by Dr. Arno Greenleaf. In the case of the sonicated sample, the DNA in the immunoprecipitate can be assayed by Dot blot hybridization with specific DNA segments.

Immunoprecipitated DNA from cells that were heat shock induced showed a 30-fold higher concentration of hsp70 gene sequences than precipitates from cells that were uninduced. In contrast, the copia gene sequences in anti-RNA polymerase immunoprecipites is reduce 8-fold from induced cells relative to uninduced cells. These and additional crosslinking results obtained with other Drosophila genes are in agreement with changes in transcriptional activity as measured by assay of pulse-labeled RNAs (5).

Examination of specific restriction fragments in the immunoprecipitates provides a means of assessing which genes of a multigene family are transcribed. RNA polymerase II is associated with all five of the hsp70 genes, with most members of the copia gene family, but with only one of the two cytoplasmic actin genes (5,6).

Very low levels of RNA polymerase II are associated with regions immediately upstream or downstream of heat shock genes examined. The RNA polymerase II density downstream of the polyadenylation site of hsp70 and hsp28 decreases rapidly (5,7), indicating that termination sites reside close to the 3'-ends of these heat shock genes.

We find low levels of RNA polymerase II present upstream of the transcription start site of hsp83 and hsp26 (5,7). In the case of hsp26, essential regulatory elements reside upstream of position -236 (16), and are well resolved from the transcription start. Thus, these crosslinking studies provide no evidence for entry of RNA polymerase II at upstream elements.

An RNA Polymerase II Molecule Is Near the Transcription Start Site of the Uninduced Hsp70 Gene.

The photo-crosslinking method provides a direct measurement of the relative density of RNA polymerase on specific regions of DNA in induced and uninduced cells. The absolute density of RNA polymerase II on induced hsp70 gene has been estimated to be as high as 20-40 molecules per gene as calculated from the rate of accumulation of hsp70 transcript. In noninduced cells, the density of RNA polymerase is twenty-fold less and thus represents approximately one molecule of RNA polymerase per gene. Additionally, restriction digests show this RNA polymerase II molecule is predominantly downstream of a Sal I site at -12 and upstream of a Pvu II site at +65 (6). This preferential association of RNA polymerase with the 5'-end of the hsp70 gene is apparent whether crosslinking is performed by a 10 minute UV irradiation of chilled cells with mercury vapor lamps or by a 40 microsecond irradiation of cells with a xenon flash lamp (6). From the latter conditions, we argue against the trivial explanation that polymerase is recruited to the heat shock promoter during the crosslinking.

Topoisomerase I Is Present on Transcribed Regions.

Analyses of DNA sequences that are crosslinked to topoisomerase I reveal that, like RNA polymerase II, topoisomerase I is recruited to heat-shock genes during the heat-shock response, since the level of topoisomerase I crosslinked to several different heat shock genes increases dramatically following their activation (7). Topoisomerase I is present at more than one site on the transcribed region and is at much lower levels, possibly absent, from regions that flank the transcribed region.

Although both topoisomerase I and RNA polymerase II are associated with active genes, they appear to interact with the genes independently. Topoisomerase I shows strong association with the nucleolus (17) which is the site of intense transcription by RNA polymerase I and not RNA polymerase II. Moreover, even with genes trancribed by polymerase II, different ratios of topoisomerase I and RNA polymerase II can be crosslinked as seen in the highly transcribed hsp70 gene and the moderately transcribed copia genes (7).

DISCUSSION

Transcription Initiation: Implications of a High RNA Polymerase II Concentration at the Transcription Start.

In induced cells, the photo-crosslinking results reveal that RNA polymerase II is at high density over the entire hsp70 gene (5,6). In uninduced cells, the amount of RNA polymerase crosslinked to the hsp70 gene is much less, but this crosslinked polymerase is predominantly localized to a short interval at the transcription start site (6). We estimate that approximately one RNA polymerase II molecule per hsp70 gene is located between nucleotides -12 and $+65$ in uninduced cells (see Results). Thus, in uninduced cells, RNA polymerase has access to and is associated with the transcription start site of hsp70. This implies that activation of the gene requires "pushing" RNA polymerase II beyond its association with the 5' end of the gene.

The structural similarities between different transcription elements raises the intriquing possibility that RNA polymerase II interaction with the hsp70 promoter could be general. Using a nuclear transcription run-on assay, Gariglio et al. (18) detected RNA polymerase II clustered at the 5' end of the nontranscribed ß-globin gene in nuclei from mature erythrocytes. In similar assays, the dihydrofolate reductase gene in mouse cells was also shown to have a higher density of RNA polymerase near the transcription start sites (19). Using the protein-DNA crosslinking method, we also have detected a higher density of RNA polymerase II on restriction fragments containing the copia promoter than on those containing the main body of the copia gene (5). These observations suggest that RNA polymerase II may have ready access to promoters of many genes and that the rate limiting step in transcription initiation is subsequent to the polymerase association. Perhaps the generation of a productively engaged RNA polymerase (a complex that transcribes the entire gene) occurs through the interaction of bound regulatory factors, which are often upstream of the trancription start, with RNA polymerase. This interaction of regulatory element and RNA polymerase II could either be through direct protein·protein contact brought about by DNA loop-out (20) or through propagation of an altered chromatin or altered DNA structure to the bound polymerase. This contact or template alteration would then catalyze productive RNA synthesis.

Transcript Elongation: Implications of Topoisomerase Recruitment to Active Transcription Units.

The topoisomerase association with transcription units may be providing topological changes that are required for transcription (21). For example, it could provide a "swivel" on an actively transcribed DNA template and thereby alleviate any topological constraints in DNA encountered during transcription. Alternatively, perhaps topoisomerase I activity is acting in opposition to a gyrase activity to provide the appropriate level of torsional strain for transcription. In this regard, perhaps eucaryotic topoisomerase I has a role analogous to its bacterial counterpart, which is involved in regulating the degree of supercoiling of the template.

Topoisomerase I may be just one of a battery of DNA-associated proteins interacting with transcribed regions. Indeed, immunofluorescent studies of the distribution of chromosomal proteins on polytene chromosomes indicate that more than several of these proteins exist in Drosophila cells (1). Precise DNA localization of topoisomerase I and other transcription-specific proteins during the time course of transcription activation may provide clues to the mechanistic relationships of the different protein components. To this end, the photo-crosslinking method using short bursts of flash lamp irradiation holds promise (6).

ACKNOWLEDGMENTS

I thank my colleagues for providing results that are summarized and cited here in the reference list or as personal communication. I thank Ann Rougvie and Ed Wong for comments on this manuscript.

REFERENCES

1. Cartwright IL, Keene MA, Howard GC, Abmayr SM Fleishmann G and Elgin SCR (1982). Chromatin structure and gene activity: the role of nonhistone chromosomal proteins. CRC Crit Rev Biochem 13:1.
2. Eissenberg JC, Cartwright IL, Thomas GH, and Elgin SCR (1985). Selected topics in chromatin structure. Ann Rev Genet 19:485.
3. Rubin GM and Spradling AC (1982). Genetic transformation of Drosophila with trnasposable element vectors. Science 218:348.
4. Gilmour DS and Lis JT (1984). Detecting protein-DNA interactions in vivo: Distribution of RNA polymerase on specific bacterial genes. Proc Nat Acad Sci USA 81:4275.

5.   Gilmour DS and Lis JT (1985) In vivo
      interactions of RNA polymerase II with genes of
      Drosophila melanogaster. Mol Cell Biol 5:2009.
6.   Gilmour DS and Lis JT. RNA polymerase II
      interacts with the promoter region of the
      noninduced hsp70 gene in *D. melanogaster* cells.
      Submitted for publication.
7.   Gilmour DS, Pflugfelder G, Wang JC, and Lis JT
      (1986). Topoisomerase I interacts with
      transcribed regions in Drosophila Cells. Cell
      44:401.
8.   Lis JT, Simon JA, and Sutton CA (1983). New
      heat shock puffs and ß-galactosidase activity
      resulting from transformation of Drosophila with
      an hsp70-lacZ hybrid gene. Cell 35:403.
9.   Simon JA, Sutton CA, Lobell RB, Glaser RL, and
      Lis JT (1985) Determinants of heat shock-induced
      chromosome puffing. Cell 40:805.
10.   Pelham HRB and Bienz M (1982). A synthetic
      heat-shock promoter element confers heat-
      inducibility on the herpes simplex virus
      thymidine kinase gene. EMBO 1:1473.
11.   Dudler R and Travers AA   (1984). Upstream
      elements necessary for optimal function of the
      hsp70 promoter in transformed flies. Cell
      38:391.
12.   Topol J, Ruden DM, and Parker CS (1985).
      Sequences required for in vitro transcriptional
      activation of a Drosophila hsp70 gene. Cell
      42:527.
13.   Simon JA, Sutton CA, and Lis JT (1985)
      Localization and expression of transformed DNA
      sequences within heat shock puffs of *D.
      melanogaster*. Chromosoma 93:26.
14.   Cohen RS and Meselson M (1984) Inducible
      transcription and puffing in Drosophila
      melanogaster transformed with hsp70-phage lambda
      heat shock genes. Proc Natl Acad Sci USA
      81:5509.
15.   Park CS, Hillel Z, and Wu C-W (1980) DNA strand
      specificity in promoter recognition by RNA
      polymerase. Nucl Acids Res 8:5895.
16.   Cohen RS and Meselson M (1985) Separate
      regulatory elements for the heat-inducible and
      ovarian expression of the *Drosophila* hsp26 gene.
      Cell 43:737.
17.   Fleischmann G, Pflugfelder G, Steiner EK,
      Javaherian K, Howard GC, Wang JC, and Elgin SCR

(1984). *Drosophila* DNA topoisomerase I is associated with transcriptionally active regions of the genome. Proc Natl Acad Sci USA 81:6958.

18. Gariglio P, Bellard M, and Chambon P (1981). Clustering of RNA polymerase B molecules in the 5' moiety of the adult ß-globin of hen erythrocytes. Nucl Acids Res 9:2589.

19. Barsoum J and Varshavsky A (1985). Preferential localization of variant nucleosomes near the 5'-end of the mouse dihydrofolate reductase gene. J Biol Chem 260:7688.

20. Enver T and Patient R (1986). Importance of helical periodicity. Nature 319:99.

21. Wang JC (1985). Topoisomerase. Ann Rev Biochem 54:665.

Transcriptional Control Mechanisms, pages 247–258
© 1987 Alan R. Liss, Inc.

NOVOBIOCIN PRECIPITATION OF
CHROMOSOMAL HISTONES[1]

Matt Cotten, Linda Sealy and Roger Chalkley

Department of Molecular Physiology and Biophysics
Vanderbilt University
Nashville, Tennessee

ABSTRACT Novobiocin interferes with in vitro chromatin
assembly using purified histones and nucleoplasmin.
The target of inhibition is not topoisomerase II; this
assembly system lacks any topo II activity. Rather,
novobiocin interacts with histones, disrupts octamer
associations and causes histones to precipitate from
chromatin solutions.

INTRODUCTION

It has been reported that chromatin assembly is an
energy-dependent process, requiring ATP and a DNA gyrase
activity for appropriate histone-DNA interactions (Glikin
et al., 1984). The major sources of evidence that a gyrase
activity is involved have been experiments using the prokaryotic gyrase inhibitor novobiocin. This compound, at the
concentrations of 250 μug/ml and above, has been shown to
inhibit the DNA supercoiling assay used to follow chromatin
assembly. Furthermore, these concentrations of novobiocin
have been demonstrated to inhibit RNA synthesis in vitro,
suggesting that a gyrase activity is also required for certain RNA polymerase activities (Ryoji et al., 1984; Kmiec

[1]This work has been supported by grants from the
National Institutes of Health to R.C. (GM 34066, GM 28817)
and to the Diabetes and Endocrinology Research Center (AM
25295) and from the American Cancer Society to L.S.
(NP518). M.C. was supported by grant GM 07728 from the Cell
and Molecular Biology Training Program of the U.S. Public
Health Service. L.S. is a scholar of the Leukemia Society
of America.

and Worcel, 1985, Glikin and Blangy, 1986; Kmiec et al., 1986).

We have monitored the effects of the DNA gyrase inhibitor novobiocin on in vitro chromatin assembly using core histones and electrophoretically-pure nucleoplasmin. Consistent with the observations of other investigators, we found that concentrations of novobiocin greater than 250 μg/ml inhibit the assembly of chromatin. However, there is no demonstrable DNA gyrase activity present in the purified components of our assembly. Therefore, we conclude that either novobiocin must be inhibiting a gyrase activity that is reconstituted when the purified components of our assembly system are mixed, or the compound interacts with other components of the assembly system. We have found that novobiocin interacts very strongly with histones and other arginine-rich proteins and is capable of removing these proteins from solution. Furthermore, novobiocin is capable of disrupting the histone octamer which forms in 2 M NaCl and the compound is capable of removing histones from previously assembled chromatin. We note that these experiments do not test the role of topoisomerase activities in chromatin dynamics but merely demonstrate that novobiocin should be used with caution in experiments designed to associate changes in DNA topology with alterations in chromatin activity.

METHODS

Nucleoplasmin was purified from unfertilized Xenopus laevis eggs as described in Sealy et al. (1986). Core histones were extracted from hydroxyapatite-bound hyperacetylated chromatin as previously described (Cotten and Chalkley, 1985).

Chromatin assembly reaction were performed using core histones, egg nucleoplasmin and topoisomerase I (wheat germ) as described in Sealy et al., 1986. The assay measures the induction of supercoils in a relaxed circular DNA molecule (pBR322) concurrent with nucleosome formation (Germond et al., 1976). DNA was resolved on 2% agarose gels containing 36 mM Tris, 30 mM sodium phosphate, 1 mM EDTA, pH 7.7 (Shure and Vinograd, 1976) with buffer recirculation.

Novobiocin was obtained from Boehringer Mannheim, dissolved in water and stored, in the dark, at -20 °C. Additional details concerning novobiocin precipitation have been published (Cotten et al., 1986).

RESULTS

Novobiocin Interferes with Chromatin Assembly Using Nucleoplasmin and Purified Components.

Chromatin assembly, in vitro, is promoted by the Xeno-pus protein nucleoplasmin (figure 1, lanes 2 and 3). When this same reaction is performed in the presence of increasing amounts of novobiocin (figure 1, lanes 4-11), we find that the induction of supercoils (indicative of nucleosome formation) is blocked, with substantial inhibition at 500 μg/ml novobiocin (figure 1, lanes 8 and 9). This observation is of interest because the target of novobiocin action, DNA gyrase, is demonstrably absent from this chromatin assembly system. Novobiocin must be exerting its effect at another site in the chromatin assembly reaction.

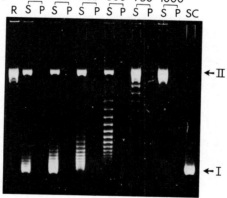

FIGURE 1. Novobiocin interferes with topoisomerase II-independent chromatin assembly in vitro. pBR322 DNA (400 ng) was relaxed with topoisomerase I (lane 1) and the mixture added to hyperacetylated HTC core histones (360 ng) and purified Xenopus nucleoplasmin (540 ng). In vitro chromatin assembly was allowed to proceed for 2 hrs at room temperature in the absence (lanes 2 & 3) or presence of novobiocin at various concentrations as indicated. At the end of the incubation period, samples were centrifuged at 15,000 x g for 10 min and separated into soluble (S) and precipitated (P) fractions before terminating the assembly reaction with SDS. DNA was then purified and analyzed on a 2% agarose gel. Ethidium bromide fluorescence of the DNA

is shown.

Novobiocin Precipitates Histones.

     In an effort to determine the site of action of novo-
biocin in this chromatin assembly reaction we have tested
the effects of the antibiotic on the activity or solubility
of the purified components of the assembly system. The
compound, in the range of 100-1000 μg/ml, has no effect of
the activity of the type I topoisomerase used in this system (data not shown). Novobiocin does not affect the solubility of nucleoplasmin (see below) and the compound does
not directly affect the topology of the DNA molecule (by
intercalation, for example; data not shown). However, at
physiological ionic strengths, the solubility of the core
histones is affected by the presence of novobiocin. Core
histone exist as H2A-H2B dimers and H3-H4 tetramers at physiological ionic strengths. When core histones are centrifuged at 15,000 x g and the material in the pellet or supernatant is resolved on a Triton-acid-urea gel we find that
these complexes are soluble (figure 2, lanes 1 and 2).
However, if core histone are incubated with novobiocin we
find that histones H3 and H4 are precipitated by concentrations of novobiocin greater than 200 μg/ml (figure 2, lanes
7 and 8). All four core histones are precipitated by novobiocin concentrations greater than 500 μg/ml.

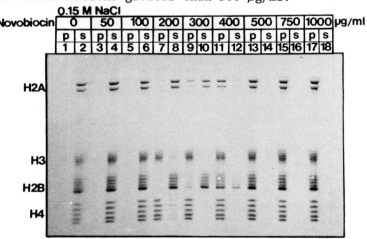

     FIGURE 2. Novobiocin precipitation of core histones in
150 mM NaCl. Histone samples (6 μg, hyperacetylated HTC
core) were incubated with the indicated concentrations of

novobiocin in 50 μl of 150 mM NaCl, 10 mM Tris, pH 8.0, for 45 minutes on ice. The samples were centrifuged for 3 minutes in an Eppendorf centrifuge (15,000 X g, room temperature) the supernatant was removed and the pellet was suspended in an equal volume of 150 mM NaCl, 10 mM Tris, pH 8.0 containing the indicated concentration of novobiocin. The sample was centrifuged again, this second supernatant was discarded and the pellet and first supernatant were resolved on a Triton-acid-urea gel as described in the Methods section. The values at the top of the figure indicate novobiocin concentration in μg/ml. P and S refer to pellet and supernatant.

Novobiocin Binds Histones

        To determine if novobiocin binds histones, increasing amount of core histones were incubated with a precipitating amount of novobiocin. The precipitated material was collected by centrifugation and the amount of novobiocin associated with the precipitated protein was determined spectrophotometrically, taking advantage of the strong absorbance of novobiocin at 300 nm. Figure 3 shows the amount of novobiocin present in the precipitate plotted as a function of the amount of histone in the sample. We find that the relationship between moles of histone present and moles of novobiocin bound is linear. The slope of this line indicates that there are approximately 15 moles of novobiocin bound per mole of core histone.

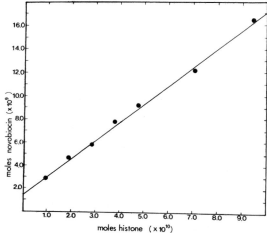

FIGURE   3.   Novobiocin binds histones. Salt extracted

core histones were incubated with 1000 μg/ml novobiocin in
150 mM NaCl, 10 mM Tris, pH 8.0, for 45 minutes on ice.
The precipitated material was collected by centrifugation
for 3 minutes in an Eppendorf centrifuge. After removing
the supernatant, the pellet was dissolved in 150 mM NaCl,
10 mM Tris, pH 8.0, 1% (wt./vol.) SDS and the absorbance at
300 nm was measured. Novobiocin has an absorbance maximum
at 300 nm with a molar absorptivity of 2.4 X 10^4 $M^{-1}cm^{-1}$ in
this buffer. Histones have virtually no absorbance at this
wavelength, therefore the absorbance of the solubilized
pellet at 300 nm was used to determine the novobiocin content. The molar concentration of histones was determined
using an average molecular weight value of 13,560 for the
four core histones. The amount of novobiocin present (in
moles X 10^9) is plotted as a function of core histone present in the sample (in moles X 10^{10}).

Novobiocin Induces Histone Precipitation During Chromatin Assembly In Vitro.

    We sought to determine if histone precipitation is
responsible for the inhibitory affect of novobiocin on the
nucleoplasmin assembly reaction. Chromatin assembly reactions were prepared as described in figure 1. The protein
present in the supernatant or pellet after exposure to various novobiocin concentrations were analyzed by SDS-PAGE
and are shown in figure 4. We find that the presence of
precipitated histones parallels the inhibition of chromatin
assembly. For example, 750 μg/ml novobiocin produces
fairly complete inhibition of nucleosome formation (figure
1, lanes 10 and 11). In the presence of this concentration
of novobiocin, nearly all of the histone H3 and H4 present
in the reaction, and half of the H2A and H2B are found in
the pellet sample after centrifugation at 15,000 x g (figure 4, lanes 11 and 12). Novobiocin inhibition of nucleosome formation in this chromatin assembly system can be
accounted for by precipitation of histones.

Novobiocin Dissociates Histones from Chromatin
Assembled In Vitro.

    Supercoiled DNA molecules relax upon introduction of
novobiocin to a chromatin assembly system. (Glikin et al.,
1984). These results have been attributed to unbalanced
topo I and topo II activities acting on the unconstrained
supercoils of dynamic chromatin. A second interpretation

could be that novobiocin is removing histones from chromatin, causing them to precipitate and the resulting naked
DNA is available to be relaxed by topo I activity in the

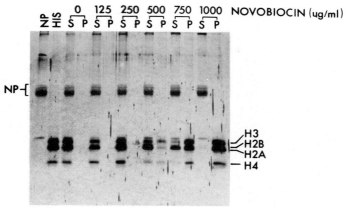

FIGURE 4. Novobiocin induces histone precipitation
during chromatin assembly _in vitro_. _In vitro_ chromatin
assembly reactions were performed with purified Xenopus
nucleoplasmin (lane 1), hyperacetylated HTC core histones
(lane 2) and topoisomerase I-treated pBR322 DNA in the
absence or presence of various concentrations of novobiocin
as indicated. After centrifugation and termination of the
assembly reations with SDS as described in the legend to
figure 1, soluble (S) and pelleted (P) fractions were analyzed directly on a 15% SDS polyacrylamide gel. Silver
staining of proteins is shown.

system. To test this possibility, chromatin was
assembled from purified components and shown to be fully
packaged (figure 5A, 0 hr). This material was then incubated with various concentrations of novobiocin (figure 5A)
for 16 hours and a portion of each reaction was centrifuged
to identify precipitated histones. A second portion of the
reaction was was treated with topoisomerase I and the DNA
was analyzed for supercoil content. We find that full supercoiling and hence, full packaging of DNA remains during
the incubation period in the absence of novobiocin (figure
5A). However, when the same chromatin was incubated with a
sufficient amount of novobiocin, the DNA became available
for relaxation (figure 5A, 1000 μg/ml novobiocin). Concom-
itant with the relaxation of DNA we find that histones are

precipitated.   (figure   5B).   The unconstrained supercoils
can be accounted for by the removal of histones from   solution   rather   than by invoking an alteration in the nucleo-

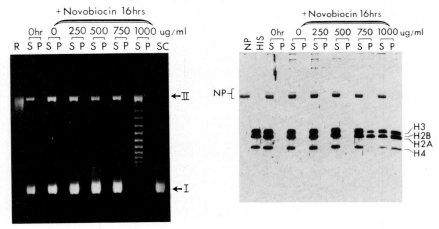

FIGURE 5. Novobiocin dissociates histones from chromatin assembled in vitro.
     A.   Chromatin was assembled in vitro from a mixture of
relaxed pBR322 DNA (lane 1), purified Xenopus   nucleoplasmin, hyperacetylated core histones and purified topoisomerase I. Supercoiled DNA isolated from the assembled   product
is   shown in lane 2.   Subsequently, this in vitro assembled
chromatin was incubated at room temperature for 16   hrs   in
the   absence   or   presence   of increasing concentrations of
novobiocin as indicated. After   additional   treatment   with
topoisomerase   I, samples were centrifuged and DNA isolated
from soluble (S) and precipitated (P) fractions for   analysis   on a 2% agarose gel.   Ethidium bromide fluorescence of
DNA is shown.
     B.   Chromatin assembly and subsequent incubation   with
novobiocin   were as described in A.   Soluble (S) and precipitated (P) fractions were analyzed directly on a   15%   SDS
polyacrylamide   gel that was subsequently stained with silver to identify histones and nucleoplasmin.

some structure.

Novobiocin Precipitates Histones in 2 M NaCl.

     We have observed that the precipitation of histones by
novobiocin   occurs   efficiently   at   high   ionic strengths.

Because the octameric association of core histones which is found in the nucleosome also occurs in 2 M NaCl (Thomas and Kornberg, 1975; Stein et al., 1979), we were interested in determining the effect of novobiocin on octameric histone complexes. One possibility is that the four core histones will begin to precipitate at the same novobiocin concentration indicating that novobiocin is incapable of disrupting the octamer. A second possibility is precipitation of histone H3 and H4 at lower concentration than histones H2A and H2B, demonstrating that the interaction of histones with novobiocin disrupts that histone octamer. The precipitation of histones in 2 M NaCl as a function of novobiocin concentration is demonstrated in figure 6. Consistent with the second possibility, histones H3 and H4 are precipitated in the presence of 100 μg/ml novobiocin while the bulk of histones H2A and H2B remain soluble (figure 6, lanes 7 and 8) indicating a disruption of the histone octamer. Complete precipitation of all four core histones occurs at concentrations of novobiocin of 300 μg/ml and greater.

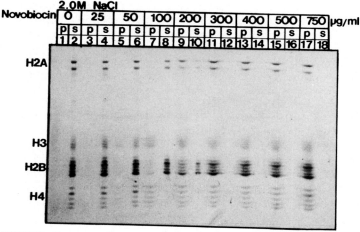

FIGURE 6. Novobiocin precipitation of histones in 2 M NaCl. Novobiocin precipitation of hyperacetylated core histones was performed as described in the legend to figure 2 except that all samples contained 2.0 M NaCl. The values at the top of the figure indicate novobiocin concentrations (μg/ml). 6 μg samples of histones were used for the precipitation. After fractionation the samples were diluted fourfold to lower the NaCl concentration to a level permissible for electrophoresis and aliquots corresponding to 1.5 μg of histone were applied to the gel. After electrophoresis and staining with Coomassie blue, the gel was stained

with silver. Note that histone H2B stains with silver much more intensely than the other core histones and thus appears to be present in lane 7 to a greater extent than H2A.

Novobiocin Disrupts the Histone Octamer.

Formaldehyde crosslinking provides a method for directly assaying the presence of histone octamers. It has been demonstrated that at physiological ionic strengths (150 mM NaCl) histones associate as H2A-H2B dimers and H3-H4 tetramers; while in 2 M NaCl, core histones associate in the octameric form that predominates in chromatin (Thomas and Kornberg, 1975; Stein et al., 1979). Because novobiocin functions to precipitate histones in the presence of high ionic strengths, and because histones H3 and H4 are precipitated at novobiocin concentrations that leave histones H2A and H2B in solution, we wondered if novobiocin might disrupt the octameric association of histones in 2 M NaCl. To test this possibility, we exposed core histones to the crosslinking agent formaldehyde. We find that in 150 mM NaCl, the crosslinked products are mainly 25 Kd and 50 Kd in size, presumably H2A-H2B dimers and H3-H4 tetramers. In the presence of 2 M NaCl, a species of approximately 110 Kd, shown by other investigators to be a histone octamer, is produced. We find that the presence of even 25 μg/ml novobiocin interferes with the formation of this histone complex, while the crosslinking of tetramers and dimers is unaffected by these low concentrations of novobiocin. Recall that the precipitation of histones H3 and H4 does not occur until concentrations of 100 μg/ml novobiocin are reached (see figure 6). Clearly novobiocin is disrupting higher order histone interaction at concentrations that do not interfere with histone solubility.

DISCUSSION

We have demonstrated that novobiocin precipitates histones at concentrations normally used to inhibit topoisomerase II activity. The interactions between novobiocin and core histones appear to disrupt the octameric associations of the histone molecules. Thus, in 2 M NaCl, where the predominant state of histones is the heterotypic octamer (Thomas and Kornberg, 1975) 100 μg/ml novobiocin precipitates H3 and H4 while H2A and H2B remain in solution. Furthermore, the crosslinking of histone octamers in 2 M NaCl is interupted by novobiocin concentrations that have no effect on the solubility of the proteins. It is likely that at

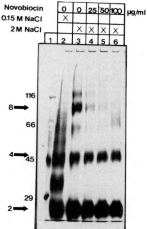

FIGURE 7. Novobiocin disrupts the histone octamer. Core histones (at a concentration of 0.5 mg/ml) were cros-slinked for 20 hours with 1% formaldehyde in 40 mM sodium borate, pH 9.7 containing either 150 mM NaCl (lane 2) or 2 M NaCl (lanes 3-6). Novobiocin was included at the indi-cated concentrations. After crosslinking, the samples were dialyzed extensively against 20 mM glycine, pH 7, resolved by SDS-PAGE and stained with silver.

physiological ionic strengths, novobiocin may inter-fere with the formation of higher order histone interac-tions which may be crucial for the formation of appropriate chromatin structure.

Novobiocin is capable of removing histones from solu-tions containing nucleoplasmin (figure 4) and from chroma-tin (figure 5) It is not surprising that an agent which can disrupt the octameric structure of core histones should also alter the interactions of these histones with DNA.

Clearly the experiments implicating DNA gyrase in both the chromatin assembly reaction (Glikin et al., 1984; Ryoji et al., 1984) and RNA synthesis (Kmiec and Worcel, 1985; Glikin and Blangy, 1986; Kmiec et al., 1986,) should be reconsidered in light of the conclusions of this paper. We have assayed for the precipitation of histones in the chro-matin assembly extract described by Glikin et al., (1984). We find that novobiocin, at concentrations up to 1000 μg/ml does not cause appreciable precipitation of histones. On the other hand, novobiocin is capable of disrupting histone-histone interactions without causing histone precipitation (figure 7) and these disruptions may interfere with chromatin assembly. This assembly extract contains high concentrations of RNA (approximately 200 μg/ml) and when this RNA is extracted and added to pure histone solutions, it allows the histones to remain soluble in 500 μg/ml novobiocin (M. C., unpublished data). Furthermore, precipitation of histones from this extract proceeds quantitatively in the presence of 2 M NaCl. This high salt concentration may function to disrupt histone-RNA interactions which allow histone solubility in the presence of novobiocin.

With regard to RNA synthesis, it has been recently reported that novobiocin inhibits an in vitro 5S RNA synthesis system, while anti-topoisomerase II antibodies and the topisomerase II inhibitor VM-26 have no inhibitory effect (Gottesfeld, 1986). These results suggest a alternate target for novobiocin, consistent with our observations.

REFERENCES

Cotten, M. and Chalkley, R. (1985) Nucleic Acids Research 13, 401-414.

Cotten, M., Bresnahan, D., Thompson, S., Sealy, L. and Chalkley, R. (1986) Nucleic Acids Res. in press.

Germond, J. E., Hirt, B., Oudet, P., Gross-Bellard, M. and Chambon, P. (1975) Proc. Natl. Acad. Sci. USA 72, 1843-1847.

Glikin, G.C., Ruberti, I. and Worcel, A. (1984) Cell 37, 33-41.

Glikin, G. C. and Blangy, D. (1986) EMBO J. 5, 151-155.

Gottesfeld, J. M. (1986) Nucleic Acids Res. 14, 2075-2088.

Kmiec, E. B. and Worcel, A. (1985) Cell 41, 945-953.

Kmiec, E. B., Ryoji, M. and Worcel, A. (1986) Proc. Natl. Acad. Sci. USA 83, 1305-1309.

Ryoji, M. and Worcel, A. (1984) Cell 37, 21-32.

Sealy, L., Cotten, M. and Chalkley, R. (1986) Biochemistry in press.

Shure, M. and Vinograd, J. (1976) Cell 8, 215-226.

Stein, A., Whitlock, J. P. and Bina, M. (1979) Proc. Natl. Acad. Sci. USA 76, 5000-5004

Thomas, J.O. and Kornberg, R.D. (1975) Proc. Natl. Acad. Sci. USA. 72, 2626-2630

Transcriptional Control Mechanisms, pages 259–274
© **1987 Alan R. Liss, Inc.**

ESTROGEN REGULATION OF VITELLOGENIN mRNA STABILITY[1]

John E. Blume[2], David A. Nielsen[2] and David J. Shapiro

Department of Biochemistry, University of Illinois,
1209 W. California, Urbana, IL 61801

ABSTRACT   The estrogen induction of vitellogenin mRNA
in Xenopus liver cells is achieved both through
regulation of vitellogenin gene transcription and by
selective stabilization of vitellogenin mRNA against
cytoplasmic degradation. Our measurements of the
absolute rate of nuclear gene transcription and of the
rate of accumulation of vitellogenin mRNA in the
cytoplasm demonstrate that the efficiency with which
each intervening sequence is excised from vitellogenin
transcripts in vivo is at least 99%. The 30 fold
increase in the stability of vitellogenin mRNA in the
presence of estrogen is a specific, reversible
cytoplasmic effect of the steroid hormone. The
stabilization of vitellogenin mRNA in the presence of
estrogen does not appear to be due to the unusually
efficient translation of vitellogenin mRNA. The
abundant vitellogenin synthesizing polysomes represent
a discrete polysome peak and exhibit a size
characteristic of polysomes translating an mRNA coding
for a polypeptide of approximately 200,000 daltons.
Following withdrawal of estrogen, there is a gradual
disappearance of the vitellogenin synthesizing polysome
population with no changes in its position in a
polysome gradient. These data provide no support for
the hypothesis that changes in the translational
efficiency of vitellogenin mRNA and in the distribution

[1]This work was supported by grant #DCB 83-02451 from
the National Science Foundation. D.N. is an American Cancer
Society Postdoctoral Fellow.
[2]These workers contributed equally to the experiments
described in this report.

of ribosomes on the mRNA play a major role in
regulating its stability.

INTRODUCTION

Addition of estrogen to primary cultures of Xenopus
liver cells results in the massive induction of the mRNA
coding for the egg yolk precursor protein vitellogenin
(26). In fully induced Xenopus liver cells, vitellogenin
mRNA represents approximately half the cell's mRNA. The
induction of vitellogenin mRNA is achieved by three basic
mechanisms: (a) an increase of at least several thousand
fold in the absolute rate of vitellogenin gene
transcription, which rises from undetectable levels
($<$1 transcript/vitellogenin gene/day) to approximately 5
transcripts/gene/min, (b) an increase of approximately 40
fold in the absolute rate of total nuclear RNA synthesis,
and (c) a selective stabilization of vitellogenin mRNA
against cytoplasmic degradation. Vitellogenin mRNA is
essentially stable in the presence of estrogen, exhibiting a
half life of approximately 500 hours, and is degraded with a
half life of 16 hours when estrogen is removed from the
culture medium. The regulation of vitellogenin mRNA
stability is a reversible cytoplasmic effect of estrogen
(3,4,26)
     In this work we have examined the possibility that
changes in the efficiency with which vitellogenin mRNA is
translated play a major role either in its stabilization or
in the shift from the stable to the unstable form. Our data
indicate that vitellogenin mRNA is translated in the
presence of estrogen on a polysome of the size appropriate
for such a large mRNA, and that dense packing of ribosomes
on the mRNA is unlikely to play a significant role in its
stabilization by estrogen. Removal of estrogen from the
culture medium results in a gradual disappearance of the
vitellogenin synthesizing polysome peak but does not change
its location on polysome gradients. The data does not
support the view that the destabilization of vitellogenin
mRNA results from a dramatic failure to translate the mRNA
successfully in the absence of estrogen. Changes in
translational efficiency therefore appear unlikely to play a
significant role in the selective stabilization of
vitellogenin mRNA by estrogen.

RESULTS

In the presence of estrogen, vitellogenin mRNA is induced from much less than 1 molecule per cell to approximately 50,000 molecules per cell, at which time it represents approximately half the cell's mRNA (3). In order to identify the regulatory strategies employed by Xenopus liver cells to achieve this impressive induction, we have made direct measurements of rates of vitellogenin gene transcription in primary Xenopus liver cultures. The data presented in Table 1 summarizes our work in this area. In the absence of estrogen, there is no detectable transcription of the vitellogenin genes down to a level of <1 molecule of vitellogenin mRNA synthesized per cell per day. When estrogen is present in the culture medium, vitellogenin gene transcription is induced by at least several thousand fold, to a rate of 20 molecules of vitellogenin transcript per cell per minute. This is among the highest rates of gene transcription seen for eucaryotic cellular genes.

TABLE 1

ESTROGEN REGULATION OF VITELLOGENIN mRNA LEVELS[a]

Treatment	Vitellogenin Gene Transcription	Vitellogenin mRNA Accumulation	Nuclear RNA Synthesis	Vitellogenin mRNA Degradation
No estrogen	<1 molecule/ cell/day	0	2 pmol/10^6 nuclei/min	$t_{1/2}$=16 hr
+ Estrogen	20 molecules/ cell/min	13 molecules/ cell/min	40 pmol/10^6 nuclei/min	$t_{1/2}$=500 hr

[a]These data are derived from measurements of the absolute rates of vitellogenin gene transcription, total nuclear synthesis, and vitellogenin mRNA degradation. In general transcription rates were measured by labeling nuclear RNA in whole cells with [3]H—uridine, determining the specific radioactivity of the cellular and nuclear UTP pools by HPLC, and hybridizing the nuclear RNA to filter-bound vitellogenin cDNA clones. Vitellogenin mRNA degradation was measured in a similar way as detailed in the text (summarizes data of 3 and 4).

Our simultaneous measurements of the rate of nuclear
vitellogenin gene transcription and of the rate of
vitellogenin mRNA accumulation in the cytoplasm of these
cells has allowed us to make what is perhaps the only
quantitative estimate of the efficiency with which
intervening sequences are excised from cellular
transcripts. Each vitellogenin gene contains 32 intervening
sequences which must be removed to produce functional
cytoplasmic mRNA. Since newly synthesized vitellogenin mRNA
is stable in the cytoplasm of these cells and exhibits a
half life of approximately 500 hours, all mRNA which
successfully exits the nucleus will accumulate in the
cytoplasm. Our finding that nuclear gene transcription
produces 20 molecules of vitellogenin mRNA per minute, while
only 13 molecules appear in the cytoplasm, suggests that
approximately 35% of the newly transcribed vitellogenin mRNA
is subject to intranuclear degradation. The efficiency with
which each of vitellogenin's 32 intervening sequences are
excised can therefore not be less than approximately 99% for
each intervening sequence.

Estrogen Stabilizes Vitellogenin mRNA.

Several types of indirect evidence suggested that
vitellogenin mRNA is relatively stable in cells maintained
in estrogen. We made direct measurements of the stability
of vitellogenin mRNA by pulse labeling the RNA during
induction and hybridizing the cytoplasmic RNA to
vitellogenin DNA clones immobilized on nitrocellulose
filters. The data presented in Figure 1 indicates that in
cells maintained in estrogen, vitellogenin mRNA is
essentially stable for an extended period and exhibits a
half life of approximately 500 hours, which is about 3
weeks. The stability of vitellogenin mRNA in the presence
of estrogen is particularly striking in view of its very
large size of 6,500 nucleotides. When estrogen is removed
from the culture medium, there is a lag period of 12 to 24
hours during which residual estrogen is metabolized. Then,
vitellogenin mRNA is degraded with a half life of
approximately 16 hours, which is similar to the half life of
total cell poly(A) mRNA in both the presence and absence of
estrogen. Readdition of estradiol-17β restabilizes the
vitellogenin mRNA to a half life of 500 hours. These data
and other confirming experiments provide direct evidence for

a specific estrogen mediated stabilization of vitellogenin mRNA against cytoplasmic degradation.

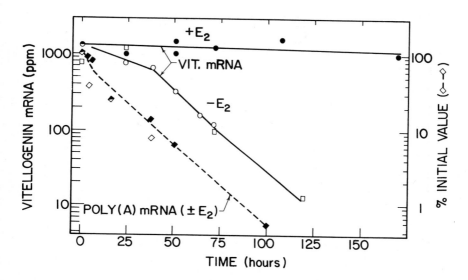

FIGURE 1.   Estrogen stabilizes vitellogenin mRNA against degradation. The solid lines show the decay of pulse labeled vitellogenin mRNA in primary liver cube cultures maintained in medium containing estrogen ($\bullet$) or lacking estrogen (o). The decay of hybridizable vitellogenin mRNA ($\square$) and of total poly(A) mRNA in the same cultures containing estrogen ($\blacklozenge$- $-$ $\blacklozenge$) and lacking estrogen ($\lozenge$- $-\lozenge$) is also shown. (Redrawn from 3.)

Although the regulation of vitellogenin mRNA by estrogen is a particularly striking example of this process, control of mRNA stability has been observed in at least 20 eucaryotic and procaryotic systems (Table 2). These data indicate that regulation of mRNA stability is a flexible and powerful control mechanism with the potential to complement transcriptional controls.

TABLE 2

SOME BIOLOGICAL SYSTEMS THAT EXHIBIT REGULATION OF SPECIFIC mRNA STABILITY

mRNA	Cell or Tissue	Regulatory Signal	Half Life		Reference
			+ Effector	- Effector	
1) Vitellogenin	Xenopus liver	Estrogen	+E$_2$; 500 hr	-E$_2$; 16 hr	(3)
2) Albumin	Xenopus liver	Estrogen	+E$_2$; 3 hr	-E$_2$; 10 hr	(32,17)
3) Vitellogenin	Avian liver	Estrogen	+E$_2$; 22 hr	-E$_2$; ~2.5 hr	(31)
4) ApO VLDL II	Avian liver	Estrogen	+E$_2$; 26 hr	-E$_2$; 3 hr	(31)
5) Casein	Rat mammary gland	Prolactin	+Pro; 92 hr	-Pro; 5 hr	(10)
6) Growth Hormone	Rat pituitary cell cultures	Dexamethasone & T$_3$	+Dex & T$_3$; 20 hr	-Dex, -T$_3$; 2 hr	(6)
7) Insulin	Rat pancreas islet cells	Glucose	+Gluc; 77 hr	-Gluc; 29 hr	(29)
8) Ovalbumin	Hen oviduct	Estrogen, Progesterone	+E$_2$; ~24 hr	-E$_2$; 2-5 hr	(21,23,16)
9) Prostatic Steroid Binding Protein	Rat ventral prostate	Androgen	$\Delta t_{1/2}$; ~30 x		(22)
10) Prolactin	Rat pituitary cell cultures	Thyroliberin (TRH)	+TRH; 27 hr	-TRH; 17 hr	(20)
11) Histone	Yeast	DNA replication	During Rep; 15 min	Post Rep; ~5 min	(15)

			During Rep; ~4 hr	Post Rep; ~15 min	
12) Histone	HeLa & mouse cell cultures	DNA replication	During Rep; ~4 hr	Post Rep; ~15 min	(5,14)
13) HSP70	Drosophila, Schneider cells	Heat shock, HSP70 autoregulation, Canavanine	Early h.s.; 6 hr	Late h.s.; 1 hr + Canavanine; 6 hr	(7)
14) E1A	Ad$_2$ infected HeLa cells	72 Kd binding protein	+72 kd; 30 min	-72 Kd; 85 min	(1)
15) E1B	Ad$_2$ infected HeLa cells	Early/late infection	+72 Kd; 30 min	-72 Kd; 150 min	(30)
16) Actin	Ad$_2$ infected HeLa cells	Viral (Ad$_2$) infection	Infected; 19 hr	Uninf; 9 hr	(18)
17) Actin	Friend's erythro-leukemia cells	DMSO induced	+DMSO; 5 hr	-DMSO; 22 hr	(19)
18) Tubulin	Friend's erythro-leukemia cells	differentiation	+DMSO; 3.5 hr	-DMSO; 7 hr	(19)
19) λ Integrase	λ infected E. coli	Early/late infection	Antitermination; 45 sec (early)	Termination: >5 min (late)	(24)
20) Gene 32	T$_4$ Infected E. coli	infected/uninfected	>20 min	< 5 min	(9)

Mechanisms Regulating Vitellogenin mRNA Stability.

In order to more precisely localize the site at which the stability of vitellogenin mRNA was regulated, we performed experiments designed to establish whether nuclear modifications of the RNA or reversible cytoplasmic events are required for its stabilization. Our data indicated that the stabilization of vitellogenin mRNA is a reversible cytoplasmic effect of estrogen which is presumably mediated by reversible protein nucleic acid interactions in the cytoplasm of these cells (3). These data, and related experiments which demonstrated that the stabilization and destabilization of vitellogenin mRNA could occur in the absence of protein synthesis, indicated that this response might not be mediated by the estrogen receptor (ER) in Xenopus liver cells, which is largely a nuclear protein (12). We have identified a cytoplasmic estrogen binding protein (EBP; 13,11) whose properties are consistent with a role in the regulation of vitellogenin mRNA stability (Table 3). The EBP is quite abundant in cells and is present at up to several hundred thousand sites per cell, which is a level sufficient to directly interact with vitellogenin mRNA and all of the mRNAs stabilized by estrogen in the cell, if that is its role. A second important property is that it has exactly the same high level of estrogen specificity as does the nuclear estrogen receptor.

TABLE 3

PROPERTIES OF XENOPUS ESTROGEN RECEPTOR AND ESTROGEN BINDING PROTEIN[a]

	K_D	Binding Specificity	Abundance Binding sites/cell	Localization +E Nuc.	Cyt.
E.R.	$4 \cdot 10^{-10}$ M	E, DES	400–2,000	>90%	
E.B.P.	$4 \cdot 10^{-8}$ M	E, DES	100,000–200,000		>90%

[a]Binding assays were carried out using Sephadex LH–20 minimal minicolumns. The table summarizes data from 11,12,13, and C. Sentman and D. Shapiro, unpublished observations. E = Estradiol, DES = Diethylstilbestrol.

The fact that the cytoplasmic estrogen binding protein has an affinity for estrogen approximately 100 fold lower than does the nuclear estrogen receptor suggested a plausible model for a flexible mechanism for regulating vitellogenin mRNA levels. We might hypothesize that in the presence of low concentrations of estradiol-17β the nuclear estrogen receptor may become saturated with hormone and initiate transcription of the vitellogenin genes. Since the cytoplasmic estrogen binding protein will not be saturated under these conditions, the mRNA will not be stabilized so that it will only accumulate to relatively modest levels in cells in the presence of low levels of the inducing hormone. In the presence of high levels of estradiol-17β, as are routinely used in our culture system, the estrogen binding protein in the cytoplasm will be saturated with estradiol and would then mediate the stabilization of vitellogenin mRNA, allowing a further increase of approximately 30 fold in the ultimate level of vitellogenin mRNA, thereby providing a great deal of flexibility in this control. The properties of the EBP also suggest that it is a nucleic acid binding protein.

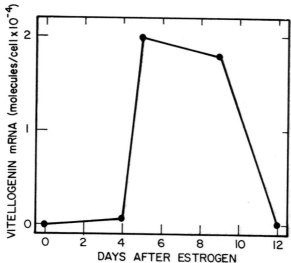

FIGURE 2.   Kinetics of vitellogenin mRNA induction in vivo. Estradiol-17β, dissolved in DMSO (1 mg/50 g) was administered on day 0.   Vitellogenin mRNA levels were determined by quantitative dot hybridization (3) and calculated as described (2).

Vitellogenin Synthesizing Polysomes During Estrogen
Stimulation and Withdrawal.

The large size of the vitellogenin polypeptides, which
have a molecular weight of approximately 200,000 daltons,
and the extremely high levels of vitellogenin mRNA in cells,
make it possible to physically isolate a population of
vitellogenin synthesizing polysomes. Through the use of
DMSO as injection vehicle, rather than propylene glycol, we
have been able to develop conditions in which reasonably
efficient induction of mRNA in the presence of estrogen is
followed by its rapid degradation in the absence of
estrogen. These data are summarized in Figure 2, which
indicates that 12 days after a single injection of
estradiol-17β, vitellogenin mRNA has declined to low
levels. These data suggested that 7 days after estrogen
administration would be an appropriate time to isolate a
population of polysomes which contains estrogen stabilized
vitellogenin mRNA and that 9 and 12 days after estrogen
administration would be appropriate times to investigate
polysomes during withdrawal. Using near physiological
conditions for the isolation of undegraded Xenopus liver
polysomes, we analyzed polysome profiles on high resolution
isokinetic sucrose gradients, as illustrated in Figure 3. A
discrete peak corresponding to vitellogenin synthesizing
polysomes is observed in the polysomes isolated at day 7.
Northern blot analysis of polysome gradient samples confirms
that the preponderant location of vitellogenin mRNA is in
this peak (Fig. 4). Comparison to other gradient profiles
indicate that vitellogenin mRNA is being translated on
polysomes containing approximately 35 ribosomes. The
vitellogenin polysome peak was significantly diminished by
day 9, and had largely disappeared by day 12. The data
suggests that the location of the vitellogenin synthesizing
polysomes in the gradients is unchanged during estrogen
withdrawal, although they can no longer be visualized as a
discrete polysome peak. These data are inconsistent with a
model for vitellogenin mRNA stabilization and degradation
based primarily on changes in the density of ribosome
loading on vitellogenin mRNA.

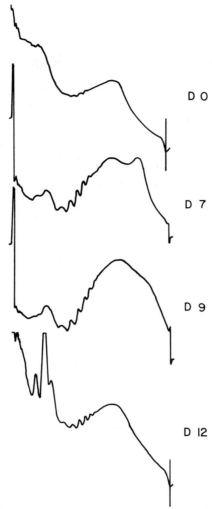

D O

D 7

D 9

D I2

FIGURE 3. Polysome profiles during estrogen stimulation and withdrawal. Total (free and membrane bound)
polysomes were isolated from <u>Xenopus</u> liver under near
physiologic conditions, at the indicated times after
estrogen administration. Total cell polysomes were isolated
on a sucrose cushion, by a modification of an earlier method
(27) and analyzed on 0.5-1.5 M isokinetic sucrose gradients.
Although the vitellogenin polysome peak at day 9 remains
substantial, its smaller size prevents its detection as a
discrete peak. Instead, a broad polysome peak is observed.

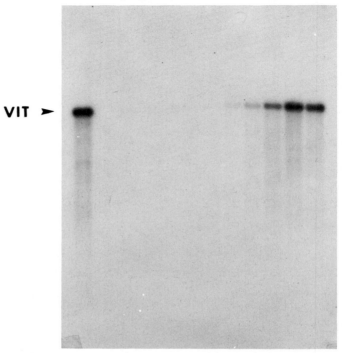

FIGURE 4. Localization of vitellogenin mRNA in polysome gradients by blot hybridization. RNA from a polysome gradient similar to the profile shown above for day 7 was isolated, fractionated by electrophoresis and analyzed by blot hybridization. The fraction with the highest level of vitellogenin mRNA corresponds to the location of the second peak in the gradient. The top of the gradient is on the left.

DISCUSSION

The 30 fold increase in the stability of vitellogenin mRNA in the presence of estrogen makes a major contribution to the induction of high levels of the mRNA. The important observation that vitellogenin mRNA, which has been pulse-labeled during induction and is undergoing rapid degradation following estrogen withdrawal, can be quantitatively restabilized by readdition of estrogen to the culture medium has focused our attention on the cytoplasmic factors

regulating vitellogenin mRNA stability. These studies and
related experiments also suggested that the nuclear estrogen
receptor was probably not a direct mediator of vitellogenin
mRNA stabilization. It seems probable that Xenopus liver
cells sense the estrogen concentration in the cytoplasm
through an estrogen binding protein we have identified. The
specificity, abundance, binding constant, localization, and
apparent affinity for the phosphate backbone for nucleic
acids all suggest a role for the cytoplasmic estrogen
binding protein in vitellogenin mRNA stabilization.

One possible model for the regulation of vitellogenin
mRNA stability in the cytoplasm of these cells is based on
translational efficiency. We might hypothesize that in the
presence of estrogen, vitellogenin mRNA is translated with
unusual efficiency in these cells. This would lead to a
high density of ribosome packing on the mRNA and might
possibly protect it against nuclease degradation. A related
and perhaps more likely hypothesis would be that following
withdrawal of estrogen from the culture medium, the
translation of vitellogenin mRNA becomes inefficient. In
the absence of the normal ribosome population on the mRNA,
it might be accessible to nuclease degradation and be
degraded more rapidly. Our data suggest that neither of
these models is likely to be correct. The translation of
vitellogenin mRNA appears to be carried out by a discrete
peak of polysomes containing on the order of 35 ribosomes.
This is roughly the size to be expected of a polysome
synthesizing a polypeptide with a molecular mass of
approximately 200,000 daltons. By comparison to other
systems such as the ovalbumin system, in which ovalbumin, a
polypeptide of molecular weight 44,000, is translated on a
12 ribosome polysome and globin, in which an 18,000 dalton
polypeptide is translated on a 4 or 5 ribosome polysome, we
would actually expect vitellogenin to be translated on a
polysome slightly larger than the one we observe. Xenopus
normally live at approximately 18°, and this may account for
a lower translational efficiency relative to 37° organisms.
Since the half life of vitellogenin mRNA in the presence of
estrogen is 500 hours, or approximately 3 weeks, translation
control mechanisms appear incapable of accounting for the
fact that vitellogenin mRNA in the presence of estrogen
appears to be essentially exempt from normal cellular
mechanisms of mRNA degradation.

In these studies we have examined the total polysome
population of Xenopus and have not distinguished between

free and membrane bound polysomes. In the presence of
estrogen, vitellogenin mRNA, which codes for a secreted
polypeptide, is translated on polysomes bound to the
endoplasmic reticulum. Our data cannot exclude the
possibility that, following withdrawal of estrogen,
vitellogenin synthesizing polysomes dissociate from the
endoplasmic reticulum. This change could potentially render
the mRNA susceptible to normal cellular degradative
processes. However, association of mRNAs with the
endoplasmic reticulum is insufficient to account for
vitellogenin mRNA stabilization. Albumin mRNA, which is
translated on a membrane bound polysome, is destabilized in
the presence of estrogen and exhibits a half life
approximately 3 fold lower than it does in the absence of
the hormone (17,32).

Our data fails to support a hypothesis for vitellogenin
mRNA stabilization based on selective translation of the
message and suggests that it may be more appropriate to
focus on the properties and structure of vitellogenin mRNP
particles in the presence and absence of estrogen.

ACKNOWLEDGEMENTS

We are grateful for the skillful assistance of Ms. T.
Houghland in typing and assembling the manuscript.

REFERENCES

1. Babich, A and Nevins, JR (1981). Cell 26:371-379.
2. Baker, HJ and Shapiro, DJ (1977). J Biol Chem 252:8434-
    8438.
3. Brock, ML and Shapiro, DJ (1983a). Cell 34:207-214.
4. Brock, ML and Shapiro, DJ (1983b). J Biol Chem
    258:5449-5455.
5. DeLisle, AJ, Graves, RA, Marzluff, WF, and Johnson,
    L.F. (1983) Mol Cell Biol 3:1920-1929.
6. Diamond, DJ and Goodman, HM (1985). J Mol Biol 81:41-
    62.
7. DiDomenico, BJ, Bugaisky, GE, and Lindquist, S (1982).
    Cell 31:593-603.
8. Goldberg, AL and St John, AC (1976). Ann Rev Biochem
    45:747-803.

9. Gorski, K, Roch, J-M, Prentki, P, and Krisch, HM (1985). Cell 34:461-469.

10. Guyette, WA, Matusik, RJ, and Rosen, JM (1979). Cell 17:1013-1023.

11. Hayward, MA, Brock, ML, and Shapiro, DJ (1982). Am J Physiol 243:(12) C1-C6.

12. Hayward, MA, Mitchell, TA, and Shapiro, DJ (1980). J Biol Chem 255:11308-11312.

13. Hayward, MA and Shapiro, DJ (1981). Devel Biol 88:333-340.

14. Heintz, N, Sive, HL, and Roeder, RG (1983). Mol Cell Biol 3:529-550.

15. Hereford, LM, Bromley, S, and Osley, MA (1982). Cell 30:305-310.

16. Hynes, NE, Groner, B, Sippel, AE, Jeep, S, Wurtz, T, Ngu, T-H, Giesecke, K, and Schutz, G (1979). Biochemistry 18:616-624.

17. Kazmaier, M, Buning, E, and Ryffel, GU (1985). EMBO J 4:1261-1266.

18. Khalili, K and Weinman, R (1984). J Mol Biol 180:1007-1015.

19. Krowczynska, R, Yenofsky, R, and Brawerman, G (1985). J Mol Biol 181:231-239.

20. Laverriere, JN, Morin, A, Tixier-Vidal, A, Truong, AT, Gourdji, D, and Martial, JA (1985). EMBO J 2:1493-1499.

21. McKnight, GS and Palmiter, RD (1979). J Biol Chem 254:9050-9058.

22. Page, MJ and Parker, MG (1982). Mol Cell Endocrinol 27:343-355.

23. Palmiter, RD and Carey, NH (1974). Proc Natl Acad Sci USA 71:2357-2361.

24. Schmeissner, U, McKenney, K, Rosenberg, M, and Court, D (1984). J Mol Biol 176:39-53.

25. Shapiro, DJ (1982). CRC Crit Rev Biochem 12:187-203.

26. Shapiro, DJ and Brock, ML (1985). In G. Litwack (ed.): "Biochemical Actions of Hormones" Vol. XII New York: Academic Press, p 139-172.

27. Shapiro, DJ, Taylor, JM, McKnight, GS, Palauos, R, Gonzalez, C, Kiely, ML, and Schimke, RT (1974). J Biol Chem 249:3665-3671.

28. Weinberg, ES, Hendricks, MB, Hemminki, K, Kuwabara, PE, and Farrelly, LA (1983). Devel Biol 98:117-129.

29. Welsh, M, Nielsen, DA, MacKrell, AJ, and Steiner, DF (1985). J Biol Chem 260:13590-13594.

30. Wilson, MC and Darnell, JE Jr (1981). J Mol Biol
     148:231-251.
31. Wiskocil, R, Bensky, P, Dower, W, Goldberger, RF,
     Gordon, JI, and Deeley, RG (1980). Proc Natl Acad Sci
     USA 77:4474-4478.
32. Wolffe, AP, Glover, JF, Martin, SC, Tenniswood, MPR,
     Williams, JL, and Tata, JR (1985). Eur J Biochem
     146:489-496.

Transcriptional Control Mechanisms, pages 275–293
© **1987 Alan R. Liss, Inc.**

MECHANISMS INVOLVED IN THE MULTIHORMONAL REGULATION
OF TRANSCRIPTION OF THE PEPCK GENE

David Chu, Kazuyuki Sasaki and Daryl Granner

Department of Molecular Physiology and Biophysics
Vanderbilt University Medical School
Nashville, TN 37232

ABSTRACT We used a nuclear RNA elongation assay to show that
cAMP analogs and dexamethasone (Dex) cause a selective
increase in transcription of the P-enolpyruvate carboxykinase
(PEPCK) gene in H4IIE hepatoma cells. Maximally effective
concentrations of Dex and cAMP give an additive response when
combined, suggesting that each acts through separate mechanisms on the PEPCK gene. Insulin and phorbol esters inhibit
PEPCK gene transcription and override the stimulatory effects
of cAMP and Dex. Each of these effects is specific, since
none of the effectors alter total RNA transcription. Insulin
does not cause attenuation of PEPCK gene transcription and
most probably affects the transcription of the PEPCK gene at
or near the site of RNA polymerase II initiation.

INTRODUCTION

Studies of various animal and cell models have contributed considerably to an understanding of the mechanism of
hormonal regulation of gene expression. The goal of our
laboratory is to understand how hormones increase or decrease
the transcription of specific genes. We have concentrated
our efforts on the regulation of the hepatic enzyme phosphoenolpyruvate carboxykinase (PEPCK) in cultured rat H4IIE
hepatoma cells. The activity of PEPCK, a rate-limiting glu-
coneogenic enzyme, is altered by a number of hormones
involved in the regulation of this metabolic process. There
is no known post-translational modification of PEPCK, thus
changes of activity are entirely due to alterations of the
rate of synthesis of the enzyme. Glucagon (via cyclic AMP)
and glucocorticoids increase PEPCK synthesis and gluconeogenesis; insulin decreases PEPCK synthesis and gluconeogenesis
(1-3). These changes could be exerted at the post-transcriptional level, but, by coupling measurements of PEPCK synthe-

sis with assays capable of quantitating both mRNA^{PEPCK}
activity and amount, we were able to exclude the following
possibilities as sites of hormonal control: 1) the transport
of mRNA^{PEPCK} from the nucleus; 2) the rate of degradation of
mRNA^{PEPCK} in the cytoplasm or nucleus; and 3) mRNA^{PEPCK}
translational efficiency (2, 4-6). Our interpretation of
these studies was that cAMP, glucocorticoids and insulin must
affect mRNA^{PEPCK} synthesis. The purpose of the studies
described in this paper was to determine whether each of
these molecules regulated PEPCK synthesis at the level of
PEPCK gene transcription, and to explore how these effectors
acted in concert.

Steroid hormones regulate metabolic process by affecting
the rate of transcription of specific genes. It is generally
thought that steroids induce a structural alteration of their
intracellular receptors that increases the receptor's affinity for a specific DNA binding site. This then somehow regulates the transcription of specific genes. Until recently
hormones that bind to plasma membrane receptors were not
though to regulate gene expression, but evidence that such is
the case is rapidly accumulating. In most cases the internalization of the plasma membrane hormone-receptor complex is
not a prerequisite for action. This implies that another
molecule or process couples the hormone-receptor interaction
to the intracellular process. Several hormones activate
adenylate cyclase which, in turn, increases cellular cAMP and
induces specific gene products including tyrosine aminotransferase, PEPCK, lactate dehydrogenase, prolactin, albumin,
alkaline phosphatase, haptoglobin and discoidin (see Ref. 7
for details).

Insulin regulates the synthesis of proteins in a variety
of tissues, but relatively little is known about the regulatory mechanisms involved, and the intracellular mediator of
this hormone has not been identified. Insulin was thought to
regulate the synthesis of proteins by affecting ribosomes,
translation factors, or both (8, 9), but it is difficult to
explain selective effects by this mechanism, especially since
stimulatory and inhibitory responses to the hormone have been
noted. Another possibility is that insulin regulates protein
synthesis at the transcriptional level, an hypothesis
supported by the observations that insulin stimultes RNA
polymerase activity and RNA synthesis (10, 11). More direct
observations support this. Insulin results in specific and
selective changes of the mRNAs that code for albumin, amylase, pyruvate kinase, fatty acid synthetase, tyrosine aminotransferase, ovalbumin, casein, and PEPCK (see Ref. 7 for
details). Although these changes could result from a post-
transcriptional effect of insulin, in many cases control at
the level of gene transcription has been demonstrated (22).
PEPCK represents the most completely studied example, and is

unique in that it represents an inhibition of gene transcription by insulin (6).

Recent studies indicate that the hydrolysis of phosphatidyl inositides play a central role in the action of several hormones (13), and there is evidence that such products are involved in insulin action (14). Inositol phospholipid hydrolysis produces transient increases of the level of diacylglycerol, which may activate protein kinase C by increasing the affinity of this enzyme for substrate (15). The tumor-promoting phorbol esters, such as phorbol-12-myristate-13-acetate (PMA), directly activate protein kinase C in vivo and in vitro, probably by substituting for diacylglycerol (15), thus are used as probes to investigate the role of protein kinase C in various metabolic processes. Since specific gene expression has been correlated with the ability of hormones to activate phosphatidyl inositide turnover and protein kinase C (16, 17), and because of the link between phosphatidyl inositide metabolism and insulin action, we incorporated studies of the tumor-promoter, PMA, into our investigations.

In this report we present an analysis of the individual and combined effects of cyclic AMP analogs, dexamethasone, phorbol esters, and insulin on transcription of the PEPCK gene in H4IIE hepatoma cells. We show that cAMP and glucocorticoids increase PEPCK gene transcription, whereas insulin and phorbol esters inhibit this process. When combinations of hormones are used, the dominant effect is exerted by the inhibitory agents. We also present evidence that the modulation of mRNA^{PEPCK} synthesis by insulin and cAMP is a consequence of altered initiation of transcription of the PEPCK gene.

EXPERIMENTAL PROCEDURES

The procedures employed have been described in detail in previous publications. Only a brief description is included in this paper.

Materials.

Swim's S-77 culture medium, hormones, chemicals, enzymes and radionuclides were obtained from standard sources and were used as described previosuly (7).

Cell Culture.

Monolayers of H4IIE cells were grown to confluency in Corning 150 cm^2 flasks. Culture and medium conditions were as described in detail in previous publications (7).

Quantitation of $mRNA^{PEPCK}$ and PEPCK Gene Transcription.

Total cell poly (A)$^+$ RNA was isolated and quantitated by the dot blot hybridization technique described previously (7). The cDNA probe employed was a 1300 base pair Sma I-Sph I fragment isolated from pPC116, a cloned recombinant DNA plasmid (18), labeled by nick translation. The results were analyzed by a densitometer scan of an autoradiogram. To quantitate PEPCK gene transcription cells were treated with hormones or effectors, nuclei were isolated, and the elongation of nascent RNA transcripts on the PEPCK gene was measured by a modification (7) of a procedure described previously (19).

RESULTS

Effect of Cyclic Nucleotide Analogs on PEPCK Gene Transcription.

We first assessed the effect dibutyryl cyclic AMP (Bt_2cAMP) has on PEPCK gene transcription (Table 1). The basal rate of incorporation of isotope into elongating PEPCK-RNA transcripts was 93 ppm, but within 20 min after adding Bt_2cAMP there was an 8-fold increase of PEPCK gene transcription. The maximal rate, a 10-fold increase, was achieved within 40 min after the addition of the cyclic nucleotide.

TABLE 1
EFFECT OF Bt_2cAMP ON PEPCK GENE TRANSCRIPTION

Treatment	Total RNA Synthesis (cpm x 10^{-6})	$mRNA^{PEPCK}$ Synthesis (ppm)
Bt_2 cAMP · 0 min	18.1 + 0.4	93 + 8
· 20 min	18.0 + 1.0	765 + 51
· 40 min	18.4 + 1.1	1004 + 175

The effect of Bt_2cAMP on PEPCK gene transcription was quantitated as described in Experimental Procedures. The data expressed reflect total RNA synthesis and mRNA PEPCK synthesis in H4IIE cells incubated for 0, 20 or 40 min in the presence of a 0.5 mM concentration of Bt_2 cAMP. Adapted from Granner, et al. (1983).

It is important to note that there is no effect of Bt_2cAMP on total RNA synthesis in these cells, thus the stimulatory effect of the nucleotide on $mRNA^{PEPCK}$ synthesis is quite specific.

Effect of a Glucocorticoid Hormone on PEPCK Gene Transcription.

We next assessed the effect of the synthetic glucocorticoid dexamethasone (Dex) on PEPCK protein and $mRNA^{PEPCK}$ synthesis (see experiment I, Table 2).   Dex at 50 nM caused a

TABLE 2
REGULATION OF PEPCK GENE TRANSCRIPTION BY DEXAMETHASONE

	Treatment	PEPCK (nmols/mg protein)	$mRNA^{PEPCK}$ Synthesis (ppm)
I	None	0.71 + 0.03	108 + 5
	Dexamethasone	2.52 + 0.11	605 + 1C
II	None	–	123
	50 nM Dexamethasone	–	305
	500 nM Dexamethasone	–	640

The effect of dexamethasone on PEPCK gene transcription in H4IIE cells was quantitated in two separate experiments.   In experiment I 50 nM dexamethasone was added to H4IIE cells for 48 hours and PEPCK protein, expressed in nmols/mg total protein, was quantitated using a specific radioimmunoassay. Transcription elongation was quantitated as described above using parallel cultures of H4IIE cells incubated in 500 nM dexamethasone for 30 min.   In experiment II the effect of 50 or 500 nM dexamethasone on transcription was quantitated as described above.

4-fold increase of PEPCK protein in H4IIE cells, and there was a corresponding increase of PEPCK gene transcription (from 108 to 605 ppm).   The half-maximal effect of Dex on PEPCK gene transcription was achieved at 50 nM, a concentration similar to that required for half-maximal occupation of the glucocorticoid receptor in these cells (Expt. II, Table 2).   Not shown is the fact that this maximal rate is achieved within 30 min, and that the hormone caused no change in total RNA transcription.   Therefore Dex enhances PEPCK protein production by selectively and rapidly increasing transcription of the PEPCK gene.

Effects of Insulin on mRNA[PEPCK] Synthesis.

Although Bt_2cAMP is an effective inducer of mRNA[PEPCK] synthesis, the results varied considerably between experiments, thus the non-metabolizable analog 8-(4-chlorophenyl-thio)cAMP (8-CPT-cAMP) was used in the following experiments. (Actually, a number of analogs have been tested, and all induce this process. Since the active moeity is cAMP, we will henceforth refer to it as such). The maximal rate of transcription was obtained 30 min after the addition of cAMP (see Figure 1A). This rate was maintained for 3-4 h then there was a gradual decline to a new steady state which was maintained at approximately 2-3 times the basal rate for at least 72 h (7). Addition of 5 nM insulin 30 min after cAMP was added resulted in a prompt decrease of mRNA[PEPCK] synthesis. By 5 min the process was substantially reduced and by 60 min after adding insulin the rate of mRNA[PEPCK] transcription was reduced to or below the basal level in spite of the fact that the cells still contained a maximally effective amount of cAMP. The action of insulin was exerted rapidly and completely overrode the effect of cAMP. This insulin-mediated effect is quite specific, since no detectable effect on total RNA transcription was observed.

H4IIE cells are remarkably sensitive to physiologic concentrations of insulin (12), and the addition of 1 pM insulin caused a substantial inhibition of mRNA[PEPCK] synthesis (Fig. 1B). Half-maximal suppression occurred between 2-5 pM, and 10 nM insulin reduced transcription to the level measured in untreated cells. Proinsulin also decreased PEPCK gene transcription, but approximately 30-50 fold higher concentrations were needed to achieve an inhibition equivalent to that obtained with insulin. These insulin concentration effects are virtually identical to those obtained when either the rate of synthesis of PEPCK or mRNA[PEPCK] are measured in H4IIE cells (6, 12). This provides additional evidence in support of the hypothesis that transcription is the primary site at which insulin regulates mRNA[PEPCK] amount, and subsequent enzyme synthesis. The 30-50 fold difference in sensitivity to insulin and proinsulin supports our previous contention that this effect is mediated through the insulin receptor (12).

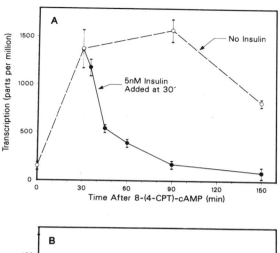

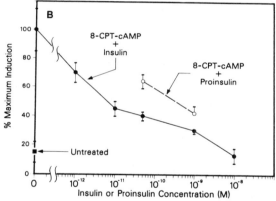

Figure 1.   Effects of 8–CPT–cAMP, Insulin and Proinsulin on mRNA[PEPCK] Transcription as a Function of Concentration and Time.   In the experiment illustrated in Panel A, nuclei were isolated from H4IIE cells that had been exposed to 0.1 mM 8–CPT–cAMP for the times indicated, or from cells to which 5 nM insulin was added 30 min after the cyclic nucleotide.   The transcriptional rate of the PEPCK gene was assayed as described and the results represent the average + S.D. of triplicate assays.   In the experiment illustrated in Panel B, H4IIE cells were exposed to 0.1 mM 8–CPT–cAMP for the 90 min and to various concentrations of insulin or proinsulin for the final 60 min.   Each point represents the mean of three transcription assays.   The rate of transcription is expressed as the percentage of maximum induction.   Details are described in ref. 7, from which this figure was reprinted, with permission.

Effect of a Phorbol Ester on mRNA^{PEPCK} Synthesis.

In studies of possible effects of calcium and phospholipid metabolites on PEPCK gene transcription, we discovered that phorbol myristate acetate (PMA) is a potent inhibitor of this process. As in Figure 2, H4IIE cells were treated with or without cAMP for 3 h with different concentrations of PMA. cAMP increased the amount of mRNA^{PEPCK} in basal and cAMP-induced cells. The half-maximal effect of PMA occurred at about 50 nM and maximal suppression of mRNA^{PEPCK} was achieved with 1 μM PMA.

The time course of this response was examined using 1 μM PMA to achieve maximal suppression of mRNA^{PEPCK} (Figure 3). mRNA^{PEPCK} increased rapidly, and remained elevated for at least 6 h in H4IIE cells exposed to cAMP. The addition of PMA 3 h after the cyclic nucleotide resulted in a rapid decrease of mRNA^{PEPCK}. Levels of mRNA^{PEPCK} decreased within 30 min after PMA treatment and a 50% reduction occured within one hour.

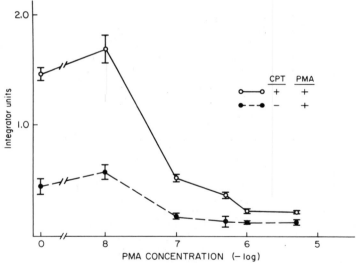

Figure 2. Phorbol Myristate Acetate Represses mRNA^{PEPCK} in H4IIE Cells. H4IIE cells were treated for 3 h with or without 0.1 mM 8-CPT-cAMP and with various concentrations of PMA. mRNA^{PEPCK} levels were determined as described above. The relative integrator units represent the mean $\pm$ SD of two samples in duplicate.

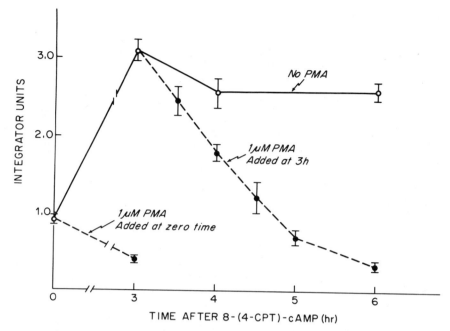

Figure 3. Rapidity of the Effect of PMA on mRNA[PEPCK].
Culture conditions, isolation of extracts, and cytodot
hybridization were as described above. Cytoplasmic extracts
were isolated from H4IIE cells that had been exposed to
0.1 mM 8-CPT-cAMP for the times indicated, or from cells to
which 1 μM PMA was added 30 min after the cyclic nucleotide
treatment. The results were expressed as the mean ± SD of
two samples in duplicate.

Effect of Hormone Combination on PEPCK Gene Transcription: The Dominant Role of Insulin and Phorbol Esters.

It has been suggested that cAMP is the primary regulator of PEPCK synthesis in rat liver (20) and that insulin
decreases PEPCK synthesis by activating phosphodiesterase,
which in turn reduces the intracellular concentration of cAMP
(21). Thus, using a poorly metabolized cyclic nucleotide, or
an agent other than cAMP to induce PEPCK, transcription
should be unaffected by insulin if the hormone acts only by
reducing cAMP. We therefore investigated insulin action in
H4IIE cells in which mRNA[PEPCK] was induced by either 8-CPT-
cAMP, an analog that is very slowly hydrolyzed, or by dexamethasone, a glucocorticoid that induces mRNA[PEPCK] in a cAMP-
independent manner (22). The experiment summarized in Table
3 illustrates the effects various hormone combinations have
on mRNA[PEPCK] transcription. Dex increased transcription by

TABLE 3
EFFECT OF HORMONE COMBINATIONS ON PEPCK GENE TRANSCRIPTION:
THE DOMINANT ROLE OF INSULIN.

	mRNA^{PEPCK} Synthesis in ppm	
Primary Treatment	No Insulin	Insulin
Control	108 + 3	70 + 6
Dexamethasone	606 + 3	58 + 23
cAMP	1355 + 43	365 + 6
Dexamethasone + cAMP	1619 + 131	520 + 10

H4IIE cells were treated for 60 min with 0.1 mM 8-CPT-cAMP,
0.5 µM dexamethasone, or the combination of these two, in the
absence or presence of 5 nM insulin. Nuclei were then isolated and transcription was quantitated as described above.
Results represent the mean + S.D. of triplicate assays.
Reprinted with permission, from Sasaki, et al, 1984.

6-fold, 8-CPT-cAMP by 13-fold, and the combination of Dex and
8-CPT-cAMP resulted in an additive induction, indicating
independent mechanisms of action. In the presence of 5 nM
insulin, mRNA^{PEPCK} synthesis was abolished in basal and dexa-
methasone-treated cells, and insulin significantly inhibited
transcription in cells treated with 8-CPT-cAMP, or the combination of dexamethasone plus 8-CPT-cAMP. Thus, in these
cells, insulin exerted the dominant effect on PEPCK gene
transcription; an effect achieved in the presence of maximally effective concentrations of inducers.

We also tested the combined effects of the inducing hormones and PMA on PEPCK gene transcription (see Table 4).
Dexamethasone increased the rate of transcription of PEPCK by
6-fold, 8-CPT-cAMP resulted in a 10-fold increase, and the
combination of both increased the transcription of PEPCK even
further. A 1 µM concentration of PMA totally abolished
mRNA^{PEPCK} transcription in the presence of 8-CPT-cAMP, dexamethasone, or the combination. As with insulin, there was no
observable change in the total RNA transcribed in the
presence of the different effectors, so the inhibitory efect
was quite specific.

TABLE 4

EFFECT OF HORMONE COMBINATIONS ON PEPCK GENE TRANSCRIPTION:
THE DOMINANT ROLE OF PMA.

Primary Treatment	mRNA^{PEPCK} Synthesis in ppm	
	No PMA	PMA
Control	113 + 23	NT
Dexamethasone	686 + 51	20 + 7
cAMP	1120 + 30	23 + 1
Dexamethasone + cAMP	1328 + 91	17 + 6

Nuclei were isolated from H4IIE cells that had been treated
for 60 minutes with 0.1 mM 8-CPT-cAMP, 0.5 µM dexamethasone,
or the combination of both, in the presence or absence of 1
µM PMA. Results represent the mean + SD of triplicate
assays.

Mechanism of Action of Insulin on PEPCK Gene Transcription.

The processes involved in the transcription of class II
genes in mammalian cells are poorly understood. The general
steps involve initiation, elongation and termination of the
transcript. Several factors are probably involved in each of
these processes, thus there are several possible sites at
which a hormone could exert a regulatory influence. Procaryotic cells regulate transcription by an additional process
called attenuation, in which the nascent transcript is
blocked at a certain point on the gene, thus preventing further elongation of the transcript. We conducted experiments
using the nuclear elongation assay, and probes directed
against various regions of the gene, to analyze whether insulin caused attenuation, retarded transcript elongation, or
inhibited transcript initiation. An analysis of the action
of cAMP on each of these processes was part of these studies.

Insulin Does Not Attenuate Transcription.

We first assessed whether transcription proceeds uniformly across the PEPCK gene in response to insulin and cAMP.
As illustrated in Figure 4, the standard nuclear elongation
assay was performed except that, in addition to the full size
genomic probe, four different fragments (see Figure 4 for
description) representing the 5' end, two middle regions, and
the 3' end of the PEPCK gene were used to probe for labeled
PEPCK transcripts isolated from nuclei of control, 8-CPT-
cAMP, and 8-CPT-cAMP plus insulin treated cells. If insulin

caused transcript attenuation the radioactivity incorporated into mRNA^{PEPCK} measured using the 5′ region probes (R–K or S–S) should exceed that detected by 3′ region probes (S–C or C–R). This is not the case, as transcription is uniformly decreased across the entire gene in insulin-treated cells (Figure 4). It is also apparent from this study that cAMP increases transcription across the entire gene. Since this assay reflects the polymerase complexes engaged in transcription, we conclude that cAMP increases, and insulin decreases, the number of polymerase complexes on the PEPCK gene.

INSULIN AND cAMP AFFECT TRANSCRIPTION INITIATION

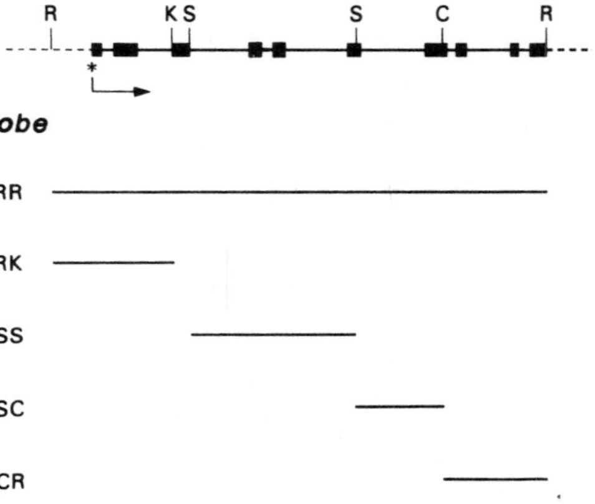

Probe		Transcription (ppm)		
		Control	8-CPT	8-CPT + Insulin
RR	———————————————	118	1031	171
RK	———	120	862	73
SS	———	233	1385	301
SC	———	132	1108	77
CR	———	57	1006	137

Figure 4. Structure of PEPCK Gene and Effects of 8-CPT-cAMP and Insulin on PEPCK Gene Transcription. The structure of the PEPCK gene is shown at the top of the Figure. Exons are indicated as the solid bars and the connecting lines are the introns. Restriction enzyme sites are abbreviated as R=EcoR I, K=KpnI, S=Sma I, C=Cla I, B=Bgl II. H4IIE cells were incubated with 0.1 mM 8-CPT-cAMP for 30 min then 5 nM insulin was added to the culture medium. Cells were harvested 30 min later and transcription was quantitated utilizing the genomic DNA fragments as shown in solid lines. Results are expressed in ppm corrected for variations of fragment length.

Insulin Retards Transcript Elongation.

A similar approach was used to assess the effect insulin treatment has on transcript elongation. H4IIE cells were treated with hormones for various times then the standard nuclear transcription run-on assay was performed using probes representing the extreme 5' (R-K) and 3' (B-S) regions of the PEPCK gene. The time required for the rate of transcription to change at the 3' end, compared to the 5' end, reflects the rate of elongation (the transit time of the PEPCK mRNA transcript). As shown in Figure 5A, incorporation at the 5' end of the gene increases immediately after the addition of cAMP and a similar change is seen at the 3' end about 5 min later. In cAMP-induced H4IIE cells, the addition of insulin caused an immediate reduction of isotope incorporation at the 5' end while the affect was not detected at the 3' end until 12 min had elapsed. Insulin thus slows down mRNA[PEPCK] transcript elongation by a factor of 2 or 3 as compared to cAMP.

This effect, while interesting, is not large enough to represent the major mechanism involved in the regulation of PEPCK gene transcription by the hormone. The short 5' region probe employed reflects the number of transcripts recently engaged, therefore we conclude that insulin inhibits, and cAMP stimulates, the initiation of mRNA[PEPCK] transcription.

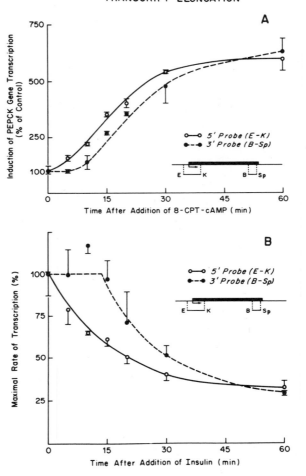

Figure 5.   Effects of 8–CPT–cAMP and Insulin on PEPCK
Gene Transcript Elongation.   Cells were incubated for 3 hr at
32°C.   Panel A: 8–CPT–cAMP was added to the medium and cells
were harvested at the indicated times.   Panel B: Insulin was
added to the medium after 1 hr of pretreatment with 8–CPT–
cAMP and cells were harvested at the indicated times.   Transcription at 5' (-0-) and 3' (-●-) regions were measured with
fragments (R-K) and (B-S), respectively.   Restriction enzyme
sites are R=EcoR I (-0.46kb), K=Kpn I (+0.88kb), B=Bgl II
(+5.02 kb), S=Sph I (+5.70kb).

DISCUSSION

We have defined a system with which we can study the individual and coordinate actions of a number of different kinds of hormones on transcription of the PEPCK gene. Our studies show that cAMP and glucocorticoids exert a rapid stimulatory effect while insulin and phorbol esters inhibit transcription of PEPCK gene. The inhibitory agents are both dominant as they each override the stimulatory actions of cAMP and glucocorticoids. All of these effects are relative-ly specific, since none of the hormones has a detectable effect on total RNA transcription. The extent of these effects is sufficient to account for the action these com-pounds have on PEPCK enzyme accumulation, and presumably on gluconeogenesis.

The major focus of our recent research has been to define the mechanism involved in the regulation of PEPCK gene transcription by insulin. Previous studies showed that the inhibitory effect of insulin on PEPCK gene transcription is: (1) achieved at physiologic concentrations of the hormone; (2) mediated through the insulin receptor; (3) specific; (4) seen in the absence of on-going protein synthesis; (5) readi-ly reversible; and (6) dominant over inducers (6, 7, 12).

These studies of the regulation of mRNA[PEPCK] synthesis provided the first evidence of an effect of insulin on gene transcription. In recent years several additional examples of the regulation of specific mRNA metabolism by insulin have been reported and these involve many different kinds of pro-teins (23). Several of these effects appear to be exerted at the level of transcription (23). It appears that the regula-tion of PEPCK synthesis represents a prototype of an impor-tant regulatory role of insulin.

Exactly how a hormone regulates the transcription of a specific gene is a question that perplexes all investigators involved in this aspect of hormone action. Unfortunately, the general mechanism of class II gene (mRNA) transcription in mammalian cells is poorly understood, but this is an extremely active area of investigation and basic principles should be elucidated in the near future. There are probably many factors that control the rate and fidelity of transcript initiation; elongation and termination may be equally com-plex. The studies reported in this paper are an approach to this general area, for insulin affects both transcript elon-gation and initiation. The latter appears to be the major role of insulin, just as it is the step affected by cAMP, albeit in the opposite direction.

The rapidity of these effects, and the fact that pro-tein synthesis is not required, suggests that these actions of insulin, cAMP, glucocorticoids, and phorbol esters may involve the covalent modification of one or more proteins

that somehow selectively regulate the transcription of PEPCK
gene. It is possible that insulin, cAMP, and phorbol esters,
(and perhaps glucocorticoids) act through a common mechanism,
perhaps a protein phosphorylation-dephosphorylation system,
that requires specific protein kinases and phosphoprotein
phosphatases.

A model that incorporates these ideas is illustrated in
Figure 6. The effectors discussed in this paper, and their
presumed mediators, are shown in panel A. Phorbol esters
(PMA) and diacylglycerol (DAG) presumably activate protein
kinase C, insulin may act through the tyrosine kinase that is
an integral part of its receptor, cAMP probably acts through
protein kinase A, and glucocorticoids may act by combining
with specific intracellular receptors. Recent evidence
suggests that the glucocorticoid receptor can be phosphorylated (24) and perhaps this modification alters the activity
of the receptor. The combined action of these four unique
classes of effectors is illustrated, as a working hypothesis,
in Figure 6B. We postulate that a protein(s) involved in

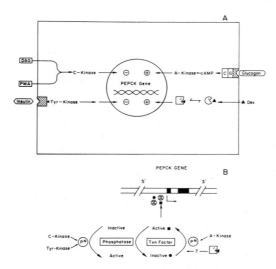

Figure 6. Multihormonal Regulation of PEPCK Gene Transcription: An Hypothetical Model. Panel A illustrates how
four different hormones, each presumably acting by a different mechanism, might regulate transcription from the PEPCK
gene. Panel B illustrates a highly speculative possibility,
namely that all of these hromones might converge on a single
protein involved in PEPCK gene transcription and alter its
activity state. Reprinted, with permission, from Granner, et
al., 1986.

PEPCK gene transcription exists in multiple phospho- and dephospho- forms and that its activity is related to the extent of phosphorylation. The phosphoprotein formed by the action of cAMP would facilitate the action of polymerase II, a process that is enhanced by glucocorticoids. The dephospho-protein would be formed by the action of a phosphatase, which in turn is activated by insulin (through tyrosine kinase) and phorbol esters (through protein kinase C). The model accounts for the individual action of each effector and explains how insulin (or phorbol esters) exert a dominant effect on transcription of the PEPCK gene.

ACKNOWLEDGEMENTS

The work discussed in this paper was supported by USPHS grants AM 35107 (to Daryl K. Granner) and AM 07061 (the Vanderbilt Diabetes Research and Training Center). We thank D. Caplenor for her assistance in the preparation of this manuscript.

REFERENCES

1. Gunn J, Tilghman S, Hanson R, Ballard F (1975) Effects of cyclic adenosine monophosphate, dexamethasone and insulin on phosphoenolpyruvate carboxykinase synthesis in Reuber H-35 hepatoma cells. Biochemistry 14:2350.
2. Beale E, Katzen C, Granner DK (1981) Regulation of rat liver phosphoenolpyruvate carboxykinase (GTP) messenger ribonucleic acid activity by N^6, O^2-dibutyryladenosine 3', 5'-monophosphate. Biochemistry 20:4878.
3. Beale E, Andreone T, Koch S, Granner M, Granner D (1984) Insulin and glucagon regulate cytosolic phosphoenolpyruvate carboxykinase (GTP) mRNA in rat liver. Diabetes 33:328.
4. Beale EG, Hartley JL, Granner DK (1982) N^6, O^2-dibutyryl cyclic AMP and glucose regulate the amount of messenger RNA coding for hepatic phosphoenolpyruvate carboxykinase (GTP). J Biol Chem 257:2022.
5. Chrapkiewicz NB, Beale EG, Granner DK (1982) Induction of the messenger ribonucleic acid coding for phosphoenolpyruvate carboxykinase in H4IIE cells. J Biol Chem 257:14428.
6. Granner D, Andreone T, Sasaki K, Beale E (1983) Inhibition of transcription of the phosphoenolpyruvate carboxykinase gene by insulin. Nature 305:545.
7. Sasaki K, Cripe TR, Koch SR, Andreone T, Petersen DD, Beale EG, Granner DK (1984) Multihormonal regulation of phosphoenolpyruvate carboxykinase gene transcription: The dominant role of insulin. J Biol Chem 259:15242.
8. Jefferson LS (1980) Role of insulin in the regulation of protein synthesis. Diabetes 29:487.

9. Korc M, Iwamoto Y, Sankaran H, Williams JA, Goldfine ID (1981) Insulin action in pancreatic acid from strepto-zotocin-treated rats. I. Stimulation of protein synthesis. Am J Physiol 240:656.

10. Horvat A (1980) Stimulation of RNA synthesis in isolated nuclei by an insulin-induced factor in liver. Nature 286:906.

11. Griswold MD, Merryweather J (1982) Insulin stimulates the incorporation of ^{32}Pi into ribonucleic acid in cultured sertoli cells. Endocrinology 11:661.

12. Andreone TL, Beale EG, Bar RS, Granner DK (1982) Insulin decreases phosphoenolpyruvate carboxykinase (GTP) mRNA activity by a receptor-mediated process. J Biol Chem 257:35.

13. Berridge MH (1984) Inositol trisphosphate and diacylglycerol as second messengers. Biochem J 220:345.

14. Farese RV, Barnes DE, Davis JS, Standarest ML, Pollet RJ (1984) Effects of Insulin and protein synthesis on phospholipid metabolism diacylglycerol levels, and pyruvate dehydrogenase activity in BC3H-1 cultured myocytes. J Biol Chem 259:7094.

15. Nishizuka Y (1984) The role of protein kinase C in cell surface signal transduction and tumor promotion. Nature 308:693.

16. Barsoum J, Varshavsky A (1983) Mitogenic hormones and tumor promoters greatly increase the incidence of colony-forming cells bearing amplified dihydrofolate reductase genes. Proc Natl Acad Sci USA 80:5330.

17. Coughlin SR, Lee WMF, Williams PW, Giels GM, Williams LT (1985) c-myc Gene expression is stimulated by agents that activate protein kinase C and does not account for the mitogenic effect of PDGF. Cell 43:243.

18. Beale EG, Chrapkiewicz NB, Scoble H, Metz RJ, Noble RL, Quick DP, Donelson JE, Biemann K, Granner DK (1985) Structure of the rat cytosolic phosphoenolpyruvate carboxykinase protein, messenger RNA, and gene. J Biol Chem 260:10748.

19. McKnight GS, Palmiter RD (1979) Transcription regulation of the ovalbumin and conalbumin genes by steroid hormones in chick oviduct. J Biol Chem 254:9050.

20. Cimbala MA, Lamers WH, Nelson K, Monahan JE, Yoo-Warren H, Hanson RW (1982) Rapid changes in the concentration of phosphoenolpyruvate carboxykinase mRNA in rat liver and kidney. J Biol Chem 257:7629.

21. Loten EG, Sneyd JGT (1970) An effect of insulin on adipose-tissue adenosine 3',5'-cyclic monophosphate phosphodiesterase. Biochem J 120:187.

22. Granner DK, Chase L, Aurbach GD, Tomkins GM (1968) Tyrosine aminotransferase: Enzyme induction independent of adenosine 3',5'-mono phosphate. Science 162:1018.

23. Granner DK, Andreone TL (1985) Insulin modulation of
    gene expression. In DeFronzo R (ed.): "Diabetes-Metabol-
    ism Reviews" New York: John Wiley & Sons, Inc. p. 139.
24. Miller-Diener A, Schmidt TJ, Litwack G (1985) Protein
    kinase activity associated with the purified rat hepatic
    glucocorticoid receptor. Proc Natl Acad Sci USA 82:4003.
25. Granner DK, Sasaki K, Andreone TL, Beale E (1986) Insulin
    regulates expression of the phosphoenolpyruvate carboxy-
    kinase gene. In Greep RO (ed.): "Recent Progress in
    Hormone Research" New York: Academic Press, p. 111.

Transcriptional Control Mechanisms, pages 295–311
© 1987 Alan R. Liss, Inc.

HORMONAL AND ENVIRONMENTAL CONTROL OF
METALLOTHIONEIN GENE EXPRESSION[1]

Michael Karin[2], Masayoshi Imagawa[2], Richard J. Imbra[2],
Robert Chiu[2], Adriana Heguy[2], Alois Haslinger[3],
Tracy Cooke[2], Sundramurthi Satyabhama[2],
Carsten Jonat and Peter Herrlich[4]

[2]Division of Pharmacology, M-013 H, School of Medicine
University of California, San Diego
La Jolla, CA 92093

ABSTRACT The expression of metallothionein (MT) genes
in animal cells is regulated at the transcriptional
level by a large number of hormonal and environmental
cues. Among these are steroid hormones (glucocorticoids
and progesterones), interferon, interleukin 1, serum
factors, heavy metal ions, phorbol esters and ionizing
radiation. In addition to the complex regulation by a
large number of inducers, the expression of the MT gene
family is further complicated by the presence of a
large number of genes which not only respond differently to induction, but also differ in their basal
levels. To fully understand the mechanisms which control both the basal and induced transcription of the MT
gene family, it is essential to characterize in detail
the various cis and trans acting genetic elements
involved in these regulatory circuits. To this end, we

[1] This work was supported by grants from the National
Institute of Environmental Health Sciences, the Environmental Protection Agency, the Department of Energy, the
Margaret Early Foundation for Medical Research, and the
Chicago Community Trust. M.K. is a Searle Scholar, R.H.C.
is supported by a Public Health Service postdoctoral
fellowship, and A.H. was supported by an EMBO long-term
fellowship.
[3] Present address: Institute of Biochemistry,
University of Vienna, Vienna, Austria.
[4] Permanent address: Institute for Genetics and Toxicology, Karlsrule University, Karlsrule, West Germany

have used an integrated approach based on mutagenesis, gene-transfer and in vitro protein-DNA binding probed by DNAse I.

The results indicate that diversity in expression of the hMT gene family is caused by the presence of distinct promoter elements on each of its members. While some of the control signals are common to all MT promoters (MRE's) others are more specific (GRE's). Similar elements can be organized in a different fashion on each promoter. These results indicate the emergence of a regulatory code, composed of short regulatory signals which serve as binding sites for distinct trans acting factors, whose activity is modulated by different hormonal and environmental factors.

INTRODUCTION

Metallothioneins (MT's) are a group of low molecular weight, cysteine-rich, heavy-metal-binding proteins. While their exact physiological roles are not clear, it is most likely that they are involved in the regulation of intracellular zinc and copper metabolism, protection of cells against toxic heavy metal ions and, possibly, protection against oxidative damage by free oxygen radicals. By controlling the intracellular level of zinc ions, MT's can indirectly contribute to the regulation of various zinc-requiring enzyme systems, some of which might be involved in replication, transcription and DNA repair (see ref. 1 for review). As appropriate for proteins that occupy a central role in regulation of cellular metabolism, the expression of MT genes itself is subject to complex and intricate control. Thus their transcription is induced in response to heavy metal ions, steroid hormones, interferons, interleukin 1, serum factors, tumor promoters and clastogenic agents. In addition, some of the MT genes may be expressed in a tissue-specific manner. Due to their complex regulation, small size and relative abundance, MT genes can serve as an excellent experimental system for studying the mechanisms by which mammalian cells respond to various hormones and environmental stress. Toward this end we have isolated the various members of the human MT gene family, determined their structure, and analyzed in detail the cis and trans-acting genetic elements controlling their expression.

REGULATION OF THE hMT-II$_A$ GENE

The human genome contains at least twelve different MT genes. While some of these are non-functional pseudogenes (2), at least six or seven of them are functional (2-5) and give rise to distinct protein species (6,7). One of the genes, the unique MT-II type gene, hMT-II$_A$, accounts for expression of at least 50% of the MT's expressed in cultured human cells or liver (8). The rest of the MT variants are contributed by various members of the MT-I group. As appropriate for its being the major MT gene, the hMT-II$_A$ gene is subject to the most intricate regulation as compared to the other MT genes.

Addition of heavy metal ions or glucocorticoid hormones to cultured mammalian cells leads to a rapid increase in hMT-II$_A$ mRNA (9) due to the increased transcription rate of the gene (10,11). Similarly, treatment of cultured cells with interferon also stimulates hMT-II$_A$ transcription (12). However, this effect is somewhat slower and of lower magnitude than the induction observed with metal ions. Another peptide hormone with immunoregulatory activity is interleukin 1 (IL-1) which mediates the acute phase response (13). IL-1 is capable of inducing hMT-II$_A$ mRNA only in a small number of human cell types including primary fibroblasts, hepatoma and promyelocytic leukemia cells (14); other cell types probably do not respond to IL-1 because they lack the appropriate receptors.

More recently we found that in primary foreskin fibroblasts and in HepG2 cells, a differentiated hepatoma cell line, the level of hMT-II$_A$ mRNA is regulated by factors present in serum. The level of MT mRNA decreases to 10-30% of the basal level within 24 hrs of switching the cells from growth in medium containing 10% fetal calf serum to 0.5% serum. This is a specific effect as the levels of several other mRNA's, including the ones coding for α-tubulin and cAMP-dependent protein kinase R subunit, do not change to a significant extent. On the other hand, the levels of hMT-II$_A$ mRNA in HeLa cells are not affected by the serum concentration in the growth medium, probably because these cells are highly transformed and growth factor-independent. Upon refeeding of serum-starved fibroblasts or hepatoma cells with serum, hMT-II$_A$ returns to its normal basal level within 8 hr.

To determine whether the decrease in MT mRNA upon serum starvation is due to a block of cell division, we examined MT mRNA levels in wild-type CHO cells and a temperature-

sensitive proliferation mutant under permissive and
non-permissive conditions. The results indicated that the
decrease in MT mRNA after serum starvation is not simply due
to blocking DNA synthesis, because arresting cell division
by a temperature up-shift of the ts mutant had no significant effect on MT mRNA. On the other hand, serum starvation
of the wild-type and the mutant cells led to a similar
decrease in MT mRNA as observed before in human cells. We
therefore conclude that MT mRNA synthesis is specifically
regulated by some factors present in serum (15). We are
currently trying to identify these factors.

Since in several other systems the effects of serum on
cell division and gene expression can be mimicked by treatment of cells by phorbol ester tumor promoters, such as
tetradecanoate phorbol acetate (TPA), we examined the effect
of TPA on MT mRNA expression in a number of cell lines. In
most cases, low doses of TPA (10-100 ng/ml) lead to rapid
but transient increase in MTmRNA synthesis. Like the other
agents mentioned above, TPA also acts as a primary inducer,
as its stimulatory effect on MTmRNA synthesis does not require de-novo protein synthesis. Nuclear run-off experiments
indicated that TPA stimulates hMT-II$_A$ gene transcription
(15). TPA is known to exert many of its effects by the activation of protein kinase C, a cellular enzyme whose normal
activator is diacylglycerol, generated during phospholipid
breakdown (16). To test whether protein-kinase C activation
is indeed responsible for increased hMT-II$_A$ mRNA synthesis,
we have examined whether DiC8 (1,2-dioctanoylglycerol, a
water-soluble derivative) will have an effect similar to
that of TPA. Incubation of HepG2 cells with 10-100 nM of
DiC$_8$ led to a significant increase of MT mRNA, that also was
independent of de-novo protein synthesis. These results
suggest that activation of protein kinase C is indeed
responsible for the increased transcription of the hMT-II$_A$
gene after treatment of cultured cells with TPA and possibly
with serum.

Besides being able to induce MT synthesis, TPA has many
other effects, among them the induction of the UV response.
The UV response is a pleiotropic response, characterized by
induction of a set of proteins and generation of a UV
resistant state, after treatment of cultured mammalian cells
with UV, X-rays, mitomicin C (MMC) and TPA. One of the
proteins induced during the UV response is EPIF, a secreted
protein that, when added to unstimulated cells, is able to
induce the same spectrum of proteins as induced by UV, MMC
or TPA (17). By differential screening of cDNA libraries of

UV-treated human fibroblasts, it became evident that hMT-II$_A$ is one of the genes whose expression is induced during the UV response (18). More recently it was shown that UV, X-rays, MMC and EPIF all lead to primary induction of hMT-II$_A$ mRNA (19).

STRUCTURE OF THE hMT-II$_A$ PROMOTER/REGULATORY REGION AND ITS ACTIVATION BY HEAVY METAL IONS AND GLUCOCORTICOIDS

Because of its unusual responsiveness to a large number of hormonal and environmental cues, the hMT-II$_A$ gene is an excellent system for analyzing the molecular mechanisms involved in signal transduction to the transcriptional machinery. The approach we have taken for analyzing these responses consists of mapping the cis-elements involved in regulation of the hMT-II$_A$ gene, using in-vitro mutagenesis and gene transfer experiments, and the identfication of trans-acting factors that bind to these elements, by the DNaseI footprinting procedure (20). These experiments have led to a detailed understanding of the organization of the hMT-II$_A$ promoter and its recognition by different trans-acting factors.

The basal level of hMT-II$_A$ promoter activity is controlled by three different elements, a GC box, located between nucleotides -68 to -57, and two basal level enhancer elements present between positions -203 to -156 and -126 to -80 (21). The GC box is a binding site for the specific transcription factor SP1 (22). In comparison to the GC box, the enhancers are complex and are composed of at least three different transcription control signals (21) that serve as binding sites for distinct regulatory proteins (22; M. Imagawa and M. Karin, unpublished results). Interspersed between the elements important for basal promoter activity, including the TATA box, are four metal-responsive elements (MRE's) (21). The MRE's are highly conserved, short, (12 bp in length) repetitive sequences that mediate MT gene induction by heavy metal ions (21,23-25). We have attempted to reveal the presence of MRE binding proteins by preparing extracts from HeLa cells incubated for 1-2 hr in either the absence or the presence of CdCl$_2$; this incubation period is sufficient to increase the transcription rate of the hMT-II$_A$ gene to its maximal level. After incubation of these extracts with an hMT-II$_A$ 5' probe, and DNase I footprinting, no differences in the protection patterns were observed. The GC box and the enhancers were extensively protected from

DNaseI digestion due to the presence in the extracts of specific trans-acting factors recognizing these elements (Fig. 1). In fact, the only upstream elements over which no protection was observed were the MRE's.

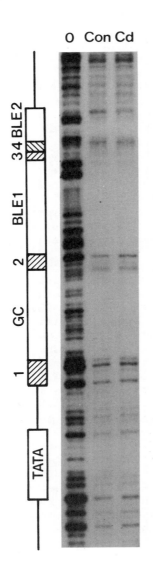

FIGURE 1. Footprinting of the hMT-II$_A$ gene with extracts from Cd-induced and uninduced HeLa cells.

Whole cell extracts were prepared from either Cd-induced (2 hrs with 5 μM Cd^{++}) or uninduced HeLa cells. The extracts were fractionated on a heparin-agarose column and the fraction which eluted between 0.2 M and 0.4 M KCl was used for the foot-printing experiments. Approximately 2 ng of an hMT-II$_A$ 5' probe labelled at position +75 was incubated with 10 μg of either extract and subjected to DNaseI footprinting as described in ref. 22.

The glucocorticoid regulatory element (GRE) is located upstream to the second enhancer, between positions -270 to -240. The GRE, defined genetically by deletion mutagenesis, coincides with the binding site for the purified rat liver glucocorticoid receptor, as defined by in-vitro binding DNaseI footprinting and DMS methylation protection (23). An 80 bp fragment containing the GRE behaves as a hormone-inducible enhancer element (A. Haslinger and M. Karin, unpublished results), but it is not clear if the gluco-corticoid receptor is the only protein which binds to this enhancer, as most enhancers contain binding sites for several factors. However, footprinting experiments with HeLa cell extracts failed to reveal the binding of any other trans-acting factors to this fragment (M. Imagawa and M. Karin, unpublished results).

Deletion or substitution of the GC box decreases the transcriptional activity of the hMT-II$_A$ promoter by one order of magnitude without affecting the extent of induction by Cd^{++}. Therefore the GC box is not directly required for metal induction. In contrast, the enhancers are far more complex and contain at least three binding sites for distinct regulatory proteins. In addition, one or two MRE's are present adjacent to each of the basal level enhancers and, upon addition of heavy metal ions, they stimulate the activity of the enhancers. The relationship between the MRE's and the enhancer has been examined in detail. Deletion of the MRE's eliminates metal-inducibility of the enhancer, but has no effect on its basal level of activity. On the other hand, a linker scanning mutation that inactivates the basal enhancer by destroying the binding site for two of the enhancer binding factors also prevents metal induction, even though the MRE's are left intact by the mutation. These results indicate that on their own the MRE's are not capable of functioning like independent enhancer elements, i.e., they can not confer metal inducibility upon heterologous promoters from a distance. Only when they are adjacent to an already active enhancer are the MRE's capable of acting from a distance by regulating the activity of the enhancer (21). This is different from what was observed for the GRE. In that case, a small fragment containing only the glucocor-ticoid receptor binding site was capable of conferring hormone-inducibility upon a heterologous promoter at various distances, regardless of its orientation.

How does the MRE act upon the enhancer? This question was answered by in-vivo competition experiments which demonstrated that, in the presence of heavy metal ions, the hMT-II$_A$ enhancer is able to compete with the SV40 enhancer more efficiently for binding of a cellular factor. Results obtained from the analysis of a series of deletion mutants for competition activity, and the absence of any MRE-like activity in the SV40 enhancer, suggest that the MRE does not act directly, as the target site for binding of the limiting factor, but indirectly, by increasing the ability of the basal level enhancer to bind the cellular factor. It is likely that the positively-acting factor that binds to the MRE in the presence of heavy metal ions is acting by stabilizing, via protein-protein interactions, the binding of the limiting trans-acting factor which binds to a site within the adjacent enhancer (26).

REGULATION OF THE hMT-II$_A$ PROMOTER BY TPA

As mentioned earlier, treatment of cultured human cells with TPA leads to increased expression of the hMT-II$_A$ gene. To elucidate the mechanism by which TPA affects transcription of that gene, different in-vitro constructed deletion mutants of the hMT-II$_A$ promoter and various constructs containing its enhancer elements in combination with the SV40 early promoter were introduced into HeLa cells either transiently or stably, and their expression was determined by CAT assays and S1 nuclease protection. The results of these experiments indicate that separate cis elements are responsible for TPA and Cd induction. For example, a deletion of hMT-II$_A$ to position -50 eliminated metal induction, while a small (twofold) but consistent response to TPA was still observed. In addition, constructs containing either the proximal or the distal hMT-II$_A$ enhancer elements upstream to the SV40 early promoter are responsive to both Cd and TPA, and enhancer deletion mutants that eliminate metal induction are still responsive to TPA. In addition to dissociating the Cd and TPA induction mechanisms, these results indicate that at least three regions within the hMT-II$_A$ promoter contain TPA-responsive cis-acting elements: -206 to -150. -130 to -69 and -50 tc +70. Computer analysis of these three regions has pointed out a short repeat common to these regions that may be the putative target for the TPA effect (19). Further analysis of point mutants and short deletions introduced into these regions will be required to prove whether it is the actual TPA-responsive element.

TPA is known to act by activation of protein kinase C (16). Recently we have shown that another protein kinase C

activator, DiC8, also induces hMT-II$_A$ expression (15). To
further examine the role of protein kinase C in the
regulation of hMT-II$_A$ by TPA. We examined the effects of a
specific protein kinase C inhibitor on the induction
response. Treatment of transfected cells with H-7 (1-5-iso-
quinolinsulfonyl)-2-methylpiperazine dihydrochloride), a
protein kinase inhibitor with marked preference for the C
kinase (27,28), prevented induction of phMT-II$_A$-CAT by TPA
but not by Cd or dexamethasone. In contrast, HA1004 (N-2-
guanidinoethyl)-5-isoquinolinesulfonamide hydrochloride), a
weak protein kinase C inhibitor (29), had no effect on
hMT-II$_A$-CAT expression. These results support the premise
that the induction of hMT-II$_A$ gene expression by TPA is
mediated via the activation of protein kinase C.

TWO SEPARATE PATHWAYS FOR HORMONAL REGULATION OF GENE EXPRESSION

The induction of hMT-II$_A$ expression by glucocorticoids
and other steroid hormones on one hand, and by TPA, EPIF and
IL-1 on the other, are mediated by two different mechanisms.
Upon entering into the cells, glucocorticoids bind to and
activate a specific cytoplasmic receptor and lead to its
migration into the nucleus. Once in the nucleus, the
receptor binds to a specific site on steroid-responsive
genes (see ref. 30 for a review of this model). This site
is the GRE (23). After binding to the GRE, the receptor
probably interacts with other trans-acting factors already
bound to the gene and increases the rate of transcription
(see Fig. 2).
TPA, while not a hormone itself, may represent a class
of activators that includes EPIF, IL-1, and serum factors,
all of which may act through protein kinase C to induce the
transcription of hMT-II$_A$ and other genes. A possible model
explaining the effect of these agents on gene expression is
shown in Fig. 2. According to this model, these activators,
directly or indirectly, increase the activity of protein
kinase C. The activated kinase then can either directly
phosphorylate its target or trigger a phosphorylation
cascade by activating another kinase, the end result being
increased phosphorylation of a trans-acting factor that
binds to the hMT-II$_A$ promoter. If this factor is present in
low levels in the cell, and its non-phosphorylated form has
lower affinity towards DNA, then it probably will not bind
to the hMT-II$_A$ promoter unless its affinity is increased as

Two Pathways of Hormone Mediated Gene Activation

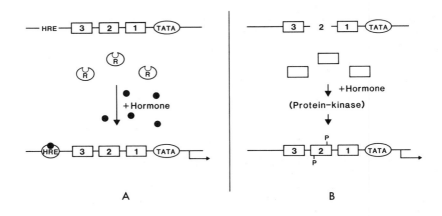

A   B

FIGURE 2. Two separate pathways of hormone-mediated gene induction.

Pathway A represents induction of gene expression by steroid hormones which bind and activate a receptor (R) that binds to a specific hormone-responsive element (HRE) on the target gene. Boxes 1, 2 and 3 represent binding sites for other trans-acting factors responsible for the basal expression of the gene.

   Pathway B represents the model of induction by activators of protein kinase C (or any other hormone-responsive kinase). In this case one of the factors responsible for the basal level of expression (factor 2) does not bind to its recognition site until it is phosphorylated as a consequence of a protein kinase activation.

a consequence of its phosphorylation. Unlike steroid receptors, this factor is not specific for a particular class of hormones, but may be modulated by many different agents affecting protein kinase C activity. In addition, it also is one of the factors that normally recognize and bind to the hMT-II$_A$ promoter, and therefore is responsible for its basal activity. In fact, we have not been able to dissociate between the cis elements required for basal expression and those required for TPA induction in contrast to the separate elements that mediate steroid hormone and heavy metal action. As shown in Fig. 2, this factor inter-acts with other trans-acting factors to stimulate hMT-II$_A$ expression. Another possibility, not shown in the figure, is that the factor is bound to the promoter at all times. However, in its non-phosphorylated state it binds to DNA but does not stimulate transcription very efficiently. After it is phosphorylated it becomes more active in stimulating transcription, possibly by increased ability to interact with other transcription factors. Footprinting and in vitro transcription experiments with extracts prepared from TPA-treated and untreated cells should help in elucidating the exact mechanisms involved in the TPA effects on hMT-II$_A$ gene expression.

EFFECTS OF INDUCERS ON CHROMATIN STRUCTURE

Probably the most important question concerning trans-criptional activation is "What are the intermediates between the binding of trans-acting regulatory proteins and the observed increase in transcription rate?" Since we have already demonstrated the binding of several trans-acting factors to the hMT-II$_A$ promoter, we would like to know what happens during the time between their binding and the observed increase in hMT-II$_A$ gene transcription.

We have begun to characterize some of these events by examining the effect of dexamethasone and Cd^{++} on the structure of hMT-II$_A$ chromatin. We found that addition of either Cd^{++} or dexamethasone to HeLa cell cultures elevated hMT-II$_A$ gene transcription in isolated nuclei so that it reached a maximum after two hours. Therefore, we exposed HeLa cells to either inducer for 2 hr, isolated the nuclei, and incubated them with different concentrations of S1 nuclease. The digestion was terminated after 10 minutes, genomic DNA was prepared and digested with HindIII and sepa-rated on a 1% agarose gel. After transfer to nitrocellulose

the blot was hybridized with a probe specific for the 3' end
of the hMT-II$_A$ gene. This indirect end-labelling procedure
led to the detection of an inducible S1 nuclease hypersensitive site in the immediate 5' flanking region of the hMT-II$_A$
gene (Fig. 3). While the exact location of the site has not

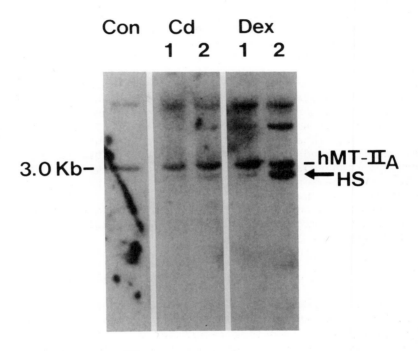

FIGURE 3. Cd^{++} and dexamethasone induce an S1 nuclease
hypersensitive site in the hMT-II$_A$ 5' flanking region.

        Nuclei were isolated from either control (con) HeLa
cells or cells incubated for 2 hrs with 5 μM of Cd^{++} (Cd) or
1 μM of dexamethasone (DEX). The isolated nuclei were incubated with S1 nuclease for 10 min, at the end of which the
genomic DNA was extracted, digested with Hind III, separated
on a 1% agarose gel and transferred onto nitrocellulose.
The filter was probed with a probe specific for the 3' end
of the hMT-II$_A$ gene. The size of the Hind III fragment on
which that gene resides is 3.0 Kb. HS indicates the fragment
which starts at the hypersensitive site and ends at the 3'
Hind III site.

yet been accurately mapped, both Cd and dexamethasone seem to induce the same hypersensitive site, even though they are known to act via different cis elements (23). Therefore, we believe that this inducible hypersensitive site is more likely to reflect a general alteration in the chromatin structure of the hMT-II$_A$ gene common to both inducers rather than a specific alteration in chromatin conformation after binding of each regulatory protein. It is not clear whether the observed change in chromatin structure precedes or follows the increased transcription of the gene.

MECHANISMS OF INDUCTION BY HEAVY METAL IONS

As discussed earlier, treatment of cultured cells with heavy metal ions such as Cd^{++} leads to rapid induction of MT gene expression. This transcriptional activation is retained in isolated nuclei as shown by the nuclear run-off assay (11). Also, the isolated nuclei conserve the effect of Cd^{++} on hMT-II$_A$ chromatin, as shown by the presence of the S1 nuclease hypersensitive site (see Fig. 3). However, when isolated nuclei, or even whole cells, are used for preparation of extracts, the binding of a Cd^{++}-activated factor to the MRE can not be detected (see Fig. 1) even though such a factor should exist. Some insight into the possible mechanism by which Cd^{++} affects the transcription of the hMT-II$_A$ gene is afforded by the results of the in-vivo competition experiments (26) and the mutational analysis of the metal-responsive enhancer element (21) described earlier. In addition, the positions occupied by the MRE's within the hMT-II$_A$ control region may give us a hint on the possible mechanism of metal induction, as the MRE's are present between all the different cis elements responsible for the basal expression of the promoter. The unique location of the MRE's and the results mentioned above lead us to suggest the following model, illustrated in Fig. 4, to explain how heavy metal ions induce MT gene transcription. It is assumed that heavy metal ions bind to an intracellular trans-acting factor and lead to its activation. This (putative) metal-responsive factor (MRF) then binds to the MRE's and, because of their unique location, contacts other factors that bind to the adjacent cis-elements, which are responsible for the basal activity of the promoter. This interaction stabilizes the binding of these factors to DNA and leads to formation of a large and stable protein-DNA complex spanning the promoter region. In some way this complex is a more efficient

substrate for transcription by RNA polyermerase II. In the
absence of heavy metal ions, the MRF does not bind to DNA,
and the only way in which a large protein-DNA complex can
form is if the unoccupied DNA between the various cis
elements "loops out" or bends to allow the different factors
bound to these elements, which are responsible for basal
activity, to contact each other. This process is likely to
be less efficient than the one stimulated by the MRF and,
therefore, less of the active complex would be formed or the
formed complex would be thermodynamically less stable,
resulting in a lower transcription rate.

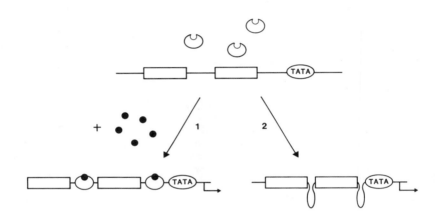

FIGURE 4. A model for induction of metallothionein
gene expression by heavy metal ions.

Heavy metal ions (solid circles) bind to a metal
responsive factor (MRF, symbolized by the ellipsoid) which
binds to the MRE and leads to formation of a large, stable
complex with other factors which bind to the cis elements
which flank the MRE's and are responsible for the basal expression of the gene. In the uninduced state a complex can
form between the factors responsible for basal expression by
bending or looping-out of the DNA which contains the MRE's.
This process is less efficient, resulting in a lower rate of

REFERENCES

1. Karin M (1985). Metallothioneins: Proteins in search of function. Cell 41:9-10.

2. Karin M, Richards RI (1982). Human metallothionein genes: Primary structure of the metallothionein-II gene and a related processed pseudogene. Nature 299:797-802.

3. Richards RI, Heguy A, Karin M (1984). Structural and functional analysis of the human metallothionein I_A gene: Differential induction by metal ions and glucocorticoids. Cell 37:263.

4. Heguy A, West A, Richards RI, Karin M (1986). Structure and tissue-specific expression of the human metallothionein I_B gene. Mol Cell Biol, in press.

5. Schmidt CJ, Jubier MF, Hamer DH (1985). Structure and expression of two human metallothionein-I isoform genes and a related pseudogene. J Biol Chem 280:7731-7737.

6. Klauser S, Kagi JHR, Wilson KJ (1985). Characterization of isoprotein patterns in tissue extracts and isolated samples of metallothioneins by reversed-phase high pressure liquid chromatography. Biochem J 209:71-80

7. Hunziker PE, Kagi JHR (1983). Isolation of five human isometallothioneins. 15th Meeting of the Federation of European Biochemical Societies, Burssels, 1983. Abstracts p. 217.

8. Karin M, Herschman HR (1980). Characterization of the metallothioneins induced in HeLa cells by dexamethasone and zinc. Eur J Biochem 107:395-401.

9. Karin M, Anderson RD, Slater E, Smith K, Herschman, HR (1980). Metallothionein mRNA induction in HeLa cells in response to zinc or dexamethasone is a primary induction response. Nature 286:295-297.

10. Karin M, Haslinger A, Holtgreve H, Cathala G, Slater E, Baxter JD (1984). Activation of a heterologous promoter in response to dexamethasone and cadmium by metallo-thionein gene 5'-flanking DNA. Cell 36:371-379.

11. Imbra RJ, Wight PA, Smith GD, Spindler SR (1986). Transcriptional regulation of human metallothionein gene expression. Manuscript submitted.

12. Friedman RL, Stark GR (1985) α-Interferon-induced transcription of HLA and metallothionein genes containing homologous upstream sequences. Nature 314:367-639.

13. Kushner I (1982). The phenomenon of the acute phase response. Ann NY Accad Sci 389:39-45.
14. Karin M, Imbra RJ, Heguy A, Wong G (1985). Interleukin 1 regulates human metallothionein gene expression. Mol Cell Biol 5:2866-2869.
15. Imbra RJ, Chiu RH, Karin M (1986). Serum and activators of protein kinase C regulate MT gene expression. Manuscript submitted.
16. Nishizuka Y (1984). The role of protein kinase C in cell surface signal transduction and tumor production. Nature 308:693-700.
17. Schorpp M, Mallick U, Rahmsdorf HJ, Herrlich P (1984). UV-induced extracellular factor from human fibroblasts communicates the UV response to nonirradiated cells. Cell 37:861-868.
18. Angel P, Poting A, Mallick U, Rhamsdorf HJ, Schorpp M, Herrlich P (1986). Induction of metallothionein and other mRNA species by carcinogens and tumor promoters in primary human skin fibroblasts. Mol Cell Biol 6:1760-1766.
19. Herrlich P, Jonat C, Rahmsdorf HJ, Angel P, Haslinger A, Imagawa M, Karin M (1986). Signals and sequences involved in the UV and TPA-dependent induction of genes, in Growth Factors, Tumor Promoters and Cancer Genes. Liss Inc., New York.
20. Galas D, Schmitz A (1978). DNase footprinting: A simple method for the detection of protein-DNA binding specificity. Nucl Acids Res 5:3157-3170.
21. Karin M, Haslinger A, Heguy A, Deitlin T, Cooke T (1986) Multiple interspersed promoter elements control the expression of the human metallothionein-II$_A$ gene. Mol Cell Biol, submitted.
22. Lee W, Haslinger A, Karin M, Tjian R (1986). Identification of factors that bind and activate the human metallothionein II$_A$ promoter. Submitted.
23. Karin M, Haslinger^{A}A, Holtgreve H, Richards RI, Krauter P, Westphal H, Beato M (1984). Characterization of DNA seuqences through which cadmium and glucocorticoid hormones induce human metallothionein-II$_A$ gene. Nature 308:513-519.
24. Carter AD, Felber BK, Walling MJ, Jubier MF, Schmidt CJ, Hamer DH (1984). Duplicated heavy metal control sequences of the mouse metallothionein-I gene. Proc Natl Acad Sci USA 81:7392-7396.
25. Stuart GW, Searle PF, Chen HY, Brinster RI, Palmiter RD (1984). A 12 base-pair motif that is repeated several

times in metallothionein gene promoters confers metal regulation to a heterologous gene. Proc Natl Acad Sci USA 81:7318-7322.

26. Scholer H, Haslinger A, Heguy A, Holtgreve H, Karin M (1986). In vivo competition between a metallothionein regulatory element and the SV40 enhancer. Science 232:76-80.

27. Hidaka H, Inagaki M, Kawamoto S, Sasaki Y (1984). Isoquinolinesulfonamides, novel andpotent inhibitors of cyclic nucleotide dependent protein kinase and protein kinase C. Biochemistry 23:5036-5047.

28. Kawamoto S, Hidaka H (1984) 1-(5-isoquinolinesulfonyl)-2-methylpiperazine (H-7) is a selective inhibitor of protien kinase C in rabbit platelets. Biochem Biophys Res Commun 125:258-266.

29. Asano T, Hidaka H (1984) Vasodilatory action of HA1004 [N-(2-guanoethyl)-5-isoquinolinesulfonamide], a novel calcium antagonist with no effect on cardiac function. J Pharmacol Exp Ther 231:141-151.

30. Rousseau GG, Baxter JD (1979). "Glucocorticoid hormone action." Springer-Verlag: Heidelberg.

Transcriptional Control Mechanisms, pages 313–323
© 1987 Alan R. Liss, Inc.

DNA SEQUENCE ELEMENTS REGULATING CASEIN GENE EXPRESSION[1]

C.A. Bisbee and J.M. Rosen

Department of Cell Biology, Baylor College of Medicine,
Houston, Texas 77030

ABSTRACT Caseins are abundant milk proteins. Casein
gene expression is controlled by the developmental and
hormonal states of the mammary gland. Tissue- and
hormone-specific elements regulating casein gene transcription are being identified by using genomic sub-
clones containing both 5'-flanking and internal
sequences to direct chloramphenicol acetyltransferase
(CAT) gene transcription in the expression vector
pSV_0cat. Cells are transfected on hydrated collagen
gels because maintenance of mammary epithelial cells on
collagen gels is an essential prerequisite for the
expression of differentiated function. In transiently
transfected mammary epithelial cells, constructions
containing exon I and part of intron A consistently
show 10- to 20-fold greater CAT activity when compared
with constructions containing only 5'-flanking
sequences. Casein gene promoter activity appears weak
and no hormonal regulation of these constructions can
be demonstrated. The difference in CAT activity seen
between constructions with and without internal
sequences is not as prominent in NIH-3T3 cells. Both a
strong mammary-specific inhibition by proximal 5'-
flanking sequences and a weaker mammary-specific
enhancement by internal sequences are seen when CAT
activity for each vector construction is normalized to
pSV_2cat levels in the appropriate cell type. These
data suggest casein gene expression is controlled by a
complex interaction between both positive and negative
regulatory elements.

[1]This work was supported by NIH grant CA16303.

INTRODUCTION

Caseins are the predominant calcium-binding milk proteins that provide essential amino acids and both calcium
and phosphate for mammalian neonates. Casein secretion in
the adult mammary gland is the end result of a multifaceted
developmental process (1-4). Milk protein gene expression
in mammary epithelial cells is regulated by a complex interaction and synergism among several peptide and steroid
hormones (1,4). Maximal expression of casein mRNA is
observed in the presence of insulin (I), hydrocortisone (F),
and prolactin (M) in both organ and cell culture (1,4,5).
Prolactin causes a 10- to 250-fold increase in casein mRNA
levels 24 hours after addition to rat mammary gland explant
cultures maintained in insulin and hydrocortisone (4). This
effect of prolactin is exerted primarily at the post-
transcriptional level, presumably by affecting the rate of
nuclear RNA processing (6). In mammary epithelial cell
cultures, casein gene expression is additionally dependent
on cell-substratum interactions as indicated by increased
casein mRNA levels in cells maintained on hydrated collagen
gels (5).

The objective of the described studies is to elucidate
the mechanisms regulating the developmental and multihormonal control of casein gene expression. The **cis**-acting
regulatory sequences required for tissue- and hormone-
specific control of casein gene expression are being identified by the use of DNA-mediated gene transfer (7). These
studies have focused on the conserved elements that have
been identified in the 5'-flanking and exon I sequences of
the casein genes (8,9) in order to characterize regions
responsible for the maintenance of promoter activity. In
addition, we have attempted to determine the elements
responsible for hormonal and cell-substratum effects on
casein gene transcription. The characterization of these
cis-acting DNA sequences is a necessary prerequisite to
isolating the inducible **trans**-acting factors that interact
with these regulatory sequences. Presumably, one or more of
these factors will be, or will lead to the identification of
the still unknown, second messenger(s) involved in prolactin
action. The major problems encountered in our studies have
been the low level of casein gene promoter activity and the
selection of suitable culture and cell systems in which
casein gene expression is maintained and induced. Other
studies in this laboratory in which the entire β-casein gene
or casein minigenes are transfected into mammary epithelial

cells are directed at elucidating the sequences involved in post-transcriptional regulation (8).

RESULTS AND DISCUSSION

Transient gene transfer experiments have been employed primarily to determine the sequences that are important for the regulation of casein gene expression. Casein gene promoter sequences have been inserted 5' to the coding sequence of CAT in the SV40-based expression vector pSV_0cat (10). Since this enzyme is not expressed in mammalian cells (7, 10), it is a useful "reporter" enzyme in experiments assaying for casein gene control sequences.

The general structure of pSV_0cat along with the casein genomic sequences inserted as promoters in this expression vector are shown in Figure 1. Two β-casein-CAT fusion plasmids were constructed containing 1100 bp or ~500 bp of 5'-flanking DNA upstream from the mRNA CAP site. An additional β-casein construction contained the above 500 bp of 5'-flanking DNA plus the proximal 500 bp of internal sequence which included exon I and 500 bp of intron A. A γ-casein construction was employed which contained 1100 bp of 5'-flanking DNA and 650 bp of internal sequence. Again, the latter included exon I and part of intron A. Note that the γ-casein has a non-canonical TATA homology sequence (TTTAAAT). Several putative hormone binding sites are present in these 5'-flanking DNA sequences (8). Extensive sequence conservation has been observed between members of the rat casein gene family extending 200 bp upstream, and between the rat and bovine α-casein genes extending 550 bp upstream of the mRNA CAP site (8,9).

These constructions were transfected into COMMA-1D cells, a mammary epithelial cell line (11), which had been plated onto hydrated collagen gels. Functional differentiation of primary mammary epithelial cell cultures requires that cells be plated onto attached collagen gels and that the gel with adherent cells subsequently be released to float in the medium (5,12,13). The cell shape changes that occur upon release of the collagen gel are a prerequisite for functional differentiation of mammary epithelial cells (5,12,13). β-casein mRNA levels respond to hormonal and substratum influences in COMMA-1D cells in a manner analogous to that seen in primary mammary epithelial cells. (unpublished observations). Figure 2 shows the CAT assay results from a representative experiment. CAT activity is

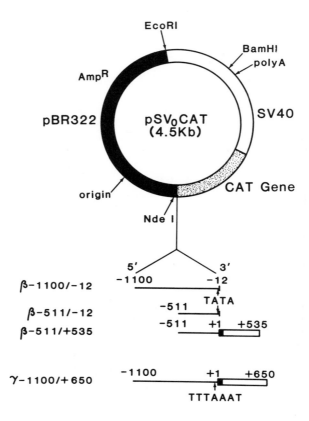

Figure 1. Construction of casein-CAT fusion genes. The structure of pSV0cat (7,10,14) is shown. Numbering is relative to the casein mRNA CAP site designated as +1. Structural gene sequences are shown with exons in black and introns in white.

high in cells transfected with pSV2cat in which CAT gene expression is driven by the SV40 promoter and enhancer. In contrast, the constructions containing casein gene sequences as promoters showed reduced CAT activity. However, there was orientation specificity of this activity as demonstrated by a greatly reduced level of CAT activity using constructions in which casein gene promoter sequences were inserted in the 3'→5' direction (15).   No hormonal or substratum regulation of CAT expression was seen in any of these constructions.   However, both the β- and γ-casein inserts containing internal sequences directed a higher level of CAT activity than those which contained only 5'-flanking sequences.   Similar results have been observed in several other experiments using this cell line as well as in primary rat mammary epithelial cells (unpublished observations) and stable, neomycin-resistant COMMA-1D cells obtained by cotransfection with pSV2neo and the appropriate casein-CAT fusion gene followed by G418 selection (6).

   These experiments lead to the conclusion that the 5'-flanking regions of the β- and γ-casein genes are weak pro-

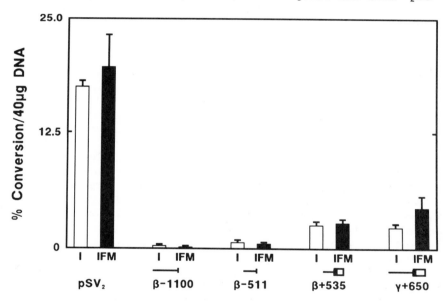

Figure 2.   Assay of casein-CAT fusion genes in COMMA-1D cells. Transfections and CAT assays were performed according to standard methods (7), except that cells were transfected while on hydrated collagen gels.   Cells were harvested by placing cultures into 5 ml of 0.1% collagenase in Hanks' Basic Salt Solution (HBSS) for 30 minutes at 37°C.   Conversion to acetylated chloramphenicol derivatives was expressed per 40 μg DNA, the approximate amount of DNA per 100 mm Petri dish harvested.   Extracts (25-50% of the total from each dish) were assayed in a 3-hour incubation at 37°C. Symbols and abbreviations are as indicated previously.

moters since they direct CAT expression at a level of 1% to 10% of that observed with the SV40 promoter and enhancer present. Also, the presence of a non-canonical TATA homology sequence in the γ-casein construction has no obvious effect on function. In fact, the presence of an internal sequence in this and one of the β-casein constructions appears of primary importance in obtaining increased CAT activity. There could be at least two reasons for the importance of this internal sequence. First, the 5' noncoding exon I may be providing an efficient eukaryotic translational signal (16) for the bacterial CAT gene. Second, the unusual conservation of the 5' noncoding region of casein mRNAs (6,8,9) suggests that the increased activity of the construction might reflect a post-transcriptional role for these sequences, as has been recently reported for c-myc and histone mRNAs (17,18). Exon I encodes the 5' leader sequence of casein mRNAs. The start codon for protein synthesis is further downstream in exon II in these genes. Sequence comparisons between the two rat casein genes have revealed homology in exon I and 5'-flanking DNA sequences, but little conservation in intron A sequences (6,8,15). In fact, the γ-casein gene contains a large repeated sequence element in intron A which is absent in the β-casein gene. This conservation in exon I and lack of it in intron A suggest the potential importance of exon I in eliciting the increased activity of casein-CAT fusion genes containing these structural gene regions. CAT mRNA levels currently are being measured to determine if this increased level of CAT enzyme is mediated by increased mRNA levels. Additional studies are also underway to identify the site of initiation in the β- and γ-casein gene constructions containing exon I and intron A.

The failure to observe any hormonal effect on casein-CAT fusion genes suggests that important cis-acting regulatory sequences lie elsewhere in the casein genes. Several studies have shown hormone regulatory sequences (19-22) and tissue-specific control elements (23-25) reside in proximal 5'-flanking DNA sequences. In fact, sequence analysis has shown that the casein genes contain putative steroid hormone-binding sequences in this region (8). However, more recent work has shown that both types of regulatory sequences can occur up to 7000 bp 5' or even internally to the mRNA CAP site (26,27). In light of our data and the other studies mentioned above, we are constructing vectors that contain additional 5'-flanking, as well as internal and

3'-flanking, DNA in order to identify hormone regulatory sequences.

Cell- and tissue-specificity of promoter function have been observed in a number of systems (24-28). After identifying the regions of casein genomic DNA that are requisite to promoter function, the cell-specificity of this function was tested. To this end, NIH-3T3 fibroblasts were cultured on hydrated collagen gels and transfected using the same procedures as for COMMA-1D cells. 3T3 cells were chosen because they are non-mammary and non-epithelial in origin. In using cells derived from the same species, we expected any differences in activity would reflect cell-type and not species-specific effects. Figure 3 shows the results of this experiment. It can be seen that pSV2cat expression is 6 times higher in 3T3 cells, which most likely reflects a greater transfection efficiency of the 3T3 cells. However, an interesting picture emerges when CAT activities from casein-CAT fusion plasmids are normalized to those seen with pSV2cat and ratios of these normalized activities are calculated for each vector construction. The β-casein construction containing internal sequences shows a small mammary-specific enhancement of CAT activity as tabulated in the last column where activity ratios are listed. However, the β-casein constructions containing only 5'-flanking sequences show a dramatic and specific inhibition of CAT activity.

	COMMA-1D	**NIH-3T3**	**C1D/3T3**
pSV₂	1.000±0.069(27*)	1.000±0.130(125*)	--
β-1100/-12	0.002±0.001	0.039±0.003	0.05
β-511/-12	0.008±0.002	0.028±0.006	0.29
β-511/+535	0.110±0.006	0.088±0.011	1.25
γ-1100/+650	0.036±0.006	0.039±0.005	0.92

* % Conversion ¹⁴C-chloramphenicol/40µg DNA

Figure 3. Mammary-specific regulation of expression of casein-CAT fusion genes. Only 25% of the pSV2cat-transfected 3T3 cell extract was assayed in order to maintain enzymatic conversion within the linear range. Symbols and abbreviations are as indicated previously.

Other mammary and non-mammary cell types are being tested to determine the generality of this phenomenon. This result is in contrast to most other genes characterized to date in which sequences residing in the proximal few hundred bp of 5'-flanking DNA are necessary for cell-specific, positive regulation (24-28).

The control of gene expression in eukaryotes must be achieved by a complex interaction of positive and negative regulatory elements. We appear to have identified a mammary-specific, positive regulatory element, most likely in conserved exon I, as well as a negative regulatory element localized in the proximal 5'-flanking DNA region. The existence of negative regulatory elements has recently been reported in several systems (29-32). However, in the present case, the negative element is fully operable in differentiated mammary epithelial cells, in contrast to other systems where the negative control occurs in uninduced or non-homologous cells (30-32). More detailed deletion and mutagenesis experiments are required to precisely localize these presumptive regulatory sequences. Presumably, other regulatory regions either further 5' of, or internal to, the mRNA CAP site interact with this negative element to overcome its effect. These interactions could be with internal elements as has been seen in the immunoglobulin genes (27), which interestingly also appear to be regulated primarily at the post-transcriptional level (33). Genes like the caseins that do not need to be acutely regulated may not require strong cell-specific promoters. The observed post-transcriptional regulation of casein mRNA synthesis (6) may be the more important regulatory pathway. Clearly, this is only the beginning of the definition of such complex regulatory mechanisms for casein gene expression.

ACKNOWLEDGMENTS

We would like to acknowledge the excellent technical assistance of Craig Couch and Sara Rupp. Secretarial assistance was ably provided to us by Patricia Kettlewell.

REFERENCES

1. Topper, YJ, Freeman, CS (1980). Multiple hormone interactions in the developmental biology of the mammary gland in vitro. Physiol Rev 60:1044.

2. Kratochwil, K (1969). Organ specificity in mesenchymal induction demonstrated in the embryonic development of the mammary gland of the mouse. Dev Biol 20:46.

3. Sakakura, T, Sakagami, Y, Nishizuka, Y (1982). Dual origin of mesenchymal tissues participating in mouse mammary gland embryogenesis. Dev Biol 91:202.

4. Hobbs, AA, Richards, DA, Kessler, DJ, Rosen, JM (1982). Complex hormonal regulation of casein gene expression. J Biol Chem 257:3598.

5. Lee, EY-HP, Lee, W-H, Kaetzel, CS, Parry, G, Bissell, MJ (1985). Interaction of mouse mammary epithelial cells with collagen substrata: Regulation of casein gene expression and secretion. Proc Natl Acad Sci USA 82:1419.

6. Rosen, JM, Rodgers, JR, Couch, CH, Bisbee, CA, David-Inouye, Y, Campbell, SM, Yu-Lee, L-Y (1986). Multi-hormonal regulation of milk protein gene expression. In "Metabolic regulation: Application of recombinant DNA techniques," New York: New York Academy of Sciences, in press.

7. Gorman, C (1985). High efficiency gene transfer into mammalian cells. In Glover, DM (ed.): "DNA Cloning: A Practical Approach," Volume II, Oxford: IRL Press, p 143.

8. Rosen, JM, Jones, WK, Rodgers, JR, Compton, JR, Bisbee, CA, David-Inouye, Y, Yu-Lee, L-Y (1986). Regulatory sequences involved in the hormonal control of casein gene expression. In "Endocrinology of the Breast: Basic and Clinical Aspects," Volume 464, New York: New York Academy of Sciences p 87.

9. Yu-Lee, L-Y, Richter-Mann, L, Couch, CH, Stewart, AF, Mackinlay, AG, Rosen, JM (1986). Evolution of the casein multigene family: Conserved sequences in the 5'-flanking and exon regions. Nucl Acids Res 14:1883.

10. Gorman, CM, Moffat, LF, Howard, BH (1982). Recombinant genomes which express chloramphenicol acetyltransferase in mammalian cells. Mol Cell Biol 2:1044.

11. Danielson, KG, Oborn, CJ, Durban, EM, Butel, JS, Medina, D (1984). Epithelial mouse mammary cell line exhibiting normal morphogenesis in vivo and functional differentiation in vitro. Proc Natl Acad Sci USA 81:3756.

12. Emerman, JT, Pitelka, DR (1977). Maintenance and induction of morphological differentiation in dissociated mammary epithelium on floating collagen membranes. In Vitro 13:316.

13. Bisbee, CA, Machen, TE, Bern, HA (1979). Mouse mammary epithelial cells on floating collagen gels: Transepithelial ion transport and effects of prolactin. Proc Natl Acad Sci USA 76:536.

14. Subramani, S., Southern, PJ (1983). Analysis of gene expression using simian virus 40 vectors. Anal Biochem 135:1.

15. Rosen, JM, Jones, WK, Campbell, SM, Bisbee, CA, Yu-Lee, L-Y (1985). Structure and regulation of peptide hormone responsive genes. In Czech, MP, Kahn, CR (eds.): "Membrane Receptors and Cellular Regulation," UCLA Symposia on Molecular and Cellular Biology, Volume 23, New York: Alan R. Liss, p 385.

16. McGarry, TJ, Lindquist, S (1985) The preferential translation of Drosophila hsp70 mRNA requires sequences in the untranslated leader. Cell 42:903.

17. Rabbitts, PH, Forster, A, Stinson, MA, Rabbitts, TH (1985). Truncation of exon 1 from the c-myc gene results in prolonged c-myc mRNA stability. EMBO J 4:3727.

18. Morris, T, Marashi, F, Weber, L, Hickey, E, Greenspan, D, Bonner, J, Stein, J, Stein, G. (1986). Involvement of the 5'-leader sequence in coupling the stability of a human H3 histone mRNA with DNA replication. Proc Natl Acad Sci USA 83:981.

19. Camper, SA, Yao, YAS, Rottman, FM (1985). Hormonal regulation of bovine prolactin promoter in rat pituitary tumor cells. J Biol Chem 260:12246.

20. Dean, DC, Knoll, BJ, Riser, ME, O'Malley, BW (1983). A 5'-flanking sequence essential for progesterone regulation of an ovalbumin fusion gene. Nature 305:551.

21. Karin, MA, Haslinger, A, Holtgreve, H, Cathala, G, Slater, E, Baxter, JD (1984). Activation of a heterologous promoter in response to dexamethasone and cadmium by metallothionein gene 5'-flanking sequences. Cell 36:371.

22. Wynshaw-Boris, A, Short, JM, Hanson, RW (1986). The determination of sequence requirements for hormonal regulation of gene expression. BioTech 4:104.

23. Hanahan, D (1985). Heritable formation of pancreatic β-cell tumours in transgenic mice expressing recombinant insulin/simian virus 40 oncogenes. Nature 315:115.

24. Khillan, JS, Schmidt, A, Overbeek, PA, deCrombrugghe, B, Westphal, H (1986). Developmental and tissue-specific expression directed by the α2 type I collagen promoter in transgenic mice. Proc Natl Acad Sci USA 83:725.27.

25. Ohlsson, H, Edlund, T (1986). Sequence-specific interactions of nuclear factors with the insulin gene enhancer. Cell 45:35.

26. Godbout, R, Ingram, R, Tilghman, S (1986). Multiple regulatory elements in the intergenic region between the α-fetoprotein and albumin genes. Mol Cell Biol 6:477.

27. Grosschedl, R, Baltimore, D (1985). Cell-type specificity of immunoglobulin gene expression is regulated by at least three DNA sequence elements. Cell 41:885.

28. Grichnik, JM, Bergsma, DJ, Schwartz, RJ (1986). Tissue restricted and stage specific transcription is maintained within 411 nucleotides flanking the 5' end of the chicken skeletal actin gene. Nucl Acids Res 14:1683.

29. Brand, AH, Breeden, L, Abraham, J, Sternglanz, R, Nasmyth, K (1985). Characterization of a "silencer" in yeast: A DNA sequence with properties opposite to those of a transcriptional enhancer. Cell 41:41.

30. Nir, U, Walker, MD, Rutter, WJ (1986). Regulation of rat insulin 1 gene expression: Evidence for negative regulation in nonpancreatic cells. Proc Natl Acad Sci USA 83:3180.

31. Goodbourn, S, Burstein, H, Maniatis, T (1986) The human β-interferon gene enhancer is under negative control. Cell 45:601.

32. Hen, R, Borrelli, E, Fromental, C, Sassone-Corsi, P, Chambon, P (1986). A mutated polyoma virus enhancer which is active in undifferentiated embryonal carcinoma cells is not repressed by adenovirus-2 E1A products. Nature 321:249.

33. Gerster, T, Picard, D, Schaffner, W (1986). During B-cell differentiation enhancer activity and transcription rate of immunoglobulin heavy chain genes are high before mRNA accumulation. Cell 45:45.

Transcriptional Control Mechanisms, pages 325–332
© 1987 Alan R. Liss, Inc.

HORMONAL REGULATION OF GENE EXPRESSION IN YEAST[1]

Scott W. Van Arsdell and Jeremy Thorner

Department of Biochemistry, University of California
Berkeley, California 94720

ABSTRACT   The mating process in yeast is initiated
when haploid cells of opposite mating type (MATa and
MATα cells) exchange oligopeptide pheromones. These
intracellular signal molecules cause specific physiological changes in their target cells, including the
modulation of gene expression. We report here that
a conserved DNA sequence motif, which may be a cis-
acting regulatory site that plays a role in the pheromonal induction of transcription, is located upstream
of several pheromone-induced genes.

INTRODUCTION

Peptide hormones, growth factors, and other extracellular signal molecules play an essential role in the regulation of cell growth and differentiation in eukaryotic organisms. In the unicellular eukaryote Saccharomyces cerevisiae,
haploid cells of opposite mating type (MATa and MATα cells)
must exchange oligopeptide pheromones before the mating process can occur, to yield diploid (MATa/MATα) cells. Each
mating pheromone (MATa cells secrete a-factor and MATα cells
secrete α-factor) elicits specific physiological changes in
its target cell which are required for conjugation, including
arrest of cell growth in the G1 phase of the cell cycle,
synthesis of cell surface agglutinins, and characteristic
changes in cell shape (for review, see references 1 and 2).
The yeast mating pheromones also cause rapid and dramatic changes in gene expression in their target cells (3,4).
We have characterized a yeast genomic clone that encodes

[1]This work was supported by NRSA Postdoctoral Fellowship
GM09499 to S.W.V. and by NIH Research Grant GM21841 to J.T.

a 650-base polyA+ RNA which is induced 20-50 fold in MATa
cells that have been treated with α-factor (5). The 650-
base RNA is transcribed from a cluster of repetitive
sequences containing both full-length and truncated sigma
and delta elements adjacent to a tRNA^{trp} gene (5). Yeast
delta elements are the 333-337 bp repeated sequences which
exist either as isolated "solo" elements or as direct repeats
flanking the 5.3 kb transposable element, Ty (6). None of
the λScG7 delta elements are associated with Ty sequences.
Members of the sigma family of repeated sequences are 339-341
bp in length and have many of the structural characteristics
of eukaryotic transposable elements (7,8). The most unusual
feature of sigma is that every element that has been analyzed
at the nucleotide sequence level to date is located 15 to 18
bp upstream of a tRNA gene (7-10). The single full-length
sigma element in the genomic insert of λScG7 is located
15 bp upstream of a tRNA^{trp} gene (5).

Transcript mapping studies (5) indicate that transcription of the 650-base α-factor-induced RNA initiates within
the full-length sigma element of λScG7 (which we have called
sigma-650), proceeds in the direction away from the tRNA^{trp}
gene, and terminates within or distal to a full-length delta
element. In addition to the 650-base transcript, MATa cells
contain two other abundant α-factor-inducible polyA+ RNA
species (500 and 5,300 bases) that are homologous to the
same strand of sigma, but are transcribed from other locations in the genome (5). The three sigma-related transcripts
also accumulate in MATα cells that have been treated with a-
factor; none of these RNAs is present at significant levels
in diploids (G. Stetler and S. Van Arsdell, unpublished results). Induction of the sigma-related RNAs is rapid, requires a functional STE2 gene product (the putative α-factor
receptor, see 14), and does not require new protein synthesis, suggesting that stimulation of sigma transcription is a
primary response to the pheromone (5).

We have used gene fusion techniques to show that the
pheromone-responsive promoter of λScG7 is located within
the sigma-650 element. A recombinant plasmid was constructed
in which the sigma-650 element was inserted just upstream of
the coding sequence and terminator of a yeast invertase
(SUC2) gene which lacked its own promoter. In MATa cells
transformed with this construction, a new prominent polyA+
RNA that hybridizes to both sigma and SUC2 probes appears in
response to α-factor, and intracellular invertase activity
is induced 10-fold (5). Primer extension studies indicate
that synthesis of the hybrid transcript initiates uniquely

within sigma sequences (5). Deletion analysis of the sigma-SUC2 fusion gene has shown that upstream sequences which are essential for α-factor induction are located entirely within the sigma element (11). These results suggest that at least the sigma-650 element, and perhaps others of the 25-30 sigma sequences in the yeast genome (9), may function as hormone-inducible promoters that activate the expression of downstream genes during the mating process. However, the role, if any, of the three major pheromone-inducible sigma-related RNAs in the mating process is not yet known.

The expression of several mating-specific genes is induced by α-factor in MATa cells (11-13), including: the STE2 gene, which encodes a structural component of the α-factor receptor (14); the MFa1 and MFa2 genes, which encode a-factor precursors (15); and BAR1 (also called SST1), which encodes an extracellular protease that degrades α-factor (V. MacKay, personal communication). Expression of the STE3 gene, which is presumed to encode the a-factor receptor (16), is induced in MATα cells by a-factor (4). Recent evidence indicates that a-factor and α-factor modulate gene expresssion in their target cells by generating a common intracellular signal (17). In this paper we report that short blocks of DNA sequence homology are located in the 5'-flanking regions of pheromone-induced genes. We propose that these conserved sequence motifs may be cis-acting regulatory sites that play a role in the pheromonal induction of transcription.

RESULTS AND DISCUSSION

The DNA sequence of the sigma-650 element, which has been shown to contain a pheromone-responsive promoter (5), is compared to the DNA sequences of several sigma elements in Figure 1. The length of these sigma elements varies from 339 bp (sigma-650) to 340 bp (pFD2) to 341 bp (pFD12, pFD17, SUQ5, and SUP2). The sigma-650 sequence is ⩾97% homologous to each of the other sigma sequences. Thus, with the exception of two truncated sigma elements (5,7), the lengths and sequences of sigma elements appear to be highly conserved.

We have compared the DNA sequence upstream of the major transcription initation site of sigma-650 (marked by an asterisk in Figure 1) to the 5'-flanking sequences of other pheromone-inducible genes (STE2, BAR1, MFa1, MFa2, and STE3). A nearly perfect 20 bp homology was found between the upstream regions of sigma-650 and STE2 (see Figure 2). This homologous sequence is located within the region of sigma-650

```
                        --------                              --------  --------
sigma-650  TGTTGTATCTCAAAATGAGATATGTCAGTATGACAATACGTCACCCTGAACGTTCATAAAACACATATGAAACAACCTTATAACA

pFD2                                                            T    A

pFD12                                                           T    A

pFD17                                                           T

SUQ5                                                            T

SUP2                                 C                          T    A


                        --------
sigma-650  AAACGAACAACATGAGACAAAACCCGACCTTCCCTAGCTGAACTACCC-AAAGTATAAATGCCTGAACAATTAGTTTAGATCCGG

pFD2                          T         T              -    A                                    A

pFD12                         T         G                   A                                    A

pFD17                                   T                   A                                    A

SUQ5                                    T                   A                                    A

SUP2                                    T                   A                                    A


                                                                           *
sigma-650  GATTCCCCGCTTCCACCACTTAGTATGATTCATATTTTATATAATATATAAGATAAGTAACATTCCCGTGAATTAATCTGATAAAC

pFD2                          A

pFD12

pFD17             T                              G

SUQ5              T

SUP2


sigma-650  TG-TTTGACAACTGGTTACTTCCCTAAGACTGTTTATATTAGGATTTTCAAGACACTCCGGTATTACTCGAGCCCGTAATACAACA

pFD2        T                         C              G

pFD12       T                                        G

pFD17       T  A                                     G

SUQ5        T  A                                     G

SUP2        T                                        G    A
```

FIGURE 1. Comparison of the nucleotide sequence of sigma elements. The nucleotide sequence (in the 5'-to-3' direction) is shown for the sigma-650 element (5). In the lines below are indicated those sites at which the other sigma elements (7,8) differ from the sequence of sigma-650. The homology between sigma-650 and STE2 (see Figure 2) is overlined. The 8 bp consensus sequence (see Figure 3) is overlined with a dashed line. The major start of sigma-650 transcription is indicated by an asterisk.

sigma   -167 ATaAAACACATATGAAacAA -148

STE2   -166 ATgAAACACATATGAAgaAA -147

FIGURE 2. A nucleotide sequence upstream of the transcription initiation site of sigma-650 (5) is homologous to
a sequence located in the 5'-flanking region of STE2 (12,19).
Base mismatches in the 20 bp region of homology are indicated
by lower case letters. The numbers refer to the distance
between the first or last base of the homologous region and
the site of transcription initiation.

that is required for pheromone induction (11) and is found at
almost precisely the same distance upstream of the transcription initiation sites of sigma-650 (-148 to -167) and STE2
(-147 to -166). This 20 bp region of homology between STE2
and sigma-650 contains within it two nearly perfect copies of
the octanucleotide, ATGAAACA, in direct repeat, separated by
the trinucleotide, CAT. Regions homologous to this 8 bp sequence motif are found at other sites upstream of the sigma-
650, STE2, BAR1, MFa1, MFa2, and STE3 genes (see Figure 3).
      The 20 bp homology between sigma-650 and STE2 is loca-
ted in a highly conserved part of sigma. The sequences of
all six sigma elements compared in Figure 1 are identical in
this region. This high degree of conservation may reflect
some functional significance. Perhaps the 20 bp homology
serves as an upstream activating sequence (UAS) (18) which
renders downstream promoters pheromone-inducible. However,
it is not known whether any of the sigma elements shown in
Figure 1, aside from sigma-650, is transcriptionally active.
Indeed, it is suprising that there are only three prominent
hormone-induced polyA+ RNAs in haploid cells which are homologous to sigma (5) since the yeast genome contains approximately 30 sigma elements (9). One possible explanation is
that only a subset of the total population of sigma elements
has retained all the cis-acting transcriptional signals required for pheromone induction. Alternatively, the majority
of sigma elements may promote transcription in response to
mating pheromones, but only those elements that are located
upstream of functional terminators and polyA+ addition sites
could direct the synthesis of discrete, stable polyA+ RNAs.
      It is intriguing that sequences that match the 8 bp con-
sensus are found at multiple sites proximal to both α-factor-
and a-factor-inducible genes. Since pheromonal induction of

```
           -208    -201   -167                      -148  -126   -119
sigma      ┌────────┐      ┌────────┐    ┌────────┐        ┌────────┐
           │ATGAgAtA│ ···  │ATaAAACA│ CAT│ATGAAACA│ A  ··· │ATGAgACA│ ··· +1
           └────────┘      └────────┘    └────────┘        └────────┘

           -292    -285   -166                    -147
STE2       ┌────────┐      ┌────────┐    ┌────────┐
           │ATGAAACA│ ···  │ATGAAACA│ CAT│ATGAAgaA│ A ················ +1
           └────────┘      └────────┘    └────────┘

           -274    -267   -184   -177                            +1
BAR1       ┌────────┐      ┌────────┐
           │cTGAAACA│ ···  │cTGAAACA│ ························· ATG
           └────────┘      └────────┘

           -199    -192   -106   -99    -72    -65              +1
MFa1       ┌────────┐      ┌────────┐    ┌────────┐
           │ATGAAACc│ ···  │ATGAAAtA│ ···│ATGAAttA│ ············· ATG
           └────────┘      └────────┘    └────────┘

           -196    -189   -178   -171  -158   -151             +1
MFa2       ┌────────┐      ┌────────┐    ┌────────┐
           │ATGAAACA│ ···  │AgGAAAaA│ ···│AgGAAAaA│ ············· ATG
           └────────┘      └────────┘    └────────┘

           -768    -761   -295   -288  -272   -265             +1
STE3       ┌────────┐      ┌────────┐    ┌────────┐
           │tTGAtACA│ ···  │ATGAAggA│ ···│tTGAAACt│ ············· ATG
           └────────┘      └────────┘    └────────┘
```

CONSENSUS   [ATGAAACA]

FIGURE 3.   An 8 bp sequence motif is found at multiple
sites upstream of several pheromone-inducible genes. Regions
homologous to the 8 bp consensus sequence 5'-ATGAAACA-3' that
are located in the 5'-flanking sequences of the sigma-650
gene (5), the STE2 gene (12,19), the BAR1 gene (V. MacKay,
personal communication), the MFa1 and MFa2 genes (15), and
the STE3 gene (12,16) are boxed. Base mismatches between
the homologous regions and the consensus sequence are indicated by lower case letters; perfect matches to the consensus
sequence are double underlined. The numbers above the sequences refer to the distance between the first or last
base of the homologous region and either the major site of
transcription initiation (sigma-650, STE2) or the translation
initiation codon (BAR1, MFa1, MFa2, STE3).

transcription seems to involve the same intracellular signal
in MATa and MATα cells (17), it is possible that all pheromone-induced genes share common upstream regulatory sites.
The conserved octanucleotide, ATGAAACA, may be such a cis-
acting sequence that plays a role in the process of pheromonal induction. This hypothesis is currently being tested
by determining whether mutations in the conserved regions
have any effect on pheromonal induction of transcription and
whether the conserved sites specifically interact with trans-
acting regulatory factors.

ACKNOWLEDGMENTS

We thank Vivian MacKay for communication of results prior to publication. The BIONET programs of Intelligenetics, Inc. were used for comparisons of nucleotide sequences.

REFERENCES

1. Thorner, J (1981). Pheromonal regulation of development in Saccharomyces cerevisiae. In Strathern, JN, Jones, EW, Broach, JR (eds): "Molecular Biology of the Yeast Saccharomyces: Life Cycle and Inheritance," Cold Spring Harbor: Cold Spring Harbor Laboratory, p 143.
2. Sprague, GF jr, Blair, LC, Thorner, J (1983). Cell interactions and regulation of cell type in the yeast Saccharomyces cerevisiae. Ann Rev Microbiol 37: 623.
3. Stetler, GL, Thorner, J (1984). Molecular cloning of hormone-responsive genes from the yeast Saccharomyces cerevisiae. Proc Natl Acad Sci USA 80: 1144.
4. Hagen, DC, Sprague, GF jr (1984). Induction of the yeast α-specific gene STE3 by the peptide pheromone a-factor. J Mol Biol 178: 835.
5. Van Arsdell, SW, Stetler, GL, Thorner, J (1986). Yeast repeated element sigma contains a hormone-inducible promoter. Cell, submitted for publication.
6. Roeder, GS, Fink, GR (1983). Transposable elements in yeast. In Shapiro, JA (ed): "Mobile Genetic Elements," New York: Academic Press, p 299.
7. del Rey, FJ, Donahue, TF, Fink, GR (1982). sigma, a repetitive element found adjacent to tRNA genes of yeast. Proc Natl Acad Sci USA 79: 4138.
8. Sandmeyer, SB, Olson, MV (1982). Insertion of a repetitive element at the same position in the 5'-flanking regions of two dissimilar tRNA genes. Proc Natl Acad Sci USA 79: 7674.
9. Brodeur, GM, Sandmeyer, SB, Olson, MV (1983). Consistent association between sigma elements and tRNA genes in yeast. Proc Natl Acad Sci USA 80: 3292.
10. Genbauffe, FS, Chisholm, GE, Cooper, TG (1984) tau, sigma, and delta. J Biol Chem 259: 10518-10525.
11. Van Arsdell, SW, Thorner, J (1986). A conserved DNA sequence motif required for pheromonal induction of gene expression in yeast, Proc Natl Acad Sci USA, to be submitted.
12. Nakayama, N, Miyajima, A, Arai, K (1985). Nucleotide

sequences of STE2 and STE3, cell type-specific sterile genes from Saccharomyces cerevisiae. EMBO J 4: 2643.

13. Hartig, A, Holly, J, Saari, G, MacKay, VL (1986). Multiple regulation of STE2, a mating-type-specific gene of Saccharomyces cerevisiae. Mol Cell Biol 6: 2106.

14. Jenness, DD, Burkholder, AC, Hartwell, LH (1983). Binding of α-factor pheromone to yeast a-cells: Chemical and genetic evidence for an α-factor receptor. Cell 35: 521.

15. Brake, AJ, Brenner, C, Najarian, R, Laybourn, P, Merryweather, J (1985). Structure of genes encoding precursors of the yeast peptide mating pheromone, a-factor. In Gething, M-J (ed) "Protein Transport and Secretion," Cold Spring Harbor: Cold Spring Harbor Laboratory, p 103.

16. Hagen, DC, McCaffrey, G, Sprague, GF jr (1986). Evidence the yeast STE3 gene encodes a receptor for the peptide pheromone a-factor: Gene sequence and implications for the structure of the presumed receptor. Proc Natl Acad Sci USA 83: 1418.

17. Bender, A, Sprague, GF jr (1986). Yeast peptide pheromones, a-factor and α-factor, activate a common response mechanism in their target cells. Cell, submitted for publication.

18. Guarente. L (1984). Yeast promoters: Positive and negative elements. Cell 36: 799.

19. Burkholder, AC, Hartwell, LH (1985). The yeast α-factor receptor: Structural properties deduced from the sequence of the STE2 gene. Nucleic Acids Res 13: 8463.

Transcriptional Control Mechanisms, pages 333–342
© 1987 Alan R. Liss, Inc.

A TRANS-ACTING FACTOR NEGATIVELY REGULATES TRANSCRIPTION AT THE MMTV LTR

Michael G. Cordingley, H. Richard-Foy,
A. Lichtler, and Gordon L. Hager.

Hormone Action and Oncogenesis Section,
Lab. of Exptl. Carcinogenesis,
Bldg. 37, Rm. 3C30, NCI, NIH,
Bethesda, MD 20892

ABSTRACT We have previously reported that in the absence of glucocorticoids the promoter of mouse mammary tumor virus is markedly refractory to activation by a constitutive transcriptional enhancer (1). Oligonucleotide directed mutations in the hormone regulatory element extensively impair the hormone response and allow effective activation of the promoter by the enhancer. This indicated that the HRE mediates negative regulation of transcription in the absence of hormone. Experiments utilizing protein synthesis inhibitors revealed that a labile protein inhibits transcription initiation at the LTR. Superinduction of transcription initiation results when protein synthesis is inhibited during hormone stimulation of the promoter. This effect is particularly dramatic when the enhancer of HaMSV is present at the upstream boundary of the MMTV LTR but is also observed in the absence of an enhancer element. Deletion of sequences which contain the HRE prevents superinduction. We conclude that a labile factor effects negative regulation through sequences at or closely associated with the HRE and that both positive and negative regulatory factors interact with closely associated sequences in the MMTV LTR.

INTRODUCTION

Expression of mouse mammary tumor virus is regulated at the level of transcription by glucocorticoid hormones (2,3). Nucleotide sequences contained entirely within the viral Long Terminal Repeat (LTR) are sufficient to confer hormone responsiveness to a heterologous promoter in gene transfer experiments (4,5). In addition sequences between 70 and 200 base pairs upstream from the cap site for transcription initiation are known to bind, with high affinity, purified glucocorticoid recptor in vitro (6,7). These hormone responsive sequences (which we shall term the hormone response element; HRE) have been reported to function to some extent independent of position and orientation and in this respect resemble the class of transcriptional elements termed enhancers (8,9). Previous experiments from this laboratory, utilizing plasmid constructions in which an exogenous enhancer (that of Harvey murine sarcoma virus; HaMuSV) was inserted directly upstream from the the MMTV LTR, indicated that in the absence of activating hormone the HRE in the MMTV LTR could function as a negative element (1). The MMTV promoter is thus refractory to activation by the upstream enhancer in the absence of hormone. Deletion of the HRE alleviates this negative effect and allows efficient activation of the promoter in the absence of hormone (10). The observation that the HRE functions as a negative element in the absence of its cognate positive transcription factors appears to distinguish it from other activating elements. Here we describe further experiments to elucidate the mechanism of negative regulation associated with the HRE. Oligonucleotide -directed mutagenesis of the HRE reveals that the negative effect associated with the HRE in the absence of hormone is attributable to specific nucleotide sequences which constitute a negative element. In addition we have used a series of cell lines in which MMTV LTR-fusion genes are mobilized on high copy number stably replicating bovine papilloma virus (BPV) vectors (11). In these cells the episomal MMTV promoter is highly

responsive to glucocorticoids whereas BPV transcription from the same episome is not (12).
Exposure of these cells to protein synthesis
inhibitors during hormone stimulation resulted in
higher levels of accumulated transcripts than with
hormone alone. This "superinduction" is shown to
act at least in part at the level of transcription
initiation and is dependent on MMTV LTR sequences.
On the basis of these observations we propose that
a labile protein is responsible for negative
modulation at the MMTV LTR and that its action is
mediated through sequences in or closely
associated with the HRE.

RESULTS

Oligo-scanning mutations in the HRE.
     We have carried out experiments to confirm
our earlier observation that, in the absence of
hormone, sequences coincident with the HRE contain
a negative element capable of inhibiting
activation by the upstream enhancer. Using a
novel approach to oligonucleotide-directed
mutagenesis termed "oligo-scanning" (13), mismatch
mutations up to 12 contiguous bases (and probably
larger) can be efficiently introduced into target
sequences cloned in an M13-derived vector.
Oligonucleotides generally employed in this
technique range in size from 25 to 30 residues,
with 10 nucleotides of homology on either side of
the target sequence. Using this methodology we
have introduced a series of mismatch mutations
into the HRE.
     One mismatch mutation was in sequences which
constitute a high affinity binding site for
purified glucocorticoid receptor in vitro (6) and
is critical for the hormone response in functional
assays of the promoter (7,14,15). This binding
site is located between position -182 and -166
with respect to the mRNA cap site and contains the
hexameric "core" motif 5'TGTTCT3' (-176 to -171)
which is highly conserved in glucocorticoid
receptor binding sites (6). Sequences between
-175 and -166 have been mutated, altering all but
1 base in the conserved "core" sequence. The
effect of this mutation on hormone induction and

on the ability of the HRE to modulate activation
of the MMTV promoter by the upstream enhancer was
assessed by transfection of Enhancer-MMTV LTR
v-ras^H fusion genes into NIH3T3 cells. Stably
transformed foci were derived with both the mutant
and wild type constructs. Transcription from the
wild type and mutant LTR was determined by
nuclease S1 analysis of RNA synthesized in the
presence and absence of glucocorticoid.

This mutation rendered the promoter almost
totally unresponsive to glucocorticoid in all
independently derived foci tested. Moreover
compared to the promoter with a wild type HRE the
level of transcription from this promoter in the
absence of hormone was highly elevated. Thus
alteration of sequences within the HRE permits, in
the absence of glucocorticoid, activation of the
promoter by the upstream enhancer. These results
confirm that specific nucleotide sequences in the
HRE mediate negative regulation of transcription
in the absence of the activated hormone-receptor
complex.

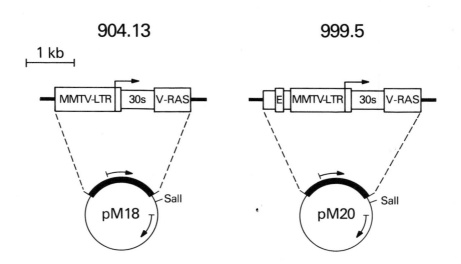

FIGURE 1. Structure of BPV-based minichromosomes in transformed C127 cell lines.

Superinduction of transcription.
    A series of C127 derived cell lines carrying
MMTV fusion genes linked to the 69% transforming
fragment of bovine papilloma virus (11) were
selected to study the effect of protein synthesis
inhibitors on induction of transcription at the
MMTV promoter. Figure 1 illustrates the structure
of the stably replicating episomal molecules in
two cell lines used in these studies. Cell line
904.13 has a BPV-minichromosome carrying an MMTV
LTR-v-ras^H fusion gene whereas cell line 999.5
contains a similar episome with an HaMuSV enhancer
inserted at the upstream boundary of the MMTV LTR.
    Transcription from the episomal LTR is highly
responsive to hormone in both cell lines (Table
1). In addition we discovered that hormone-
induced transcripts accumulated to higher levels
in the presence of the protein synthesis inhibitor
cycloheximide, than in the presence of hormone
alone (Table 1). This effect was independent of
the fusion gene product (since the same effect was
seen in 1471.1, a line carrying a BPV-mobilized
MMTV-CAT fusion gene) and also occurred in the
presence or absence of an upstream enhancer.
Cellular actin transcripts are unaffected by

TABLE 1
SUPERINDUCTION OF MMTV-INITIATED TRANSCRIPTS

Cell Line	Promoter	Induction Ratio[a] Hormone[b]	Hormone + CHX[c]
904.13	MMTV	10	30
999.5	E^d-MMTV	15	45
1471.1	MMTV	20	80

[a] Induction was determined 4 hours after hormone
 treatment by quantitative nuclease S1 analysis.
[b] Dexamethasone (10^{-6}M).
[c] Cycloheximide (100 micrograms/ml).
[d] Harvey murine sarcoma virus enhancer element.

either agent during the time course of our
experiments.

In order to determine whether superinduction
of steady-state levels of MMTV-initiated transcripts reflects increased transcription initiation or a post-transcriptional effect we carried
out _in vitro_ nuclear "run-on" assays in which
nascent RNA chains are elongated _in vitro_ in the
presence of radiolabelled ribonucleotides.   In
these experiments hybridizable counts are a
measure of the density of RNA polymerase complexes
actively transcribing the gene. Our analyses
confirm that in all cases superinduction of
transcription occurs, at least in part, at the
level of initiation. The effect is particularly
marked in the presence of the upstream HaMuSV
enhancer. Moreover we observe that while protein
synthesis inhibition in the absence of hormone has
no effect on transcription rates from the MMTV LTR
in 904.13 cells, significant induction of transcription results if an enhancer is present
upstream from the MMTV LTR as is the case in 999.5
cells. This observation is consistent with the
interpretation that depletion of a labile protein
alleviates the negative effect of the HRE
sequences. This alone is sufficient to permit
activation of the MMTV promoter by the upstream
enhancer. Induction of transcription from the
promoter which has no upstream constitutively
active enhancer, is however hormone dependent.

Further corroboratory evidence for the existence of a labile negative regulatory factor
acting upon the MMTV LTR has been obtained by
examining the ability of MMTV LTR deletion mutants
to undergo superinduction in the presence of
hormone and protein synthesis inhibitor. A
deletion mutant which retains sequences upstream
from -860 in the LTR maintains the ability to
become superinduced, however a mutant which only
retains 107 bases of upstream sequence is neither
hormone-inducible nor superinducible. These
experiments indicate that sequences between -860
and -107, which contain sequences essential for
the hormone response, also contain sequences
necessary for the negative activity of the putative labile factor.

DISCUSSION

The results presented suggest that transcription from the MMTV LTR is normally negatively regulated by a labile protein factor. Depletion of this factor results in higher levels of hormone induced initiation rates as measured by <u>in vitro</u> nuclear "run-on" assays, and is sufficient alone to allow activation of the promoter by an upstream enhancer element. Levels of transcripts from the cellular actin gene were not affected by hormone or protein synthesis inhibition during the time course of our experiments. Furthermore deletion of sequences internal to the MMTV LTR abolished superinduction. These observations strongly argue that superinduction does not result from generalized increased expression from either cellular or episomal promoters and is mediated through a sequence specific interaction of factor(s) within the MMTV promoter upstream region -860 to -107. In this respect superinduction at this promoter appears to reflect a phenomenon mechanistically relevant to the regulation of the MMTV promoter.

In experiments with oligonucleotide-directed mutants described herein we showed that a mutation in the hormone-receptor binding site (within the HRE) destroyed the negative effect associated with the HRE in the absence of hormone. On the basis of these results we propose that a labile repressor may be associated with specific recognition sequences within the HRE in the absence of hormone. Binding of the activated hormone- receptor complex must displace or negate the effect of such a factor. Although the sequences required for activity of the putative repressor appear to coincide closely with those for binding of the hormone-receptor complex <u>in vitro</u>, further mutagenesis may discriminate between negative and positive regulatory sequences in the HRE region. Present knowledge of the glucocorticoid receptor, the only characterized protein known to interact with nucleotide sequences in this vicinity, preclude the possibility that it plays a direct role in negative regulation. The unoccupied receptor is believed to be localized in the cytoplasm of the cell and has a half-life of

approximately 10 hours in the absence of hormone
(rat liver glucocorticoid receptor;16), far longer
than that proposed for the negative effector.   In
addition the receptor is known to have little
affinity for DNA in the absence of activating
hormone (17).

The precise mechanism of negative regulation
remains unclear.   The chromatin structure of the
MMTV LTR has been shown to be highly ordered into
a phased array of nucleosomes (Helene Richard-Foy,
manuscript in preparation).   Upon hormone induction sequences coincident with a nucleosome
predicted to be positioned over the HRE become
hypersensitive to DNase1 (10,18,19).   The
perturbation of nucleoprotein structure in this
region appears to result from receptor binding at
the HRE and may be central to the regulatory
mechanism.   One might envisage that a repressor
molecule may ensure the maintenence of an inactive
chromatin conformation in the absence of inducing
hormone.   Experiments are presently underway to
assess the chromatin configuration of mutant LTRs
which do not have a functional negative element,
and to biochemically characterize trans-acting
regulatory proteins bound at the HRE in the
absence of hormone.   Such proteins must surely be
candidates for a negative regulatory factor.

ACKNOWLEDGEMENTS

We wish to acknowledge the expert technical
assistance of Ron G Wolford and Diana S Berard and
to thank Michael Ostrowski for communication of
unpublished data.

REFERENCES

1.   Ostrowski MC, Huang AL, Kessel M, Wolford RG
      and Hager GL (1984). Modulation of enhancer
      activity by the hormone responsive regulatory
      element from mouse mammary tumor virus. EMBO
      J. 3: 1891.
2.   Young HA, Shih TY, Scolnick EM and Parks WP
      (1977). Steroid induction of mouse mammary
      tumor virus: effect upon synthesis and
      degradation of viral RNA. J Virol 21: 139.

3.   Ringold GM, Yamamoto KR, Bishop JM and
      Varmus HE (1977). Glucocorticoid-stimulated
      accumulation of mouse mammary tumor virus
      RNA: increased rate of synthesis of viral
      RNA. Proc Natl Acad Sci USA 74: 2879.
4.   Lee F, Mulligan R, Berg P and Ringold G
      (1981). Glucocorticoids regulate expression
      of dihydrofolate reductase cDNA in mouse
      mammary tumour virus chimeric plasmids.
      Nature 294: 228.
5.   Huang AL, Ostrowski MC, Berard D and Hager GL
      (1981). Glucocorticoid regulation of the
      Ha-MuSV p21 gene conferred by sequences from
      mouse mammary tumor virus. Cell 27: 245. 6.
6.   Scheidereit C, Geisse S, Westphal HM and
      Beato M (1983). The glucocorticoid receptor
      binds to defined nucleotide sequences near
      the promoter of mouse mammary tumour virus.
      Nature 304: 749.
7.   Pfahl M, McGinnis D, Hendricks M, Groner B
      and Hynes NE (1983). Correlation of
      glucocorticoid receptor binding sites on MMTV
      proviral DNA with hormone inducible
      transcription. Science 222: 1341.
8.   Chandler VL, Maler BA and Yamamoto KR (1983).
      DNA sequences bound specifically by
      glucocorticoid receptor in vitro render a
      heterologous promoter hormone responsive in
      vivo. Cell 33: 489.
9.   Ponta H, Kennedy N, Skroch K, Hynes NE and
      Groner B (1985). Hormonal response region in
      the mouse mammary tumor virus long terminal
      repeat can be dissociated from the proviral
      promoter and has enhancer properties. Proc
      Natl Acad Sci USA 82: 1020.
10.   Hager GL, Richard-Foy H, Kessel M, Wheeler D,
      Lichtler AC and Ostrowski MC (1984). The
      mouse mammary tumor virus model in studies of
      glucocorticoid regulation. Recent Prog Horm
      Res 40: 121.
11.   Law M-F, Lowy DR, Dvoretsky I and Howley PM
      (1981). Mouse cells transformed by bovine
      papilloma virus contain only extrachromosomal
      viral DNA sequences. Proc Natl Acad Sci 78:
      2727.
12.   Ostrowski MC, Richard-Foy H, Wolford RG,

Berard DS and Hager GL (1983).
Glucocorticoid regulation of transcription at
an amplified, episomal promoter. Mol Cell
Biol 3: 2047.

13.   Lichtler AC and Hager GL, manuscript
submitted to Gene (1986)

14.   Lee F, Hall CV, Ringold GM, Dobson DE, Luh
J, and Jacob PE (1984). Functional analysis of
the steroid hormone control region of mouse
mammary tumor virus. Nucl Acids Res 12: 4191.

15.   Majors J, and Varmus HE (1983). A small
region of the mouse mammary tumor virus long
terminal repeat confers glucocorticoid
hormone regulation on a linked heterologous
gene. Proc Natl Acad Sci USA 80: 5866.

16.   McIntyre WR, and Samuels HH (1985).
Triamcinolone acetonide regulates
glucocorticoid-receptor levels by decreasing
the half life of the activated
nuclear-receptor form. J Biol Chem 260: 418.

17.   Wrange O, Carlstedt-Duke J and Gustafsson J-A
(1979). Purification of the glucocorticoid
receptor from rat liver cytosol. J Biol Chem
254: 9284.

18.   Zaret KS and Yamamoto KR (1984). Reversible
and persistent changes in chromatin structure
accompany activation of a glucocorticoid
dependent enhancer. Cell 38: 29.

19.   Peterson DO (1985). Alterations in
chromatin structure associated with
glucocorticoid-induced expression of
endogenous mouse mammary tumor virus genes.
Mol Cell Biol 5: 1104.

Transcriptional Control Mechanisms, pages 343–355
© **1987 Alan R. Liss, Inc.**

REQUIREMENT OF AN ADDITIONAL FACTOR FOR THE
GLUCOCORTICOID RESPONSIVENESS OF THE RAT
ALPHA1-ACID GLYCOPROTEIN PROMOTER IN HTC CELLS

Elliott S. Klein,[1] Rosemary Reinke,[2,3]
Philip Feigelson,[2] and Gordon M. Ringold[1]

[1]Department of Pharmacology, Stanford University
Stanford, California 94305
[2]Institute of Cancer Research, Columbia University
New York, New York 10032

ABSTRACT Alpha1-acid glycoprotein is induced several
hundred fold by glucocorticoids in HTC rat hepatoma
cells. This induction has previously been shown to
require ongoing protein synthesis thereby implicating
either a pre-existing protein of short half-life or
an intermediary inducible protein in the glucocorticoid stimulation of AGP gene expression. Nuclear
transcription studies carried out previously have
indicated that the regulated expression of α_1-acid
glycoprotein by glucocorticoids in HTC cells might be
at a post-transcriptional step. Using gene transfection assays to investigate the role of 5'-flanking
sequences in the hormonal induction of the AGP gene,
we find that the region comprising nucleotides −763
to +1 confers glucocorticoid inducibility on heterologous genes such as chloramphenicol acetyltransferase and rabbit β-globin. The induction of rabbit
β-globin is inhibited by blockade of protein
synthesis whereas that of chloramphenicol acetyltransferase is not. Evidence is provided which
suggests the latter is due to stabilization of
chloramphenicol acetyltransferase mRNA by cycloheximide. Pretreatment with cycloheximide leads to a
diminution of the subsequent dexamethasone induction

[3]Present address: Department of Biological Chemistry,
University of California, Los Angeles, California 90024

of α_1-acid glycoprotein. Thus it appears that the hormone inducible property of the AGP gene resides within the AGP promoter and that a pre-existing labile protein is required for this induction to proceed efficiently.

INTRODUCTION

Regulation of transcription initiation plays a prominent role in controlling levels of gene expression. A useful model to analyze transcriptional regulation in higher eukaryotes is the altered expression of specific genes in response to steroid hormones (1). Many of the physiological effects of glucocorticoids are mediated by their interaction with specific receptor proteins (2). This interaction results in an increased affinity of glucocorticoid-receptor complexes for specific DNA sequences, and a subsequent increase in the transcription rate of associated coding sequences (3).

Alpha$_1$-acid glycoprotein (AGP) is a serum protein whose level is elevated by glucocorticoids (4) and during the acute phase response (5). AGP can be induced in vivo by either dexamethasone (Dex) administration (4) or by subcutaneous injection of turpentine (5), and in cultured HTC rat hepatoma cells by treatment with Dex (6). Although its physiological role is uncertain, candidate functions of AGP include its actions as an inhibitor of platelet aggregation (7) and/or involvement in the immune response (8).

The Dex mediated induced expression of the AGP gene in rat hepatoma HTC cells (6) and the cloned AGP gene transfected into L-cells (9) requires on-going protein synthesis. This inhibition is contrasted by the non-attenuated Dex induction of mouse mammary tumor virus (MMTV) RNA during blockade of protein synthesis with cycloheximide. Although the reason for this differential sensitivity of glucocorticoid responsive genes to cyclo-heximide is not known, the Dex induction of AGP could either be due to the requirement of a short-lived protein or the induction of another Dex inducible protein which mediates AGP induction. Experiments described herein support the former hypothesis.

Analyses of AGP gene transcription rates in nuclei from HTC cells indicate that the AGP gene is transcribed at high levels constitutively (i.e., in the presence or absence of hormone). This and the several hundred-fold in vitro elevation of AGP cytoplasmic mRNA levels by Dex has led to the proposal that expression of AGP is regulated at a post-transcriptional level (10). In contrast, however, similar studies using nuclei derived from livers of rats demonstrate that enhanced transcription of the AGP gene is associated with the AGP induction by glucocorticoids in vivo (11).

To further investigate the regulated expression of AGP, we have constructed fusion genes containing the 5'-upstream region of the rat AGP gene. In contrast to the proposal that hormone regulated post-transcriptional events mediate AGP mRNA induction we find that the AGP promoter itself is responsive to glucocorticoids. In addition, evidence is presented which implicates the AGP promoter in the cycloheximide mediated attenuation of AGP gene expression.

RESULTS

The AGP Promoter is Responsive to Glucocorticoids

AGP mRNA is induced several hundred fold by glucocorticoids in HTC rat hepatoma cells. Nuclear transcription assays however, have indicated that the AGP gene is transcribed at a high level constitutively and is relatively unaffected by glucocorticoids. To test whether this result accurately reflects the in vivo situation we explored whether the AGP promoter will direct expression of a heterologous sequence in a constitutive or an inducible fashion. A fusion gene (pAGPCAT) containing the 5'-upstream portion (-763 to +20) of the AGP gene, derived from the genomic clone previously described (9), was attached to the chloramphenicol acetyltransferase (CAT) gene (12). The plasmids pAGPCAT and pSV2neo (13) were co-transfected into the MMTV infected HTC cell line JZ.1 (14) by the method of Graham and van der Eb (15). Stable transformants resistant to the neomycin analog G-418 were screened for CAT activity (12). Figure 1 shows the results obtained from three different clonal isolates, one of which is negative for CAT activity.

346   Klein et al

J2   J3   J4

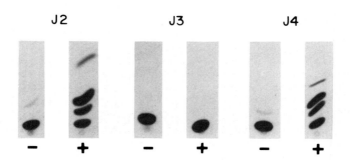

FIGURE 1. Screening for pAGPCAT stable transformants of JZ.1 via assay of CAT activity in the absence (-) and presence (+) of 1 µMolar Dex. Duration of Dex treatment was fifteen hours. After selection of G-418 resistant colonies, CAT activity was assayed as described (12) using clonally derived stable transformants.

Of the eight independently isolated G418 resistant clones, six were positive for CAT activity. All six positive clones exhibited Dex inducible CAT activity.
The plasmid pAGPCAT includes the first 20 nucleotides of the AGP primary RNA transcript. These 20 nucleotides could conceivably confer a Dex mediated induction of CAT RNA by providing a signal for stabilization of the transcript. Thus we removed these 20 b.p. generating the plasmid pAGPCAT Δ20.1. Analysis of this deletion of the plasmid pAGPCAT is shown in Figure 2.
Removal of the region +1 to +20 of the AGP gene results in essentially no change in the hormonal inducibility of CAT activity imparted by the AGP promoter. Although the relative fold induction of CAT activity from both pAGPCAT and pAGPCAT Δ20.1 is greater than 20-fold, precise comparison of these inductions is difficult as variables such as gene copy number and the sites of integration have not been controlled. In addition, preliminary evidence suggests the region -260 to +1 is sufficient for conferring hormone responsiveness to CAT activity (data not presented).

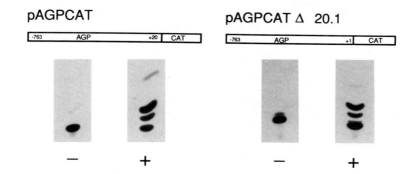

FIGURE 2. CAT activity of pAGPCAT and pAGPCAT Δ20.1 stable transformants of JZ.1 in the absence (-) and presence (+) of 1 μMolar Dex. Duration of Dex treatment was fifteen hours.

Induced Expression of RNA from the AGP Promoter: Evidence for the Requirement of a Preexisting Protein

The induction of AGP RNA by Dex is inhibited when protein synthesis is blocked with cycloheximide (6,9). We sought to determine if this property of the AGP induction process is maintained after fusion of the AGP promoter to heterologous sequences. Poly(A) enriched RNA was isolated from stable transformants containing either pAGPCAT or pAGP-β-globin (16), both of which contain the AGP 5'-upstream region from -763 to +20. As seen in Figure 3, like AGP RNA, β-globin RNA is induced very poorly by Dex when protein synthesis is inhibited. Thus the inhibition of Dex-inducible AGP RNA accumulation by cycloheximide appears to involve the 5'-flanking region.

In contrast to the results with AGP and β-globin RNAs, the abundance of CAT RNA is increased in cyclo-heximide treated cells. Additional evidence (to be presented subsequently) suggests that CAT RNA accumulates under these conditions because inhibition of protein synthesis dramatically increases CAT RNA stability.

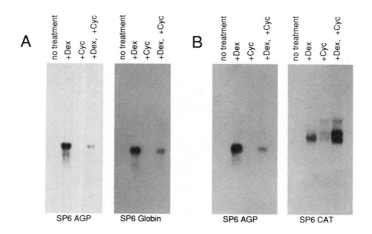

FIGURE 3. (A) Northern analysis of poly(A) selected RNA from a JZ.1 stable transformant carrying the construct pAGP-β-globin.   SP6 derived riboprobes are indicated. (B) Northern analysis of poly(A) selected RNA from a JZ.1 stable transformant (J2) carrying the construct pAGPCAT. Dex and cycloheximide treatment in both panels A and B were for fifteen hours.

Based on previous data and those with AGP-β-globin transformants (Figure 3) we considered the possibility that stimulation of the AGP promoter requires the production of a Dex-inducible protein.   Specifically, after 15 hours of concurrent Dex and cycloheximide treatment, the level of expression of AGP RNA is considerably inhibited as compared to Dex treatment alone. We were therefore interested in studying earlier durations of concurrent treatment.   As seen in Figure 4 the Dex induction of AGP RNA is not inhibited by cycloheximide after treatments of approximately two hours or less. Contrary to the notion that a Dex inducible protein mediates AGP induction, this delay in inhibition is suggestive of a preexisting and labile protein which is required for the induction process.

If the Dex induction of AGP RNA requires the involvement of such a protein, then inhibiting protein synthesis

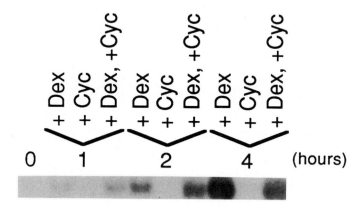

FIGURE 4. Northern analysis of poly(A) selected RNA from uninduced and Dex induced JZ.1 cells and the effect of cycloheximide at 1, 2, and 4 hours. Both Dex and cycloheximide were added to the medium at the same time. A SP6 derived AGP riboprobe was used.

prior to Dex administration should decrease the inducibility of AGP. Similarly, a relatively short cycloheximide pretreatment should have less of an effect on subsequent AGP induction than a longer one. The change in the degree of induction of AGP RNA after cycloheximide pretreatment is shown in Figure 5. Consistent with the hypothesis that a preexisting protein is required for AGP induction, an increase in the duration of cycloheximide pretreatment is associated with a decrease in the subsequent induction of AGP RNA. While the identity of this protein is unknown, the fact that the dex induced expression of MMTV RNA is not inhibited by cycloheximide even after 24 hours of combined treatment (6) would argue against a significant decrease in the level of glucocorticoid receptors in the presence of cycloheximide.

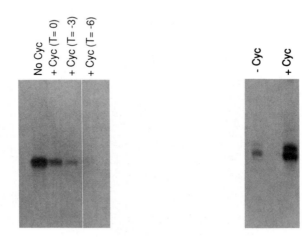

FIGURE 5. (Left panel) Northern analysis of poly(A) selected RNA from JZ.1 cells after a four hour Dex treatment and the effect of cycloheximide pretreatment. Dex was added at T = 0 hr. Cycloheximide was added at either T = -6 hr, T = -3 hr, or T = 0 hr. RNA was harvested at T = +4 hr. A SP6 AGP riboprobe was used.

FIGURE 6. (Right panel) Northern analysis of poly(A) selected RNA from untreated and cycloheximide treated JZ.1 stable transformants carrying the construct pSV2CAT. A SP6 CAT riboprobe was used.

Expression of CAT RNA from the AGP Promoter

As shown in Figure 3, cycloheximide superinduces the Dex induction of CAT RNA from the AGP promoter. The notion of a transcriptional inhibitory domain residing in the AGP promoter to explain this result is contradicted by the directly opposite effect cycloheximide has on the induction of AGP and β-globin RNAs. To examine the effect of cycloheximide on CAT RNA expression, independent of the AGP promoter, stable HTC transformants containing the plasmid pSV2CAT (12) were isolated. CAT activity in such transformants is constitutive and independent of Dex administration (data not shown). Expression of CAT RNA, however, is induced by cycloheximide treatment (Figure 6).

In both pAGPCAT and pSV2CAT stable transformants (Fig. 3 and Fig. 6, respectively) multiple species of poly(A)+ CAT RNAs are present. The two major species are approximately 1100 and 1400 b.p. in length. Both of these RNA species which vary at their 3' ends (data not shown) are generated irrespective of whether the CAT gene is under control of SV40 sequences or AGP sequences.

The HTC cell line JZ.1 used in these experiments does not express T-antigen thus the cycloheximide mediated induction of CAT RNA in pSV2CAT transformants cannot be explained by a derepression of the SV40 early region promoter. It therefore appears most likely that cycloheximide can stabilize CAT mRNAs. Such a stabilization event might account for the observed superinduction, by cycloheximide, of CAT RNA from the DEX inducible AGP promoter.

The relative abundance of CAT RNA over time in the pAGPCAT transformant J2 is shown in Figure 7.

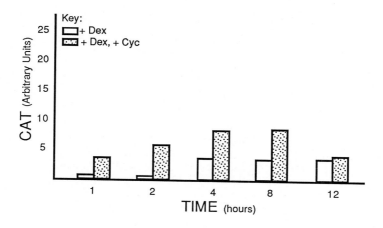

FIGURE 7. Dex induction of CAT poly(A)+ RNA in the pAGPCAT transformant J2 over time and the effect of cycloheximide. The data was normalized to expression of β-actin mRNA which was unchanged under these conditions.

Again, cycloheximide causes a superinduction of CAT RNA over that seen with Dex treatment alone, however this superinduction decreases with time. This attenuated effect of cycloheximide on CAT RNA is consistent with a decrease in the transcriptional activity from the AGP promoter observed in Dex plus cycloheximide treated cells seen in Figures 4 and 5.

DISCUSSION

AGP RNA is induced in rat HTC cells upon treatment with glucocorticoids (6). Although previous nuclear run-on assays indicated that the AGP gene is transcribed in a constitutive and hormone-independent manner in HTC cells (10), we have now shown the 5'-flanking region -763 to +20 of the AGP gene confers at least a 20-fold DEX induction of both CAT and β-globin RNAs from the heterologous constructs pAGPCAT and pAGP-β-globin, respectively. Removal of the region +1 to +20 from the construct pAGPCAT (i.e., pAGPCAT Δ20.1) had no effect on the induction of CAT RNA. Thus it seems likely that the results of the nuclear run-on assays from HTC cells are compromised by some as yet unidentified complication.

In contrast, nuclear run-on assays have shown the in vivo induction of AGP mRNA in response to either an acute phase stimulus or Dex treatment is due at least in part to an increase in hepatic AGP gene transcription (11). Although this contrasts with that seen in cultured rat HTC cells, it is consistent with the presently observed Dex induction of CAT and β-globin RNAs after transfection of AGP fusion genes.

Preliminary evidence suggests the region -260 to +1 of the AGP gene is sufficient for the Dex induction of a heterologous gene. Comparisons of the degree of inducibility of different deletions of the AGP promoter has not been possible as the copy number of individual stable transformants has not been determined. A suitable transient transfection system is currently being developed to circumvent this problem.

The Dex induction of AGP in HTC cells requires on-going protein synthesis (6). This has previously been explained as a requirement for the induction of a Dex inducible protein which is involved in the induction of AGP. Similarly, the glucocorticoid responsive expression of $\alpha 2u$-globulin in isolated hepatocytes is inhibited by cycloheximide (17,18). The induction of AGP RNA after approximately one to two hours of Dex treatment is not attenuated by concurrent cycloheximide administration, whereas longer treatments exhibit a marked cycloheximide inhibition of AGP. This would suggest that a pre-existing, labile protein is required for the Dex induction of AGP. In support of such a proposed protein is the decrease in the degree of AGP induction by Dex with increased duration of cycloheximide pretreatment.

The mechanism by which this putative regulatory protein allows the induction of AGP is unknown. Likewise, we do not know whether the induction of AGP by Dex involves the binding of the glucocorticoid-receptor complex to 5'-upstream sequences of the AGP gene. Actions of the preexisting protein might involve direct or indirect specific DNA binding in concert with the hormone-receptor complex. Alternatively, such a protein may cause an alteration of the hormone-receptor complex which is required for the induction of AGP but which is not required for the induction of MMTV.

The detection of CAT mRNA in transfected cells has proven to be relatively difficult even when expressed from a strong promoter such as that from SV40. We believe this may be due to an innate instability of CAT mRNA in HTC cells. The stability of CAT mRNA appears to be remarkably increased in HTC cells by inhibiting protein synthesis with cycloheximide. The mechanism of this stabilization event is not yet obvious. For this reason, while CAT constructions are clearly useful due to the ease and sensitivity of the enzymatic assay, we caution that their use may be inappropriate for investigating some aspects of gene regulation.

ACKNOWLEDGMENTS

We thank Karen Benight for help in preparation of this manuscript. This work was supported by grants from the NIH (GM25821) to GMR and (P01-CA22376) to PF, by NIH pre-doctoral training grants to Stanford University (GM07149) and Columbia University (AM07327), respectively. GMR is an Established Investigator of the American Heart Association.

REFERENCES

1. Ringold GM (1985). Steroid hormone regulation of gene expression. Annu Rev Pharmacol Toxicol 25:529.
2. Feigelson P, Beato M, Colman P, Kalimi M, Killewich L, Schutz G (1975) Studies on the hepatic glucocorticoid receptor and on the hormonal modulation of specific mRNA levels during enzyme induction.   Rec Prog Horm Res 31:213.
3. Yamamoto KR, Alberts BM (1976). Steroid receptors: Elements for modulation of eukaryotic transcription. Annu Rev Biochem 45:721.
4. Baumann H, Firestone GL, Burgess TL, Gross KW, Yamamoto KR, Held WA (1983). Dexamethasone regulation of α_1-acid glycoprotein and other acute phase reactants in rat liver and hepatoma cells.   J Biol Chem 258:563.
5. Koj A (1974). Acute phase reactants.   In Allison AC (ed): "Structure and Function of Plasma Proteins", London: Plenum, p. 73.
6. Vannice JL, Ringold GM, McLean JW, Taylor JM (1983). Induction of the acute-phase reactant, alpha-1-acid glycoprotein, by glucocorticoids in rat hepatoma cells. DNA 2:205.
7. Snyder S, Coodley EL (1976). Inhibition of platelet aggregation by α_1-acid glycoprotein. Arch Intern Med 136:778.
8. Schmid K, Kaufman H, Isemura S, Bauer F, Emura J, Motoyama T, Ishiguro M, Nanno S (1973). Structure of α_1-acid glycoprotein. The complete amino acid sequence, multiple amino acid substitutions, and homology with the immunoglobulins. Biochemistry 12:2711.
9. Reinke R, Feigelson P (1985) Rat α_1-acid glycoprotein: Gene sequence and regulation by glucocorticoids in transfected L-cells.   J Biol Chem 260:4397.
10. Vannice JL, Taylor JM, Ringold GM (1984).   Glucocorticoid-mediated induction of α_1-acid glycoprotein: Evidence for hormone-regulated RNA processing. Proc Natl Acad Sci USA 81:4241.

11. Kulkarni AB, Reinke R, Feigelson P (1985). Acute phase mediators and glucocorticoids elevate α_1-acid glycoprotein gene transcription. J Biol Chem 260:15386.

12. Gorman C, Moffat L, Howard B (1982). Recombinant genomes which express chloramphenicol acetyl-transferase in mammalian cells. Mol Cell Biol 2:1044.

13. Southern PJ, Berg P (1982) Transformation of mammalian cells to antibiotic resistance with a bacterial gene under control of the SV40 early region promoter. J Mol Appl Genet 1:327.

14. Vannice JL, Grove JR, Ringold GM (1983) Analysis of glucocorticoid-inducible genes in wild-type and variant rat hepatoma cells. Mol Pharmacol 23:779.

15. Graham FL, Van der Eb AJ (1973). A new technique for the assay of infectivity of human adenovirus 5 DNA. Virology 52:456.

16. The construct pAGP-β-globin was derived from the plasmid pSBeta, which was a gift from Dr. Andrew Buchmann. The AGP promoter fragment −763 to +20 was inserted into pSBeta in place of the SV40 early region promoter.

17. Chen CC, Feigelson P (1979) Cycloheximide inhibition of hormonal induction of α_{2u}-globulin mRNA. Proc Natl Acad Sci USA 76:2669.

18. Chen C-LC, Feigelson P (1980) Hormonal control of α_{2u}-globulin synthesis and its mRNA in isolated hepatocytes. Ann NY Acad Sci 349:28.

Transcriptional Control Mechanisms, pages 357–368
© 1987 Alan R. Liss, Inc.

CONTROL OF TRANSCRIPTION OF THE
OVALBUMIN GENE IN VITRO: A TRANSCRIPTION
FACTOR BINDS TO AN UPSTREAM PROMOTER ELEMENT[1]

Martine Pastorcic, Heng Wang, Sophia Y. Tsai,
Ming-Jer Tsai, and Bert W. O'Malley

Department of Cell Biology, Baylor College
of Medicine, Houston, Texas 77030

ABSTRACT We have used an in vitro system including
HeLa cell extracts to characterize the transcription
factors and promoter elements involved in initiation
of transcription of the ovalbumin gene. By 5'-deletion
mapping we demonstrate that a promoter element located
about 80 bp upstream from the cap site is required for
quantitative transcription. The enhancement of transcription conferred by the -80 region appears to be
specifically competed in trans by DNA fragments
containing this portion of upstream sequence.
Furthermore, DNase I footprinting analysis identifies
a protein binding over the -80 region in both crude
nuclear extracts and a partially purified fraction
containing a transcription factor. The upstream
control region thus defined includes a direct repeat
of the sequence GTCAAA which is homologous but not
identical to the CAAT-box consensus. We propose that
a transcription factor binds to the -80 region and
activates transcription of the ovalbumin gene.

INTRODUCTION

The efficiency of transcription of many mRNA coding
genes by RNA polymerase II depends on promoter elements

[1]This work was supported by grants from the National
Institutes of Health (HD-08188 and HD-07495 to BWO and HD-
17379 to MJT).

located upstream from the TATA-box. The evidence for such promoter elements was provided by the study of the effects of mutations in transcription assays in vivo and in vitro (1,2). These upstream elements appear to be the target sites for transcription factors (3-13). Sequence motives that are conserved upstream from many eucaryotic genes such as the GGGCGG-box (2,13) and the CAAT-box (11,12) have been found to be present in some of these promoter elements. But the functional relationships between the transcription factors binding to corresponding homologies upstream from different genes have been poorly documented.

The expression of the chicken ovalbumin gene in transfected HeLa cells is markedly affected by deletion of DNA sequences between -95 and -77 bp upstream from the cap site (14). As a first step in the molecular analysis of the mechanisms of transcription initiation of the oval-bumin gene, we describe an in vitro system including crude HeLa cell extracts which defines sequence requirements very similar to those initially identified in vivo. We present data to suggest that the "-80 region" is the binding site for a transcription factor required for the upstream sequence dependent activation of transcription.

RESULTS

Efficient Transcription of the Ovalbumin Gene In Vitro Requires DNA Sequences Upstream from the TATA-Box.

The ovalbumin gene template used in transcription assays is described in Fig. 1. Nuclear (15) or whole-cell (16) extracts were the source of RNA polymerase and transcription factors. Transcription was performed in the conditions described in Fig. 2 (17), and the RNA products were purified and analyzed by RNase A mapping (18) using the RNA probes indicated in Fig. 1 to monitor transcripts from the ovalbumin promoter and from the SV40 early promoter as an internal control.

Comparison of the template efficiency of the various 5'-deletion mutants assayed (indicated above the lanes) reveals particularly the importance of sequences between -95 and -77 in transcription with both nuclear (Fig. 2) and whole-cell extracts (Table 1). Thus we delineated an upstream promoter element between -95 and -77. This sequence interval contains one of the 2 direct repeats of the sequence GTCAAA which shares partial homology to the

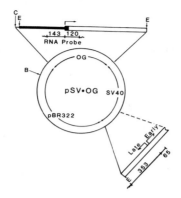

Figure 1. The ovalbumin gene and the promoter mutants. Chicken ovalbumin sequences (■) were fused to chicken β-globin sequences (□) in this ovalglobin (OG) construct (14). ⟶ direction of transcription. 5'-deletions were derived from pSV.OG by Bal 31 digestion from the Cla I site. The RNA probes used to map the products of <u>in vitro</u> transcription are indicated. They were synthesized <u>in vitro</u> using the Sp6 transcription system (17). C: Cla I, E; Eco RI, B: Bam HI.

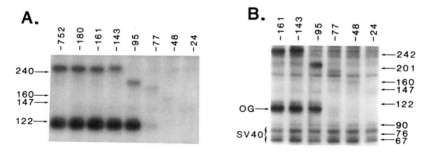

Figure 2. Functional mapping of the ovalbumin promoter <u>in vitro</u>. The end-points of the 5'-deletions assayed are indicated above the lanes. Transcription assays were carried out in 100 μl and contained 1 to 2 μg of circular DNA template, 500 μg of nuclear extract (17), 12 mM Hepes (pH 7.9), 3.75 mM $MgCl_2$, 60 mM KCl, 1 mM DTT, 0.1mM EDTA, 12% glycerol, and 500 μM each of ATP, GTP, UTP, CTP. The RNA transcripts were purified and analyzed by RNase A mapping (17). A: ovalbumin specific transcripts (120 NT). B: ovalbumin (OG) and SV40 early transcripts.

TABLE I
RELATIVE TRANSCRIPTION EFFICIENCY OF VARIOUS
5'-DELETION MUTANTS OF THE OVALBUMIN PROMOTER[a]

	-161/-143	-95	-77	-48	-24
NE	100	80	<10	ND	ND
WCE	100	60	25	15	ND

[a]Transcription was quantified by densitometric scanning of low-exposure autoradiograms. ND indicates that transcription was not detectable.

CAAT-box consensus (Fig. 6). The quantitative differences in the effects of the deletions between nuclear and whole-cell extracts are discussed below.

Specific Competition of In Vitro Transcription by DNA Fragments Containing the -80 Region.

To assess if the cis-acting element delineated by 5'-deletion mapping is the binding site for a transcription factor we carried out transcription competition assays. Competitors were blunt-ended DNA fragments of ovalbumin sequences from position -753 to various 3'-ends (-125, -89, -73, -56 or -44). The 3'-extremity of each fragment was linked to 41 bp of pBR322 (17). The presence of sequences between -89 and -73 in the competitors appears to increase significantly their ability to compete for transcription (Fig. 3A). This suggests that a transcription factor indeed binds to the -80 region. Some evidence that this transcription factor might be specifically required for the upstream sequence dependent activation of transcription comes from the following two results. First, transcription of the oval-bumin template retaining only 48 bp of upstream sequence (pSV.OG 5' Δ -48) was not differentially competed by fragments with end-points at -125 or -44 (Fig. 3B). In the same conditions, transcription of the template retaining 161 bp of flanking sequence was markedly reduced by the competitor containing sequences to -44 to a level indistinguishable from the transcription efficiency observed on pSV.OG (5' Δ −48). Second, we monitored SV40 transcription in the same assay conditions and we found that the presence of the −80 region on the competitors did not affect transcription (Fig. 3C).

If sequences which are critical for the binding of a transcription factor are located between −89 and −73, one should note that the most effective competitor contains sequences farther downstream: to −44. Thus sequences between −56 and −44 could influence the binding of the transcription factor recognizing the −80 region or contribute to the binding of another transcription factor.

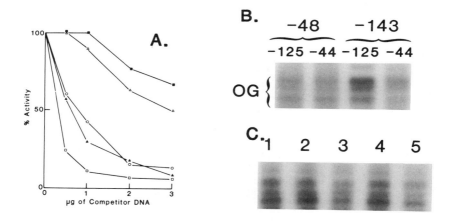

Figure 3. Competition of in vitro transcription by ovalbumin upstream sequences. A: 0.5 to 3 µg of competitor DNA fragments were incubated for 20 minutes on ice in 1.1x transcription conditions with 300 µg of nuclear extract. 1 µg of pSV.OG (5' Δ −161) template was then added and further incubated for 30 minutes at 30°C. 0.5, 1, 2, 3 ug of competitor DNA (■ :−125, Δ :−89, 0:−73, ▲ :−56, □ −44) correspond respectively to 5, 10, 20 and 30 fold molar excess of promoter sequences. B: transcription of pSV.OG (5' Δ −48) and (5' Δ −143) in the presence of 3 µg of competitor with end-point at −125 or −44. C: Competition of SV40 transcription. 0 (lane 1), 1 (lanes 2 and 4) or 3 µg (lanes 3 and 5) of competitor with end-point at −44 (lanes 2 and 3) or −125 (lanes 4 and 5) were added. Ovalbumin (A, B) or SV40 (C) transcription was monitored by RNase A mapping.

A Factor Binding at the −80 Upstream Promoter Element.

DNase I protection experiments (19) detect a DNA-binding activity in the −80 region in nuclear extracts on both coding and non-coding strands. The protected interval is located approximately from −70 to −90 (17).

We fractionated the crude nuclear extract succes-sively on DEAE-sephadex and phosphocellulose columns as previously described by Tsai et al. (20). We found most of the activity binding to the −80 region in the P1000 fraction. The binding site appears between −66 and −89 with two slightly enhanced cleavage sites at −66 and −68.

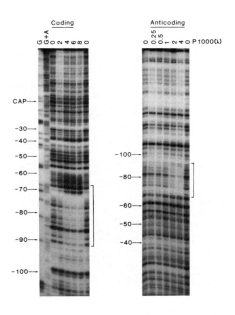

Figure 4. DNase I footprinting of the ovalbumin promoter with the P1000 fraction. Binding assays were in 1.5 x transcription conditions, at 22° for 10'. 20 µl reactions contained 400 ng of pBR322, 7 ng of end-labelled probe and various amounts of P1000 which contained 2 mg of protein/ml. A and G+A Maxam and Gilbert sequencing ladders and the corresponding positions on the ovalbumin promoter are indicated. The DNA probes −161 to +120 (coding strand) or −296 to +40 (non coding strand) were labeled by ^{32}P respectively at position −161 and +40.

In a reconstituted transcription system using partially purified fractions instead of crude extracts, transcription efficiency also depends on upstream sequences. The P1000 fraction is absolutely required for transcription. Thus transcription activity correlates with the DNA binding activity through these initial purification steps.

Competition of the −80 Binding Activity by Fragments of Ovalbumin Upstream Sequences.

We used the same fragments tested in transcription competition assays to assess the specificity of the DNase I footprint over the upstream promoter element (Fig. 5). We find that the fragments with end-points at −125 or

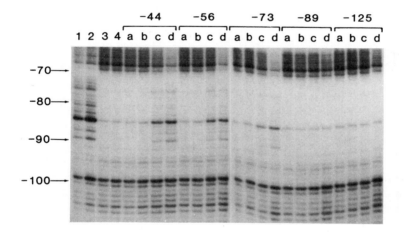

Figure 5. Competition of the DNA-binding activity over the −80 region by fragments of ovalbumin promoter. Binding was carried out in transcription conditions with 1 ng of probe, 500 ng of pBR322, 50 μg of nuclear extract and various amounts of competitors (−44, −56, −73, −89 and −125) in 20 μl reaction, lanes a: 10 ng, lanes b: 30 ng, lanes c: 100 ng, lanes d: 300 ng, lanes 1 and 2: no nuclear extract added, lanes 3 and 4: no competitor. Binding (20 minutes) and DNase I digestion (30 seconds) were carried out at 22°C. The DNA probe (−143 to +120) was end-labeled by ^{32}P on the coding strand at −143.

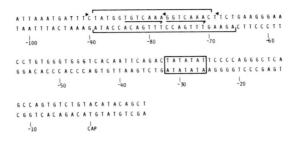

Figure 6. DNA sequence of the ovalbumin promoter and summary of the footprinting experiments. DNase I footprints are indicated by brackets. Arrowheads show barriers to exonuclease digestion on the non coding strand (27). Horizontal arrows underline the direct repeats of the CAAT-box homology (21).

-89 showed no competition, whereas fragments ending at -73, -56 or -44 were efficient competitors. This demonstrates the specificity of the DNA-binding activity to the -80 region of the promoter and correlates well with the results obtained in transcription competition assays.

Note that in the transcription assays the fragment with an end-point at -44 was a more effective competitor than the fragments with end-points at -56 or -73. However, in the footprinting assays the 3 competitors appear to have the same efficiency. This suggests the possibility of a binding site for another transcription factor between -44 and -56. The absence of differential effect of the -44 and -125 competitors on the transcription of the SV40 or the ovalbumin template retaining only 48 bp of upstream sequence suggests that this factor could also be specifically involved in the upstream sequence dependent activation of transcription of a class of genes containing the ovalbumin gene.

DISCUSSION

The data presented in this report suggest that the efficient transcription of the ovalbumin gene in vitro requires the binding of a transcription factor to an upstream promoter element.

By 5'-deletion mapping we partially delineated a

promoter element between −95 and −77 bp upstream from the cap site. Transcription competition experiments indicate that it is actually the binding site for a factor selectively required for the upstream sequence dependent activation of transcription of the ovalbumin gene.

A factor binding over the −80 region can be detected in crude nuclear extracts by DNase I footprinting. We can correlate this binding activity with transcriptional activity through two initial purification steps by chromatography on DEAE and phosphocellulose. We have recently purified further the DNA-binding activity through sephacryl-S300 and heparin sepharose. We find that the fractions containing the activity are also required for transcription in reconstituted extracts. But clearly, the unequivocal identification of the DNA-binding activity with a transcription factor requires purification of the protein to homogeneity.

Our 5'-deletion mapping indicates the importance of the −80 region for transcription in both nuclear extracts and whole-cell extracts. Although the same sequences affect transcription with both extracts, there are quantitative differences between the two assays. In particular, the effect of sequences between −143 and −95 is more apparent in whole-cell extracts than in nuclear extracts. Similarly, the data obtained by transfecting HeLa cells with the same plasmids also indicate a 20 to 40% decrease in ovalbumin gene expression upon deletion of sequences between −143 and −95. We do not know the reason for those quantitative differences. However, they might be most simply explained if the interaction of the transcription factor with the −80 region is influenced by sequences upstream from −95. The concentration of transcription factor might be high enough to obscure the effects of sequences upstream from −95 in nuclear extracts but not in whole-cell extracts or in vivo. This hypothesis is consistent with the overall reduction of transcription following deletion of sequences between −143 and −77 being greater with nuclear extracts than with whole-cell extracts.

A direct repeat of the sequence GTCAAA is present between −71 and −83. This sequence shares limited homology with the CAAT-box consensus (21). The importance of the CAAT-box in transcription assays in vivo (23,24), in vitro (12) as well as in the developmental expression of the human _Aγ- globin gene' (25,26) has been documented. The homology between the −80 region and the

CAAT-box was originally reported by Benoist et al. (21). However the GTCAAA sequence diverges significantly from the general GG(T/C)CAATC consensus. It is possible that the ovalbumin -80 region represents a novel transcription control region. It is also possible that the homology between the -80 region and more conserved CAAT-boxes reflects a relatedness between members of a larger family of regulatory proteins.

In a previous report using exonuclease footprinting we described a DNA-binding activity over the CAAT-box region of the ovalbumin gene present in the P1000 fraction from HeLa whole-cell extracts (27). The same protein fraction also contains a DNA-binding activity to the β-globin CAAT-box. The borders of the binding site identified in the ovalbumin promoter by exonuclease footprinting coincide with the data obtained by DNase I protection experiments. At present, however, we have no evidence to suggest that the ovalbumin and β-globin CAAT-box binding activity(s) are identical. To answer some of these questions we are presently analysing the effects of competitor DNA fragments from various promoters on the transcription of the ovalbumin gene and on the specific binding to the -80 region. We are also purifying the binding activity to the -80 region of the ovalbumin gene from HeLa cells. However, the transcription factor which is able to activate transcription of the ovalbumin gene in a heterologous HeLa cell system might be different from the factor present in oviduct cells where ovalbumin gene expression is hormonally regulated. For this reason, we are purifying also the -80 promoter binding activity from chicken oviduct cells.

REFERENCES

1. Manley JL (1983). Analysis of the expression of genes encoding animal mRNA by in vitro techniques. Prog Nucleic Acids Res and Mol Biol 30:195.
2. Dynan WS, Tjian R (1985). Control of eucaryotic messenger RNA synthesis by sequence-specific DNA-binding proteins. Nature (London) 316:774.
3. Miyamoto NG, Moncollin V, Wintzerith M, Hen R, Egly JM, Chambon P (1984). Stimulation of in vitro transcription by the upstream element of the adenovirus − 2 major late promoter involves a specific factor. Nucleic Acids Res 12:8779.

4. Miyamoto NG, Moncollin V, Egly JM, Chambon P (1985). Specific interaction between a transcription factor and the upstream element of the adenovirus-2 major late promoter. Embo J 4:3563.

5. Parker CS, Topol J (1985). A Drosophila RNA polymerase II transcription factor specific for the heat-shock gene binds to the regulatory site of an hsp 70 gene. Cell 37:273.

6. Topol J, Ruden DM, Parker CS (1985). Sequences required for in vitro transcriptional activation of a drosophila hsp 70 gene. Cell 42:527.

7. Sawadogo M, Roeder R (1985). Interaction of a gene-specific transcription factor with the adenovirus major late promoter upstream of the TATA box region. Cell 43:165.

8. Carthew RW, Chodosh LA, Sharp PA (1985). An RNA polymerase II transcription factor binds to an upstream element in the adenovirus major late promoter. Cell 43:439.

9. Capasso O, Heintz N (1985). Regulated expression of mammalian histone H4 genes in vivo requires a trans-acting transcription factor. Proc Natl Acad Sci USA 82:5622.

10. Heberlein U, England B, Tjian R (1985). Characterization of Drosophila transcription factors that activate the tandem promoters of the alcohol dehydrogenase gene. Cell 41:965.

11. Jones KA, Yamamoto KR, Tjian R, (1985). Two distinct transcription factors bind to the HSV thymidine kinase promoter in vitro. Cell 42:559.

12. Graves BJ, Johnson PF, McKnight SL (1986). Homologous recognition of a promoter domain common to the MSV LTR and the HSV tk gene. Cell 44:565.

13. Dynan WS, Sazer S, Tjian R, Schimke RT (1986). Transcription factor Spl recognizes a DNA sequence in the mouse dihydrofolate reductase promoter. Nature (London) 319:246.

14. Knoll BJ, Zarucki-Schulz T, Dean DC, O'Malley BW (1983). Definition of the ovalbumin gene promoter by transfer of an ovalglobin fusion gene into cultured cells. Nucleic Acids Res 11:6733.

15. Dignam JD, Lebovitz RM, Roeder RG (1983). Accurate transcription initiation by RNA polymerase II in a soluble extract from isolated mammalian nuclei. Nucleic Acids Res. 11:1475.

16. Manley JL, Fire A, Cano A, Sharp PA, Gefter ML

(1980).   DNA-dependent   transcription   of   adenovirus genes in a soluble whole cell extract.   Proc Natl Acad Sci USA 77:3855.

17. Pastorcic M, Wang   H, Elbrecht A, Tsai SY, Tsai MJ, O'Malley BW (1986).   Mol Cell Biol. In press.

18. Zinn R, Di Maio D, Maniatis T (1983).   Identification of two distinct regulatory regions adjacent to the human β-interferon gene.   Cell 34:865.

19. Galas D, Schmitz A (1978).   DNase I footprinting: a simple method for the detection of protein-DNA binding specificity.   Nucleic Acids Res 5:3157.

20. Tsai SY, Tsai MJ, Kops LE, Minghetti PP, O'Malley BW (1981).   Transcription factors from oviduct and HeLa cells are similar.   J Biol Chem 255:13055.

21. Benoist   C,   O'Hare   K,   Breathnach   R,   Chambon   P (1980) The   ovalbumin   gene   sequence   of   putative control regions.   Nucleic Acids Res 8:127.

22. Efstratiadis A,   Posakony JW,   Maniatis T,   Lawn RM, O'Connell C, Spritz RA, DeRiel JK, Forget BG, Slightom L, Blechl AE, Smithies O, Baralle FE, Shoulder CC, Proudfoot NJ (1980).   The structure and evolution of the β-globin gene family.   Cell 21:653.

23. Dierks P, van Ooyen A, Cochran MD, Dobkin C, Reiser J, Weismann C (1983).   Three regions upstream from the cap site are required for efficient and accurate transcription of the rabbit β-globin gene in mouse 3T6 cells.   Cell 32:695.

24. Grosveld GC, Rosenthal A, Flavell RA (1982).   Sequence requirements for the transcription of the rabbit β-globin gene in vivo: the -80 region.   Nucleic Acids Res 10:4951.

25. Gelinas   R,   Endlich   B,   Pfeiffer   C,   Yagi   M, Stamatoyannopoulos C (1985). $_A$G to A substitution in the distal CCAAT box of the ^{A}γ- globin gene in Greek hereditary persistence of fetal haemoglobin.   Nature (London) 313:323.

26. Collins FS, Metherall JE, Yamakawa M, Pan J, Weissman SM, Forget B (1985).   A point mutation in the ^{A}γ - globin gene promoter in Greek hereditary persistence of fetal haemoglobin.   Nature (London) 313:325.

27. Elbrecht A, Tsai SY, Tsai MJ, O'Malley BW (1985). Identification by exonuclease footprinting of a distal promoter-binding protein from HeLa cell extracts.   DNA 4:233.

Transcriptional Control Mechanisms, pages 369–378
© 1987 Alan R. Liss, Inc.

REGULATION OF NEUROENDOCRINE GENE EXPRESSION

Michael G. Rosenfeld,[1] Stuart Leff,[1] Andrew Russo,[1]
E. Bryan Crenshaw III,[1] Christian Nelson,[1]
Rodrigo Franco,[1] Sergio Lira,[1] and Ronald M. Evans[2]

[1]Eukaryotic Regulatory Biology Program
School of Medicine
University of California, San Diego

[2]Molecular Biology and Virology Laboratory
The Salk Institute
La Jolla, CA 92138

ABSTRACT The precisely restricted temporal and spatial
patterns of developmental expression of genes of the
neuroendocrine system are controlled by the cell-specific
expression of factors which involve functional activity,
in vivo and in vitro, to cis-active genomic sequences of
their cognate genes. Thus, specific sequences in the 5'
flanking region of the prolactin and growth hormone genes
confer pituitary cell type-specific expression, and each
appears to require several trans-acting factors to effect
their actions. A second type of developmental expression
is exemplified by expression of the calcitonin/CGRP gene,
where a neuron-specific factor dictates the pattern of
exon splicing and poly(A) site utilization, resulting in
a tissue-specific pattern of alternative RNA processing
and neuropeptide biosynthesis.

A. RNA PROCESSING REGULATION IN DEVELOPMENT

Molecular cloning of DNA complementary to rat calcitonin
mRNA predicted the structure of the protein precursor to the
32 amino acid calcium-regulating hormone, calcitonin (1,2).
Proteolytic processing of the precursor was predicted to
generate an 82 amino acid N'-terminal peptide and a 16 amino
C' terminal calcitonin cleavage product as well as calcitonin
in thyroid C cells. The production of a discrete calcitonin-

related mRNA was first noted during the spontaneous "switching" of serially transplanted rat medullary thyroid carcinomas (MTCs) from states of "high" to "low" or absent calcitonin production (3,4). The unexpected explanation for the "switch" was that the calcitonin gene generated a structurally distinct transcript, referred to as calcitonin gene-related product mRNAs (CGRP mRNA) which contained no calcitonin coding information (4,5). Isolation and sequence of the calcitonin genomic DNA and calcitonin and CGRP cDNAs proved that both CGRP and calcitonin mRNAs arose by differential RNA processing from a single genomic locus (4,5). This model was established by a complete analysis of the two mature mRNA transcripts and of the calcitonin gene. Examination of the sequence of the two gene products reveals that CGRP and calcitonin mRNAs share sequence identity through nucleotide 227 of the coding region, predicting that the initial 72 N' terminal amino acids of each precursor are identical, but then diverge entirely in nucleotide sequence, encoding unique C'-terminal domains (4). Protein processing signals within the C'-terminal region of CGRP predict the excision of a 37 amino acid polypeptide containing a C'-terminal amidated phenylalanine residue (4). Based upon the structure of the calcitonin-CGRP gene, production of calcitonin mRNA involves splicing of the first three exons, present in both mRNAs, to the fourth exon, which encodes the entire calcitonin/CCP sequence. Alternative splicing of the first three exons to the fifth and sixth exons, which contain the entire CGRP coding sequence and the 3' non-coding sequences, respectively, results in production of CGRP mRNA. In this case the fourth exon is excised along with the flanking intervening sequences.

CALCITONIN GENE RELATED PEPTIDE IS THE PRODUCT OF
CALCITONIN/CGRP EXPRESSION IN THE BRAIN

To establish the physiological production of CGRP, which had been predicted as a result of recombinant DNA analysis in the absence of prior knowledge of either its structure or function, an approach was used which could be applied to study of the large number of unknown and unsuspected neuropeptides which are likely to be discovered based on the application of current recombinant DNA techniques. Evidence that CGRP mRNA was physiologically expressed was initially provided by demonstration that CGRP exon-specific probes hybridized to poly(A)-rich RNA prepared from rat trigeminal ganglia and hypothalamus (5). A strategy which combined histochemical and molecular biological approaches was then employed to evaluate the potential production of the predicted peptide in the brain and to determine the precise sites of its synthesis. Using an antisera generated against a synthetic peptide corresponding to the fourteen C'-terminal amino acids of CGRP, immunoreactive CGRP was identified in a unique distribution in a large number of cell groups and pathways in the central nervous system, distinct from that of any known neuropeptide (5). S_1 nuclease protection assay, mRNA-directed cell-free translation, and cDNA clonal sequence analyses confirmed the production of bona fide CGRP mRNA in the brain, and identified the sites of its biosynthesis. Gel filtration analysis of brain immunoreactive peptide suggests that this precursor is processed in brain to generate the predicted peptide product (CGRP) (5), and primary cultures of rat trigeminal ganglia appear to secrete authentic CGRP peptide (6). Tissue specificity of the RNA processing events is suggested because virtually no calcitonin mRNA could be identified in the rat brain (5), while in thyroid C cells, calcitonin and CGRP mRNAs (and calcitonin and CGRP) are present in a ratio of approximately 95-98:1 (7). Small amounts of CGRP are present histochemically and by radioimmunoassay in thyroid C cells (7,8), and both calcitonin and CGRP can be co-produced within the identical cell (7).

The distribution of CGRP in pathways and neurons believed to serve specific sensory, integrative, and motor systems (5) suggests several possible physiological roles for the peptide. The localization of CGRP immunoreactivity in the olfactory and gustatory systems, including taste buds, the hypoglossal, facial, and vagal nuclei, and in the hypothalamic and limbic regions strongly suggests that it may have a functional effect in ingestive behavior (5). Additional studies have revealed the widespread presence of CGRP at neuromuscular junctions, including striated muscle (9) and skeletal muscles, the first peptide identified at neuromuscular junctions in mammalian species. CGRP is present in small trigeminal and spinal sensory ganglion cells, which are known to relay thermal and nociceptive information to the brainstem and spinal cord (5). In the spinal ganglia CGRP-positive cells represent 30-50% of the total population of small ganglion cells, a percentage significantly greater than that for any other neuropeptide. For example, neurons stained with substance P antisera co-localize to a subset of CGRP-containing small cells (9). CGRP is present in a subset of cells in one of the vagal motor nuclei (n. ambiguus) and is extraordinarily widely distributed in fibers in vascular musculature and in veins of virtually all organs, suggesting a role in cardiovascular homeostasis.

Finally, CGRP-containing nerve fibers are widely distributed to most other organ systems, located mainly in the adventitia around arteries and veins, and in sensory fibers of every tissue which contains sensory innervation, and these fibers derive from the sensory ganglia or dorsal root ganglia. Thus, CGRP is present in virtually all organ systems, including the genitourinary system (e.g., bladder trigone, vagina, prostate, etc.), gastrointestinal (esophagus, duodenum, ileum), pulmonary (bronchiolar cells) and dermal systems (e.g., sensory fibers penetrating epidermus, sweat glands, hair follicles). In fact, it is perhaps the most widely distributed of the known peptides. Consistent with certain features of its anatomical distribution, administration of synthetic rat CGRP produces a unique pattern of effects on blood pressure and catecholamine release in dogs and rats and produces gastric hypoacidity (10). CGRP is also widely distributed in the endocrine system, in a subset of adrenal medullary cells, in bronchiolar cells, intestinal cells, and in fiber baskets which innervate the pancreatic islets and, interestingly, in thyroid C cells (5,10).

IDENTIFICATION OF A SECOND CGRP-RELATED GENE EXPRESSED IN THE BRAIN

The application of recombinant DNA technology accelerated the process of identifying previously unsuspected neuropeptides, and revealed the existence of families of related genes encoding structurally similar, but not identical, peptide products. The existence of such gene families account, in fact, for the generation of diversity of putative regulatory peptides in the brain and endocrine system. On this basis, the possible existence of other gene products related to CGRP was investigated by screening libraries of chimaeric plasmids containing inserts complementary to mRNAs from rat medullary thyroid carcinomas with a clonal α-CGRP cDNA probe. This analysis resulted in the identification of a novel mRNA (β-CGRP mRNA) which is related to α-CGRP mRNA (11). The sequence of this mRNA reveals a 394 nucleotide open reading frame; the first 256 nucleotides encode an 82 amino acid NH_2 N'-terminal sequence common to the precursors of calcitonin and α-CGRP, with about 30% base (and amino acid) substitutions or additions when compared to α-CGRP mRNA. There are two sets of paired basic amino acid residues in this region, such that three peptides could potentially be generated from this region, although Arg.Lys sites are not as readily cleaved as Lys.Arg sites, and this might not be utilized. In contrast, there are less than 5% base substitutions in the next 114 nucleotides constituting the β-CGRP mRNA-coding domain when compared to the equivalent region of α-CGRP mRNA; the sequence predicts excision of a 37 amino acid peptide containing a cDNA terminal phenylalanine-amide, and differing by only a single amino acid (a Lys for Glu in position 35) from the primary sequence of α-CGRP. The 5' and 3' non-coding regions of α- and β-CGRP mRNAs diverge sufficiently to allow generation of mRNA-specific hybridization probes. Such probes were used to confirm the presence of two discrete genes encoding α- and β-CGRP mRNAs; the latter gene did not contain sequences hybridized to the calcitonin-coding sequences of calcitonin/CGRP gene.

The identification of β-CGRP mRNA in rat medullary thyroid tumors led to speculation concerning its potential physiological expression relative to α-CGRP. Analysis of RNA from various regions of the nervous system revealed that β-CGRP mRNA was present in the trigeminal ganglia, lateral medulla, hypothalamus and, to a lesser extent, midbrain (11). Based on this analysis, the relative distribution of β-CGRP is similar to α-CGRP, but generally present at less than 20% that of α-CGRP in the corresponding area. Similarly, β-CGRP mRNA is less than 20% α-CGRP mRNA and less than 1% calcitonin mRNA in thyroidal C cells. The detailed pattern of differentiated expression of α- and β-CGRP mRNAs was subjected to analysis using hybridization histochemistry methodology. Overall, the pattern of expre-sion corresponded to previous histochemical analysis of CGRP expression, including the third, fourth, fifth, seventh, tenth, and twelfth cranial nerves, and parabrachial and peripeduncular nuclei. While in many areas, α-CGRP probes recorded greater than 10-fold better signals than did β-CGRP probes, the hybridization signal for β-CGRP clearly exceeds that for α-CGRP in the nuclei of the third, fourth, and fifth cranial nerves, and recorded at least equivalent signals in several other areas. Based on similar distribution of the mRNAs, it is possible that they are expressed in identical neurons, but at variable levels. In some nuclei it is possible that β-CGRP might be exclusively expressed. The finding of a second rat gene related to the calcitonin/α-CGRP gene exemplifies the biological potential of expression of families of genes encoding related neuropeptides, generated as a consequence of gene duplication events. In the case of these two related genes, it will be important to determine whether they are independently regulated and/or functionally discrete, providing functional advantages for the expression of two such related gene products.

MECHANISM OF ALTERNATIVE RNA PROCESSING
IN CALCITONIN GENE EXPRESSION

Documentation that calcitonin and CGRP mRNAs shared an
identical transcriptional start (CAP) site was provided by
both S_1 nuclease protection and primer extension analyses (12).
Therefore, both RNAs are products of a single transcription
unit. These analyses identified a cryptic splice site generating a 24 nucleotide extension of the first exon in the case
of both calcitonin and CGRP mRNAs. The utilization of this
site appears to be a stochastic event; it occurs in the untranslated region of the message, and thus does not alter the
encoded protein products. The 3' ends of calcitonin and CGRP
mRNAs were determined by S_1 nuclease mapping experiments and
confirmed by identifying the appropriate polyadenylation sequences in rat calcitonin cDNA and genomic clones. The polyadenylation site of calcitonin mRNA appears to be 18 or 19
nucleotides 3' to a sequence AATAAA located 226 nucleotides
downstream of the calcitonin termination codon. CGRP mRNA
utilizes a recognition sequence, ATTAAA, and the CGRP poly(A)
addition signal and site are situated 1.9 kilobases downstream
from the analogous region for calcitonin mRNA, defining the
end of the large 3'-CGRP non-coding exon (the sixth genomic
exon). Thus, production of calcitonin and CGRP mRNAs is associated with the selective polyadenylation of transcripts at one
of two alternative poly(A) sites. Based on neuron transcription analysis, alternative RNA processing and not alternative
RNA transcriptional termination, is the regulated event.

The alternative poly(A) site choice could reflect the primary regulated event or could result from alternative splice
commitment involving exon 3 and 4 (calcitonin-coding) and 5
(CGRP-coding). In order to distinguish between these possibilities, an extensive series of mutated calcitonin/CGRP genes and
fusion genes containing the regions encompassing the poly(A)
cleavage sites were analyzed by DNA-mediated gene transfer in
cell lines making unambiguous RNA processing choices. Based
upon these analyses (S. Leff et al., unpublished), the evidence
strongly suggests that in GGRP-producing tissues, there is a
factor(s) which is required for the exon 3 to exon 5 splice
to occur, and that it is the commitment to this splicing event
which precedes and dictates the choice of poly(A) site. The
exon 3 to exon 4 splice appears to be permitted only when information beyond the calcitonin poly(A) site is removed from the
transcript. We therefore propose a model in which splice commitment dictated by a factor(s) active or/and present only in

CGRP-producing cells and some other neuronal cell types, precedes and determines the exon 3 to exon 5 splice and the poly-(A) cleavage site. Such a factor is likely to be similar or identical to that which regulates alternative splicing events in other neuroendocrine gene transcripts such as the substance P gene or in expression of other gene families. One of the most curious aspects of calcitonin/CGRP gene expression is that, although only a subset of neurons appear to possess the CGRP splice commitment machinery, the gene is expressed in the brain only in those neurons in which this machinery is extant. The biochemical nature of this neuron-specific factor(s) remains enigmatic. Introduction of a metallothionine, calcitonin fusion gene into transgenic animals further supported the model that calcitonin mRNA represents the "null" choice, with only subsets of neurons containing the splice commitment machinery being competent to produce CGRP mRNA. A fusion gene containing the mouse metallothionine I promoter CAP site, fused to the first exon of the calciotnin/CGRP gene was introduced into fertilized mouse eggs to target expression of the gene in tissue not normally expressing the endogenous calcitonin/CGRP gene. The pattern of calcitonin and CGRP mRNA expression was analyzed in the resultant transgenic offspring. Thus, in three pedigrees analyzed, calcitonin mRNA was the predominant mature transcript in all tissues except neuronal tissue. Most, but not all, of the areas in which the fusion gene was expressed in the brain made a clear processing choice, producing CGRP mRNA (E.B. Crenshaw III, A.F. Russo, et al.).

B. DEVELOPMENTAL REGULATION OF NEUROENDOCRINE GENE EXPRESSION BASED ON HERITABLE PATTERNS OF CELL-SPECIFIC GENE TRANSCRIPTION

Cis-active regulatory sequences, referred to as enhancers, in viral and cellular genes appear to stimulate the transcription of many eukaryotic promoters in a position- and orientation-independent fashion, conferring cell-specific or regulated patterns of gene expression in transfected cells. Genomic regions of several mammalian and Drosophila genes which include the putative enhancer elements can direct a restricted, apparently tissue-specific pattern of gene expression in transgenic animals. Based on footprinting and in vitro transcriptional analyses, it is suggested that enhancers may exert their effects as a consequence of binding trans-acting factors.

The anterior pituitary gland develops from a common primordium to produce phenotypically distinct cell types, each of which expresses one of at least six discrete trophic hormones.

Two of these are the structurally-related prolactin and growth
hormone genes, which are evolutionarily derived from a single
primordial gene; these related peptide hormones are produced
in discrete cell types, referred to as lactotrophs and somatotrophs, respectively, and their expression is apparently limited to the pituitary gland. Growth hormone and prolactin are
temporally the last pituitary hormones expressed. Rat growth
hormone is initially detected at fetal day 17-18, while prolactin appears at or following birth, with the percentage of
lactotrophs rising dramatically 3-5 days postpartum to constitute approximately 10% of total cells. Because it has been
reported that growth hormone (GH) and prolactin are initially
co-produced within single cells prior to the appearance of
lactotrophs, it is possible that lactotrophs may arise developmentally from somatotrophs.

In order to evaluate whether these two closely-related
neuroendocrine genes are developmentally controlled by different factors, interacting with unique structural sequences
in each gene or are controlled by a single or common factor(s),
the potential cis-active elements dictating cell-specific expression of each gene was examined. Regions of 250 and 150
base pairs were identified 2.7 and 0.2 kb 5' of the transcription start site of the rat prolactin and growth hormones,
respectively, which act in a position- and orientation-fashion
to transfer cell-specific expression to heterologous genes.
These sequences permit expression in rat cell lines of pituitary origin which expresses the endogenous prolactin and growth
hormone genes, but fail to exert any effects in any other cell
cell line tested, including those of endocrine origin (thyroid
C cells, pancreaic B cells, ovarian cells). Discrete and
separable sequences were identified in the prolactin gene enhancer, each required but insufficient alone for biological
activity. The cis-active elements appear to act developmentally in vivo, because prolactin 5' sequences target lacto-
troph-specific expression of SV40 T antigen fusion genes in
transgenic mice (13)   A lactotroph cell line derived from
these transgenic animals contains the putative trans-active
factors to permit functional action of the prolactin, but not
the growth hormone, enhancer element.

C. CONCLUSION

The developmental code for neuroendocrine gene expression depends upon cell-specific expression of trans-acting factors which bind to specific genomic sequences or regions of the primary transcript to effect the transcriptional and RNA processing regulation characteristic of the particular cell type. Defining the molecular and functional properties of these critical regulatory factors is required for further understanding of the molecular basis of tissue-specific gene expression.

REFERENCES

1. Rosenfeld, MG, Amara, SG, Roos, BA, Ong, ES, and Evans, RM (1981). Altered expression of the calcitonin gene associated with RNA polymorphism. Nature 290:63.
2. Amara, SG, Jonas, V, O'Neil, JA, Vale, W, Rivier, J, Roos, BA, Evans, RM, and Rosenfeld, MG (1982). Calcitonin COOH-terminal cleavage peptide as a model for identification of novel neuropeptides predicted by recombinant DNA analysis. J Biol Chem 257:2129.
3. Rosenfeld, MG, Lin, CR, Amara, SG, Stolarsky, LS, Ong, ES, and Evans, RM (1982). Calcitonin mRNA polymorphism: peptide switching associated with alternative RNA splicing events. Proc Natl Acad Sci USA 79:1717.
4. Amara, SG, Jonas, V, Rosenfeld, MG, Ong, ES, and Evans, RM (1982). Alternative RNA processing in calcitonin gene expression generates mRNAs encoding different polypeptide products (1982). Nature 298:240.
5. Rosenfeld, MG, Mermod, J-J, Amara, SG, Swanson, LW, Sawchenko, PE, Rivier, J, Vale, WW, and Evans, RM (1983). Production of a novel neuropeptide encoded by the calcitonin gene via tissue-specific RNA processing. Nature 304:129.
6. Mason, RT, Peterfreund, RA, Sawchenko, PE, Corrigan, AZ, Rivier, JE, and Vale, WW (1984). Release of the predicted calcitonin gene-related peptide from cultured rat trigeminal ganglion cells. Nature 308:653.
7. Sabate, MI, Stolarsky, LS, Polak, JM, Bloom, SR, Varndell, IM, Ghatei, MA, Evans, RM, and Rosenfeld, MG (1985). Regulation of neuroendocrine gene expression by alternative RNA processing: co-localization of calcitonin and calcitonin gene-related peptide (CGRP) in thyroid C-cells. J Biol Chem 260:2589.

8. Tschopp, FA, Tobler, PH, and Fisher, JA (1984). Calcitonin gene-related peptide in the human thyroid, pituitary,
and brain. Mol Cell Endocrinol 36:53.
9. Gibson, SJ, Polak, JM, Bloom, SR, Sabate, IM, Mulderry,
PM, Ghatei, MA, McGregor, GP, Morrison, JFB, Evans, RM,
and Rosenfeld, MG (1984). Calcitonin gene-related peptide
immunoreactivity in the spinal cord of man and of eight
other species. J Neuroscience 4:3101.
10. Lenz, HJ, Mortrud, MT, Rivier, JE, Brown, MR, and Vale, WW
(1984). Calcitonin gene related peptide inhibits basal,
pentagastrin, histamine. and bethanecol stimulated secretion. Gut 26:550.
11. Hammer, R, Brinster, RL, Rosenfeld, MG, Evans, RM, and
Mayo, KE (1985). Expression of human growth hormone
releasing factor in transgenic mice results in increased
somatic growth. Nature 315:413.
12. Miller, AD, Ong, ES, Rosenfeld, MG, Verma, IM, and Evans,
RM (1984). Infectious and selectable retrovirus containing an inducible rat growth hormone minigene. Science
225: 994.
13. Nelson, C, Crenshaw, EB, III, Franco, R, Lira, SA, Evans,
RM, and Rosenfeld, MG (1986). Discrete cis-active genomic
sequences dictate the pituitary cell type-specific expression of the rat prolactin growth hormone gene. Nature
322:557.

Transcriptional Control Mechanisms, pages 379–381
© **1987 Alan R. Liss, Inc.**

CELL TYPE SPECIFIC EXPRESSION OF THE RAT GROWTH HORMONE GENE

Rodrigo Franco,[1] Christian Nelson,[1] Sergio Lira,[2]
R.M. Evans,[3] and M.G. Rosenfeld[2]

[1]Department of Biology
[2]Eukaryotic Regulatory Biology Program
University of California at San Diego, M-013
La Jolla, California 92093

ABSTRACT Cell type specific expression of the rat
growth hormone gene is regulated by an enhancer
located 48 base pairs upstream of the transcriptional
CAP site.

INTRODUCTION

Several eukaryotic genes contain transcriptional
enhancer sequences which dictate a cell specific pattern of
expression in transfected cells (1,2,3,4,5). The biological
importance of enhancer sequences for proper cell type expression is suggested by the observation that genomic regions
containing putative enhancers direct a tissue specific
pattern of gene expression in transgenic animals (6,7,8,9,10,
11,12,13,14). The rat growth hormone gene (RGH) is almost
exclusively expressed in the somatotrophs of the anterior
pituitary, making this gene a useful model for the study of
tissue specific gene expression.

CHARACTERIZATION OF A CELL TYPE SPECIFIC ENHANCER
LOCATED 5' TO THE RAT GROWTH HORMONE GENE

In order to locate genomic regions involved in tissue
specific expression of the RGH gene, fusions were constructed
containing 5' flanking regions of the gene placed 5' of
the bacterial chloramphenicol acetyltransferase gene (CAT).
Transient expression of the fusion genes in various cell
lines demonstrated measurable expression when genomic

sequences located -235 to $+8$ relative to the CAP site were assayed in two lines of GH producing cells (GH_4, G/C) but not in any other endocrine or non-endocrine cell line. Deletion of the RGH sequence to -181 resulted in loss of expression of the fusion construct.

To determine if the RGH element could act in a position and orientation independent manner, constructs were made placing the region from -235 to -48 bp 5' of the CAP in either orientation 5' of the -422 bp rat prolactin promoter, or 1.4 kb 3' of the prolactin gene transcriptional CAP. The RGH element enhanced expression of the CAT reporter gene in all positions and orientations tested, demonstrating that this element has all the properties of a transcriptional enhancer.

Furthermore, the tissue specificity of the RGH enhancer is not dependent on elements in the rat prolactin or RGH promoters as shown by the tissue specificity of the enhancer when placed 5' of the Herpes virus thymidine kinase promoter.

Further deletion mapping of the 188 bp RGH enhancer did not reveal any subfragments retaining enhancer activity. The enhancer activity may therefore reside in either multiple small domains or a single large (greater than 50 bp) domain which has not as yet been tested as a separate fragment.

DISCUSSION

We have characterized a 188 bp sequence 5' of the RGH CAP which behaves in a position and orientation independent manner to enhance expression of a fused reporter gene in a specific cell type.

Current efforts are directed at fine mapping sequences necessary for RGH enhancer function using linker scanning techniques and concurrent analysis of the thyroid hormone and glucocorticoid regulatory elements present within the same 188 bp region.

ACKNOWLEDGEMENTS

We would like to thank Charles Nelson for his invaluable assistance in maintaining cell cultures.

REFERENCES

1. DeVilliers JL, Olson L, Tyndall C, Schaffner W (1982). Transcriptional enhancers from SV40 and polyoma virus show a cell type preference. Nucleic Acids Res 10:7965.

2. Spandidos DA, Paul J (1982). Transfer of human globin genes to erythroleukemic mouse cells. EMBO J 1:15.

3. Chao MV, Mellon P, Charnay P, Maniatis T, Axel R (1983). The regulated expression of beta-globin genes introduced into mouse erythroleukemia cells. Cell 32:483.

4. Kriegler M, Botchan M (1983). Enhanced transformation by a simian virus 40 recombinant virus containing a Harvey murine sarcoma virus long terminal repeat. Mol Cell Biol 3:325.

5. Ott MO, Sperling L, Herbomel P, Yaniv M, Weiss MC (1984). Tissue specific expression is conferred by a sequence from the 5' end of the rat albumin gene. EMBO J 3:2505.

6. Brinster RL, Ritchie KA, Hammer RE, O'Brien RL, Arp B, Storb U (1983). Expression of microinjected immuno-globulin gene in the spleen of transgenic mice. Nature 306:332.

7. Storb U, O'Brien RL, McMullen MD, Gollahon KA, Brinster RL (1984). High expression of cloned immunoglobulin K gene in transgenic mice is restricted to B lymphocytes. Nature 310:238.

8. Swift GH, Hammer RE, MacDonald RJ, Brinster RL (1984). Tissue specific expression of the rat pancreatic elastase I gene in transgenic mice. Cell 38:639.

9. Grosschedl R, Weaver D, Baltimore D, Constantini F (1984). Introduction of a Mu immunoglobulin gene in the mouse germ line: Specific expression in lymphoid cells and synthesis of functional antibody. Cell 38:647.

10. Chada K, Magram J, Raphael K, Radice G, Lacy E, Constantini F (1985). Specific expression of a foreign beta-globin gene in erythroid cells of transgenic mice. Nature 314:377.

11. Krumlauf R, Hammer RE, Tilghman SM, Brinster RL (1985). Developmental regulation of alpha-fetoprotein genes in transgenic mice. Mol Cell Biol 5:1639.

12. Ornitz DM, Palmiter RD, Hammer RE, Brinster RL, Swift GH, MacDonald RJ (1985). Specific expression of an elastase human growth hormone fusion gene in pancreatic acinar cells of transgenic mice. Nature 313:600.

Transcriptional Control Mechanisms, pages 383–393
© 1987 Alan R. Liss, Inc.

MOLECULAR GENETICS OF EUKARYOTIC RNA POLYMERASES [1]

C. James Ingles[2], Matthew Moyle[2], Lori A. Allison[2],
Jerry K.-C. Wong[2], Jacques Archambault[3] and James D. Friesen[3]

[2]Banting and Best Department of Medical Research and
[3]Department of Medical Genetics, University of Toronto,
Toronto M5G 1L6, Canada

ABSTRACT Using DNA encoding the largest subunit of
Drosophila RNA polymerase II, we have isolated related
DNA sequences encoding both yeast and hamster RNA
polymerase polypeptides. A family of genes appears to
encode the largest subunit of RNA polymerases I, II and
III. Nucleotide sequencing indicated that conserved
regions in the eukaryotic RNA polymerase polypeptides
are remarkably similar in their primary structure to
analogous portions of E. coli RNA polymerase. Linker
insertion mutations, introduced into the gene encoding
the largest subunit of yeast polymerase II (RPO21),
defined important functional regions of this gene.
Surprisingly, insertions as far as 1.0 kb 3' to the
coding sequence destroyed RPO21 function. The RPO21
polypeptide of both yeast and hamster has a unique
C-terminal domain composed of an unusual tandemly
repeated heptapeptide sequence. This novel domain is
required for transcription in vitro from class II
promoters. It may mediate upstream activating sequence
and enhancer function.

INTRODUCTION

A combination of biochemical and genetic approaches has
recently been applied in an attempt to understand at the
molecular level, the mechanisms by which eukaryotic
transcription is regulated. As a result, a number of

[1]Supported by MRC (Canada)

cis-acting DNA sequences which regulate the transcription of adjacent genes and proteins that recognize these specific DNA sequences have been described. Much less is known, however, about the very complex enzymes central to all RNA transcription in eukaryotes, the RNA polymerases. Here too a combined biochemical and genetic approach may yield new insight into the molecular mechanisms which govern promoter function.

The entry point for genetic studies of eukaryotic RNA polymerases was provided by the mutations conferring resistance to the mushroom toxin α-amanitin. α-Amanitin resistant cells contain altered α-amanitin resistant RNA polymerase II activities (see ref. 1 for review). We and others have used α-amantin resistance to select, identify or characterize other classes of RNA polymerase II mutations such as those with lethal (2-3) and conditional lethal phenotypes (2-4) as well as some second site revertants of certain temperature-sensitive RNA polymerase II mutations (5). Some years ago it became obvious that the genes encoding subunits of this complex enzyme would have to be isolated. Not only could more informative studies of existing mutations be initiated, but also these genes could serve in turn as stepping stones for obtaining new mutations affecting RNA polymerase function.

RESULTS AND DISCUSSION

A Family of Related Eukaryotic RNA Polymerase Genes.

Polymerase II mutations conferring -amanitin resistance in Drosophila were used to map the chromosomal location of the gene encoding an RNA polymerase II subunit polypeptide (2). Using P-element transposon tagging, this region of the X chromosome of Drosophila was cloned (6) and the gene was shown to encode the largest subunit of RNA polymerase II (7). Since the conservation of RNA polymerase II subunit structure and antigenicity in all eukaryotes had been well established it was not too surprising to find that the nucleic acid sequences encoding analogous RNA polymerase II subunit polypeptides from a wide variety of species were also remarkably conserved. Thus DNA encoding the largest subunit of RNA polymerase II in Drosophila has been used to first identify (8) and, more recently, to isolate by cross species hybridization the analogous polymerase II genes from yeast (9), hamster, mouse

(10) and human (11) cells.

The studies of yeast DNA's which are homologous to the Drosophila RNA polymerase II DNA have been particularly illuminating. In our initial Southern blot hybridization experiments, we demonstrated that three genetic loci rather than just a single locus could be detected with hybridization conditions of reduced stringency (9). One locus, RPO21, encodes, as expected the largest subunit of yeast RNA polymerase II (9). The second locus, RPO31, we later showed encodes the largest subunit of RNA polymerase III (12). The third yeast gene having homology with Drosophila polymerase II DNA appears to be the analogous largest subunit of RNA polymerase I (RPO11).

When the amino acid sequences of both yeast polypeptides, RPO21 and RPO31, were compared, extensive homology was readily apparent. We found that the two yeast RNA polymerase polypeptides have six major colinearly arrayed regions of homology, which range in identity from 43% to 71%. These same six regions are almost equally homologous to the corresponding regions of the largest subunit, β ', of the single prokaryotic RNA polymerase (13). Thus it can be concluded that the multiple forms of eukaryotic RNA polymerase arose by processes of duplication and divergence of the genes encoding subunits of a single ancestral enzyme. The conservation of amino acid sequence between prokaryotic and eukaryotic RNA polymerases is not restricted to the yeast RNA polymerases. Sequencing of DNA encoding the largest subunit of RNA polymerase II of Drosophila (14) and Chinese hamster (data not shown) has indicated that the corresponding polypeptides of higher eukaryotic RNA polymerase II also share homology with the prokaryotic β' polypeptide. Undoubtedly other subunits of the prokaryotic "core" RNA polymerase, α and β , will be shown to have their eukaryotic counterparts too.

These results raise the interesting possibility that transcription in eukaryotic cells may be more similar to transcription in prokaryotic cells than was previously suspected. If the "core" subunits of RNA polymerase II are conserved in structure, then perhaps proteins that have been shown to interact directly with these "core" subunits to affect initiation, (σ factors) or termination (nusA), may have related counterparts in eukaryotic transcription complexes. In preliminary experiments (data not shown) using the same conditions for DNA hybridization as were used in the Drosophila - yeast RNA polymerase studies, we have detected DNA sequences in the yeast genome which are

homologous to E. coli nusA DNA. These sequences could
represent yeast nusA analogs.

Genetic approaches to RPO21 function.

     Since the yeast RPO21 gene is a single copy essential
gene in haploid yeast cells (9) it is possible to use RPO21
DNA and techniques of reverse genetics to construct a
repertoire of RNA polymerase II mutations in yeast cells.
We have previously described the construction and
characterization of mutations in the RPO21 gene which confer
a temperature-sensitive for growth phenotype (15). An
analysis of these mutants and a number of second-site
suppressing mutations of these temperature-sensitive RPO21
alleles may identify other genes whose products are
additional members of an RNA polymerase II transcription
complex. For those initial studies we had mutagenized a
portion of the RPO21 gene on a URA3-containing integrating
plasmid. We have now undertaken a more systematic study to
identify important functional domains of this yeast RPO21
gene. Twelve base pair oligonucleotide linkers encoding in
phase amino acid insertions were inserted by the technique
of linker-tailing (16) at HaeIII, RsaI, AluI, DpnI, and TaqI
sites in a fully functional 9.2 kb RPO21 gene carried on a
CEN3, URA3, ARS1 shuttle vector (pJS121). Each altered
RPO21 clone was tested for its ability to complement a null
chromosomal RPO21 mutation. Our results are summarized in
Figure 1, which shows the positions of the linker insertions
relative to landmarks of the RPO21 gene, the six regions in

     Figure 1. Linker insertion mutagenesis of the yeast
RPO21 gene. The location of linker insertions which result
in lethal (▲) and non lethal (△) and yet the determined
(|) phenotypes are shown relative to the coding sequence of
RPO21 (▬) and the location of the six evolutionarily
conserved regions (☐) of the RPO21 polypeptide. Function
of the mutant RPO21 alleles was tested by transforming a
diploid RPO21/rpo21::LEU2 strain, and analysing spore
progeny for growth at 30°C after meiosis.

the RPO21 gene which have homology with the prokaryotic RNA
polymerase (13). Insertions at some sites destroyed RPO21
function in yeast cells. At other sites there was no
apparent effect of the linker insertions on RPO21 function.
Thus regions of the RPO21 polypeptide which serve important
structural or catalytic roles can be identified.
Surprisingly, some linker insertions into the DNA which
flanks the RPO21 coding sequence had a lethal effect in this
in vivo assay system. Twelve base pair insertions in RPO21
DNA as far as about 1.0 kb 3' to the RPO21 TGA stop codon
did not complement an RPO21 null mutation. Additional
evidence that this 3' region may play an important role in
RPO21 function came from the observation that deletion of
the 3' 1.7 kb EcoRI fragment from this RPO21 construct
yielded an RPO21 allele that also did not complement a
chromosomal null RPO21 mutation. This 3' truncated RPO21
gene did however complement one of the ts RPO21 alleles
(rpo21-1) described earlier (15). The DNA 3' to the RPO21
coding sequence may directly affect the level of expression
of either the RPO21 mRNA or the RPO21 polypeptide. Our
current studies of the synthesis and stability of RPO21 mRNA
and protein may provide a better understanding of the
functioning of what could be a novel 3' regulatory sequence
in this yeast gene.

Evolutionary conservation of an unusual C-terminal domain in
RPO21.

Our nucleotide sequences of the yeast RPO21 and RPO31
genes revealed that the yeast RPO21 and RPO31 and the E.
coli β' polypeptides are colinearly homologous over almost
the entire length of the RPO31 and β' polypeptide. RPO21
encodes a larger polypeptide; its longer length is almost
entirely due to the addition of a unique C-terminal domain
consisting of 26 tandem repeats of an unusual heptapeptide
sequence. A DNA hybridization probe containing only this
yeast tandemly repeated sequence has been used in Southern
blot experiments. A similar sequence appears to be present
in the RPO21 gene of other eukaryotes (Figure 2). The
3.6 kb EcoRI fragment in Drosophila DNA is the DNA encoding
the C-terminal half of the Drosophila RpII-215 (RPO21) gene
(6-8); the 12 kb and 8.6 kb EcoRI fragments in the hamster
DNA's have also been shown to be RPO21 DNA (8). In
addition, we have also detected a related sequence in the
genomic DNA of such widely divergent species as trypanosomes
and humans.

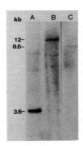

Figure 2. Higher eukaryotes have DNA sequences
homologous to DNA encoding the tandemly repeated
heptapeptide domain of yeast RPO21. 10 ug of Drosophila
melanogaster (A), 15 ug of Chinese hamster (B) and 10 ug
Syrian hamster (C) DNA were cut with EcoRI, fractionated,
transferred to nitrocellulose and hybridized under
conditions of reduced stringency as described earlier (8,9)
with a yeast RPO21 probe (nucleotides 4920 to 5404).

More convincing evidence for the evolutionary
conservation of this unusual sequence is presented in
Figure 3. Using the Drosophila RpII-215 DNA as probe, we
have isolated the RPO21 gene from genomic libraries of
Chinese hamster DNA; Corden et al. (10) have recently
reported a similar isolation of the mouse RPO21 gene. A
comparison of the yeast, hamster and mouse C-terminal amino
acid sequences revealed that the same consensus heptapeptide
sequence is present in the mammalian and yeast RPO21
polypeptides. Rather than 26 repeats as in yeast, however,
there are 52 repeats of the same consensus sequence in the
hamster (Figure 3) and mouse (10) polypeptides. Inspection
of Figure 3 indicates that the first 26 repeats in the
mammalian polypeptide, like those of the yeast polypeptide,
match the consensus sequence more closely than the distal 26
repeats. Presumably a duplication occurred during evolution
and the additional distal repeats are now more divergent.
It should be noted, however, that the hamster and mouse,
species estimated to have diverged over 60 million years
ago, have strikingly similar RPO21 C-terminal domains. The
amino acids present at each of the positions where the two
mammalian sequences differ from the consensus heptapeptide
sequence are identical. Indeed there is only a single
position in this 384 amino acid C-terminal domain where the
two mammalian sequences are not the same.

Yeast

Hamster

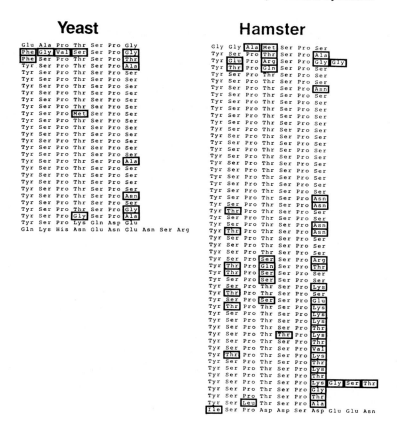

Figure 3. Amino acid sequence of the tandemly repeated C-terminal domain of yeast and Chinese hamster RPO21 polypeptides. The boxed residues indicate amino acids which differ from the consensus heptapeptide sequence.

Role of the RPO21 C-terminal domain in transcription.

The unusual C-terminal sequence appears to be in the RPO21 polypeptide of all eukaryotes. It is not present in the analogous polypeptide of RNA polymerase III, nor is it present in the β' subunit of E. coli RNA polymerase. Its remarkable evolutionary conservation indicates that it must be important for polymerase II-mediated transcription. A role for this domain is suggested by some indirect experiments. We (13) and Corden et al. (10) both have shown

independently that the proteolysis long observed to
accompany RNA polymerase II purification removes this
C-terminal domain yielding RNA polymerase IIB with its
smaller largest subunit. Using a monoclonal antibody
recognizing a determinant only present on the full length
intact RPO21 polypeptide of polymerase IIA, Dahmus and
Kedinger (17) were able to block accurately initiated in
vitro transcription of Adenovirus 2 major late DNA and other
eukaryotic gene sequences. This monoclonal antibody,
apparently recognizing the tandemly repeated heptapeptide
sequence, had no effect, however, on nonspecifically
initiated transcription in vitro on a calf thymus DNA
template. This C-terminal domain therefore likely plays an
important role in the correct initiation of transcription by
polymerase II at bona fide promoters.

     Several possibilities for a function of this unusual
RPO21 domain can be envisaged. The heptapeptide sequence
might bind a particular DNA sequence; it may interact with a
structural element in the nucleus; it could bind actin, a
putative transcription factor (18); or if phosphorylated it
may help destablize nucleosomes (10). Its unique repeating
structure also suggests that it could bind a special class
of polymerase II transcription factors. As shown in
Figure 4, a tandemly repeated structure at the C-terminus of
the RPO21 polypeptide could bind proteins at variable
(Fig. 4A) and multiple (Fig. 4B) positions. Trans-acting
sequence-specific DNA binding proteins which play an as yet
undetermined role in augmenting transcription from adjacent
start sites could therefore be positioned more flexibly
relative to polymerase II present at the start site of
transcription. Thus if the protein factor that binds to the
TATA box and mediates TATA box recognition by polymerase II
were also to interact with this RPO21 repeating domain, then
the multiple start sites of transcription seen for most
yeast genes might be in part explained. The start site of
transcription relative to the position of the TATA box in
mammalian genes is, however, more fixed. If this C-terminal
domain was involved only in mediating TATA box factor
recognition then it seems unlikely that this mammalian
domain with its 52 repeats would be twice as long in the
yeast enzyme. It seems more attractive therefore to
consider the possibility that this polymerase II domain
could also be involved in mediating upstream activating
sequence and enhancer function.

     It may be worthwhile to consider how such a domain
might have evolved in the eukaryotic enzyme. It is likely

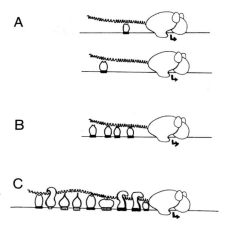

Figure 4.   A model for the function of the tandemly
repeated RPO21 C-terminal domain. RNA polymerase II with
two large and several smaller subunits is shown positioned
at the start site of transcription of a gene. A DNA-binding
protein also contacting the RPO21 tandemly repeated domain
is shown situated at variable (A) or multiple (B) positions
relative to RNA polymerase II at the start site of
transcription. Tandem duplications of this C-terminal
domain could be recognized by the variety of sequence-
specific DNA-binding proteins (C) which affect transcription
of a typical mammalian gene.

that one or more subunits of the RNA polymerase prokaryotic
"core" makes contact with sequence-specific DNA binding
proteins that activate transcription from particular
promoters (19). If such a single RNA polymerase domain,
contacted by positive-acting regulatory proteins, were to
have tandemly duplicated in an ancestral RNA polymerase,
then this positive activation might have been more
effective. There are two possibilities; now it could be
recognized by either two of the same transcription factors
or by two different transcription factors. The evolutionary
consequence of this multiple recognition would be to lock
the sequence of this ancestral tandemly repeated domain. A
change at any position in its amino acid sequence would
require at least two simultaneous compensating or
suppressing mutations in the genes encoding the
transcription factors and in the other duplicated domains in

the polymerase gene. Thus there would be a strong counter-
selection against changes in the sequence of the tandemly
duplicated domain. Further tandem duplications might have
led then to the current situation. As depicted in
Figure 4C, a much longer tandemly repeated domain could be
recognized by the whole plethora of different sequence-
specific DNA-binding transcription factors described
elsewhere in this Volume. Combinations of multiple flexibly
positioned proteins, each having the ability to contact the
RPO21 polypeptide, might result in the wide variety of
promoter stengths seen for different genes in different
states of physiological adaptation or cyto-differentiation.
It will be interesting indeed to see if this unique
C-terminal domain of RNA polymerase II provides the link
between the cis-acting DNA sequences, the sets of proteins
that recognize these sequences and eukaryotic RNA
polymerase Il.

REFERENCES

1. Ingles CJ (1985). RNA polymerases. In Gottesman MM
    (ed): "Molecular Cell Genetics", New York: John Wiley
    and Sons, p 423.
2. Greenleaf AL, Weeks JR, Voelker RA, Ohnishi S, Dickson
    B (1980). Genetic and biochemical characterization of
    mutants at an RNA polymerase II locus in D.
    melanogaster. Cell 21:785.
3. Mortin MA, Perrimon N, Bonner JJ (1985). Clonal
    analysis of two mutations in the large subunit of RNA
    polymerase II of Drosophila. Mol. Gen. Genet. 199:421.
4. Ingles CJ (1978). Temperature-sensitive RNA polymerase
    II mutations in Chinese hamster ovary cells. Proc.
    Natl. Acad. Sci. USA 75:405.
5. Wong JK-C (1985). Eukaryotic RNA polymerase II:
    mutation, identification and molecular cloning of the
    ama^R gene locus in eukaryotes. Ph.D. dissertation
    University of Toronto.
6. Searles LL, Jokerst RS, Bingham PM, Voelker RA,
    Greenleaf AL (1982). Molecular cloning of sequences
    from a Drosphila RNA polymerase II locus by P element
    transposon tagging. Cell 31:585.
7. Greenleaf AL (1983). Amanitin-resistant RNA polymerase
    II mutations are in the enzyme's largest subunit. J.
    Biol. Chem. 258:13403.

8. Ingles CJ, Biggs J, Wong JK-C, Weeks JR, Greenleaf AL (1983). Identification of a structural gene for an RNA polymerase II polypeptide in Drosophila melanogaster and mammalian species. Proc. Natl. Acad. Sci. USA 80:3396.

9. Ingles CJ, Himmelfarb HJ, Shales M, Greenleaf AL, Friesen JD (1984). Identification, molecular cloning, and mutagenesis of Saccharomyces cerevisiae RNA polymerase genes. Proc. Natl. Acad. Sci. USA 81:2157.

10. Corden JL, Cadena DL, Ahearn Jr JM, Dahmus ME (1985). A unique structure at the carboxyl terminus of the largest subunit of eukaryotic RNA polymerase II. Proc. Natl. Acad. Sci. USA 82:7934.

11. Cho KWY, Khalili K, Zandomeni R, Weinmann R (1985). The gene encoding the large subunit of human RNA polymerase II. J. Biol. Chem. 260:15204.

12. Moyle M, Hofmann T, Ingles CJ (1986). The RPO31 gene of Saccharomyces cerevisiae encodes the largest subunit of RNA polymerase III. Biochem. Cell Biol. in press.

13. Allison LA, Moyle M, Shales M, Ingles CJ (1985). Extensive homology among the largest subunits of eukaryotic and prokaryotic RNA polymerases. Cell 42:599.

14. Biggs J, Searles LL, Greenleaf AL (1985). Structure of the eukaryotic transcription apparatus: features of the gene for the largest subunit of Drosophila RNA polymerase II. Cell 42:611.

15. Ingles CJ, Himmelfarb HJ, Moyle M, Allison LA, Shales M, Friesen JD (1985). The molecular genetics of yeast RNA polymerases. In Calender R, Gold L (eds): "Sequence Specificity in Transcription and Translation, UCLA Symposia on Molecular and Cellular Biology, 30," New York: Alan R Liss, p 85.

16. Lathe R, Kieny MP, Skory S, LeCocq JP (1984). Linker tailing: Unphosphorylated linker oligonucleotides for joining DNA termini. DNA 3:173.

17. Dahmus ME, Kedinger C (1983). Transcription of adenovirus-2 major late promoter inhibited by monoclonal antibody directed against RNA polymerases II_O and II_A. J. Biol. Chem. 258: 2303.

18. Egly JM, Miyamoto NG, Moncollin V, Chambon P (1984). Is actin a transcription initiation factor for RNA polymerase B? EMBO J. 10:2363.

19. Hochschild A, Irwin N, Ptashne M (1983). Repressor structure and the mechanism of positive control. Cell 32:319.

Transcriptional Control Mechanisms, pages 395–404
© 1987 Alan R. Liss, Inc.

CHARACTERIZATION OF A
MOUSE MITOCHONDRIAL TRANSCRIPTION SYSTEM[1]

Michael W. Gray[2] and David A. Clayton

Department of Pathology, Stanford University School of
Medicine, Stanford, California 94305

ABSTRACT   We have isolated and characterized a fraction
of mouse mitochondrial enzymes that is capable of
directing specific initiation of transcription when
programmed with cloned mouse mitochondrial DNA (mtDNA).
We present evidence that accurate and efficient
initiation from the light-strand promoter (LSP) requires
the presence of at least two separable activities, a
nonspecific RNA polymerase (RNAP) and a "transcription
factor" (TF). The latter is essentially devoid of RNAP
activity but is able to confer LSP selectivity on RNAP.

INTRODUCTION

In animals, the mitochondrial (mt) genome is a compactly
organized, circular molecule of about 16.5 kbp, whose
complementary "heavy" (H) and "light" (L) strands are transcribed completely from separate promoters (HSP and LSP,
respectively) located in the D-loop region, the only
extensive non-coding region (1). Promoter-specific transcription of human mtDNA has been demonstrated in vitro in a
run-off system programmed with cloned D-loop region DNA and a
human mt extract (2). This system has made it possible to
delineate precisely those D-loop sequences that constitute
HSP and LSP (3,4), within which strand-specific initiation

[1]This work was supported by Grant GM-33088-14 from the
National Institute of General Medical Sciences to D.A.C. and
by the provision of a Medical Research Council of Canada
Visiting Scientist Award to M.W.G.
[2]Permanent address: Department   of   Biochemistry,
Dalhousie University, Halifax, Nova Scotia B3H 4H7 (Canada).

of transcription occurs. More recently, dissection of this system has shown that accurate and efficient transcription of human mtDNA requires at least two separable components: a nonspecific RNA polymerase activity (RNAP) and a "transcription factor" (TF) that is devoid of RNAP activity but confers on RNAP the capacity for promoter-specific transcription (5).

Considering that mammalian mtDNAs are organized in a very similar fashion (6-8), it has been surprising to discover that there is a pronounced species specificity in transcription of the mammalian mt genome. For example, the identified human mt LSP and HSP sequences (3) are not found at a comparable location in the D-loop region of mouse mtDNA, nor elsewhere in the mouse mt genome (6). Moreover, mouse mtDNA does not serve as a template for in vitro transcription with human mt enzymes that will actively transcribe a human mtDNA template (unpublished observations).

As a prelude to elucidating the molecular basis of this transcriptional specificity, we have isolated and characterized a fraction of mouse mt enzymes that directs LSP-specific initiation of transcription when programmed with cloned mouse mtDNA. We present evidence that, as in the human case, separate RNAP and TF activities are required to reconstitute accurate and efficient transcription of mouse mtDNA in vitro.

MATERIALS AND METHODS

Detergent lysis of gradient-purified mouse (LA9) or human (KB) mitochondria and run-off transcription assays were carried out as described in detail elsewhere (2,5). Lysates contained 10 mM Tris.HCl (pH 8.0), 0.1 mM EDTA, 0.5 mM dithiothreitol (DTT), 0.5 mM phenylmethylsulfonyl fluoride (PMSF), 7.5% glycerol (GLC), 0.5% Triton X-100 (TX), and varying concentrations (0.35 M - 1.05 M) of KCl. A high-speed supernatant (S100) was prepared by centrifugation of lysates for 1 hr at 40,000 rpm in a Beckman Ti75 rotor. Recombinant plasmids carrying D-loop inserts were linearized with an appropriate restriction endonuclease to generate templates that would support the synthesis of discrete run-off transcripts diagnostic of LSP-specific initiation. Transcripts were labeled by synthesis in the presence of $[\alpha-^{32}P]GTP$ and RNA products were analyzed by electrophoresis in 6% polyacrylamide gels. Gels were dried and used to expose Kodak XAR-5 film at -70°C with DuPont Cronex Lightning Plus intensifying screens.

RESULTS

LSP-Specific Transcription by Human and Mouse Mitochondrial Extracts.

Based on the previously mapped positions of transcription initiation sites in human (3) and mouse (9) mtDNA, the templates used here were expected to direct the synthesis of LSP-specific transcripts 416 (KB) and 210 (LA9) nucleotides long. As shown in Fig. 1, S100 fractions from KB mitochondria were very active in LSP-specific, template-dependent transcription (lanes 1-3 and 7-9), giving a strong signal on an overnight autoradiogram. (In this system, the LSP-specific transcript (▶) is by far the predominant product (2,5)). In contrast, an S100 fraction from LA9 mitochondria, prepared and assayed under the same conditions, appeared to be inactive with a mouse mtDNA template (lanes 4-6). However, upon prolonged autoradiographic exposure, a transcript of the anticipated size was detected (▷ , lanes 10-12). Compared to KB mt extracts, LA9 mt extracts consistently had very low transcriptional activity; however, what little transcription they did direct was LSP-specific and template-dependent.

Transcriptional activity was markedly enhanced when the LA9 mt S100 was prepared at higher KCl concentrations, either 0.7 M (lane 13) or 1.05 M (lane 14). This suggested that under standard extraction conditions (employing 0.35 M KCl), one or more components of the mouse mt transcriptional complex was inefficiently solubilized, and was therefore present in limiting concentration in the final extract. However, this beneficial effect of increasing KCl concentration was counterbalanced by an inhibitory effect of KCl on LSP-specific transcription, as discussed below.

Inhibition of LSP-Specific Transcription.

Complicating the assay of transcriptional activity in crude mt S100 fractions was the presence of RNA synthesis inhibitors, both endogenous and introduced. LA9 mt S100, when added together with KB mt S100, severely inhibited transcription of the human mtDNA template (not shown). The inhibitory activity was stable to heating at 95°C for 3 min and appeared to act at the level of product formation, since KB transcripts were unaffected when incubated with LA9 mt S100 added at the end of the usual 30 min assay period. Inhibition could be mimicked by addition of phenol-extracted

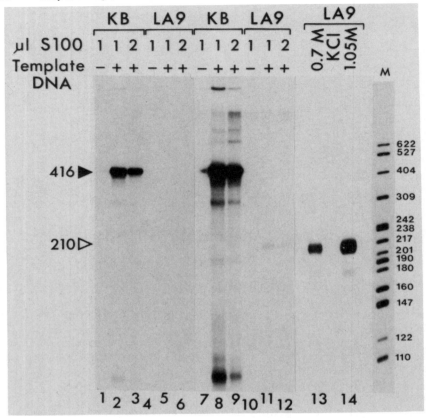

FIGURE 1. LSP-Specific transcription by human and mouse mitochondrial extracts. Triangles mark the positions of run-off transcripts diagnostic of human (▶) and mouse (▷) LSP-specific initiation, respectively. Autoradiograms were exposed for 15 hr (lanes 1-6), 150 hr (lanes 7-12) or 90 hr (lanes 13 and 14). M is a marker of ^{32}P-labeled HpaII restriction fragments of pBR322.

LA9 or KB mt nucleic acids (largely tRNA) to an active KB mt S100, or indeed by E. coli tRNA. These results indicated that endogenous nucleic acids could substantially inhibit the transcriptional activity of a crude mt S100 fraction, albeit in a non-specific manner.

Transcriptional activity was also inhibited by certain components of the mitochondrial lysis buffer itself; indeed,

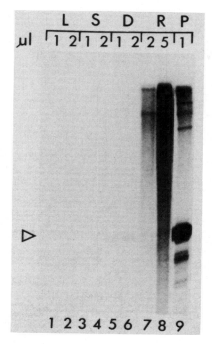

FIGURE 2. Isolation of a non-specific RNA polymerase from mouse mitochondria. Fractions were assayed for run-off transcription activity at various stages of chromatographic purification of an LA9 mt extract. L, initial lysate; S, S100; D, fraction of S100 eluted from DEAE-Sephacel at 0.3 M KCl (stepwise elution); R, fraction of S100 eluted from PC between 0.4 M and 0.5 M KCl (gradient elution); P, fraction of S100 eluted from PC with 0.7 M KCl (stepwise elution). Autoradiogram was exposed for 15 hr.

addition of a simulated lysis buffer severely reduced LSP-specific transcription catalyzed by a partially purified fraction of either mouse mt proteins (provided by D. Chang) or human mt proteins (provided by R. Fisher). When the buffer components were tested individually, it was found that KCl and TX (but not DTT, PMSF, or GLC) were inhibitory: when both KCl and TX were present at levels approximating those that might be contributed by added mt S100 (28 mM and 0.04%, respectively), profound inhibition of transcriptional activity was observed, whereas omission of both these components restored activity to control levels.

Isolation of a Non-specific RNA Polymerase from Mouse
Mitochondria.

These preliminary experiments indicated that cryptic
transcriptional activity could be unmasked by fractionation
of seemingly inactive mouse mt S100 fractions to remove
inhibitory substances, while at the same time concentrating
protein components that might be limiting in the initial
extract. In the example shown in Fig. 2, LSP-specific
transcription was not detectable in a crude lysate (lanes 1
and 2) or S100 (lanes 3 and 4) of LA9 mitochondria; however,
a run-off transcript was evident in a fraction eluted with
0.3 M KCl after the S100 had been adsorbed to DEAE-Sephacel
(lanes 5 and 6). When this fraction was further resolved on
a phosphocellulose (PC) column in the presence of a linear
gradient (0.3 M - 1.0 M) of KCl, a small peak of RNA
polymerase activity (measured by incorporation of radioactivity from [α-^{32}P]UTP in the presence of a poly(dA-dT)
template) was eluted early in the gradient (between about 0.4
M and 0.5 M KCl). After dialysis, this pooled material
(RNAP) displayed only non-specific RNA polymerase activity in
a run-off transcription assay (lanes 7 and 8), in contrast to
a fraction that had been eluted in a stepwise fashion from PC
at 0.7 M KCl (lane 9). This suggested that a factor or
factors required for LSP-specific initiation had been
separated from RNAP during PC chromatography.

Demonstration of a Mouse Mitochondrial Transcription Factor.

Fractions eluting late during PC chromatography of LA9
mt S100 showed little or no RNA polymerase activity in the
non-specific assay, and no detectable transcriptional
activity in the run-off assay (Fig. 3; 59-64, "-"). However,
LSP-specific transcriptional capacity could be reconstituted
by addition of RNAP to late-eluting column fractions (59-64,
"+"). Activity revealed in this way peaked at fraction 61.
These results indicated that a factor ("transcription
factor", TF) present in LA9 mitochondria is required, in
conjunction with a non-specific RNA polymerase activity, for
efficient LSP-specific initiation of transcription of mouse
mtDNA.

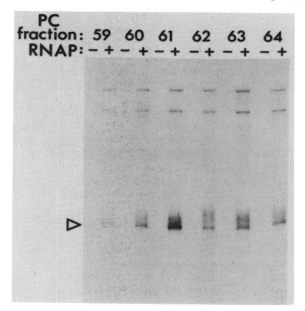

FIGURE 3. Demonstration of a mouse mitochondrial transcription factor. Aliquots (1 μl) of fractions eluting late during phosphocellulose chromatography were assayed for LSP-specific transcriptional activity in the presence (+) or absence (−) of 2 μl pooled and dialyzed PC fractions 22-35 (=RNAP). Autoradiogram was exposed for 15 hr.

Reconstitution of LSP-Specific Transcription with Separate RNAP and TF Fractions from Mouse Mitochondria.

With low levels of isolated mouse mt RNAP and relatively short autoradiographic exposures, there was no evidence of either specific or non-specific transcription in a run-off assay (Fig. 4, lane 3). The combination of RNAP + TF resulted in a strong, LSP-specific, run-off transcription signal (lanes 1 and 2). In contrast to the results with individual column fractions (see Fig. 3), a low level of LSP-specific transcriptional activity was seen with the pooled and dialyzed TF fraction (lane 4). Most likely this reflects some cross-contamination of TF with RNAP in the course of PC chromatography, although fractions constituting the two pools (RNAP, 22-35; TF, 61-64) were well separated in the elution profile. It is noteworthy that an increase in

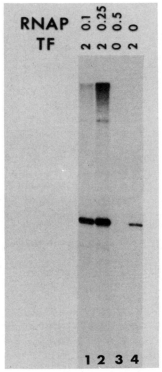

FIGURE 4. Reconstitution of LSP-specific transcription with separate RNAP and TF fractions from mouse mitochondria. Aliquots (µl) of each fraction were combined as indicated. Autoradiogram was exposed for 15 hr.

both the LSP-specific signal and non-specific background was observed with increasing RNAP concentration (compare lanes 1 and 2, Fig. 4), suggesting that TF was limiting in this experiment.

DISCUSSION

Analysis of in vivo transcripts complementary to the D-loop region of mouse mtDNA has localized the major LSP-specific initiation site to position 16,183 on the mouse mt genome (9). Since the template used in the studies reported here was truncated by cleavage at an XbaI site located at position 15,973, an LSP-specific run-off transcript would be expected to be 210 nucleotides long. As demonstrated here, a transcript of precisely this size was the major product of in vitro transcription directed by a crude or partially purified mouse mt extract. Product size varied as predicted for LSP-specific transcription in experiments in which templates were produced by cleavage with other restriction endonucleases (not shown).

As in the case of an in vitro system derived from KB mitochondria (2,5), LSP-specific transcription predominates in the LA9 mt system. Indeed, efficient HSP-specific transcription was not convincingly demonstrated in the course of this study and, as suggested in the case of KB mito-chondria (5), may require additional, unidentified factors. In contrast to KB mt extracts, LA9 mt S100 fractions consistently displayed very low transcriptional activity. This may be attributable to the combined effects of sub-optimal concentrations of one or more components of the mouse mt transcriptional complex and the presence of endogenous (mt nucleic acids) and introduced (KCl, Triton X-100) inhibitors of transcription. The results reported here emphasize the necessity of fractionating crude extracts to remove inhibitory substances, and to concentrate components of the transcriptional apparatus that may be present in limiting concentrations.

As demonstrated by PC chromatography, mouse mt extracts contain at least two separate components that must be present together to reconstitute LSP-specific transcription in vitro: a non-specific RNA polymerase (RNAP) activity that carries out RNA synthesis but is unable to initiate transcription specifically and efficienctly at LSP; and a "transcription factor" (TF) that is essentially devoid of RNAP activity but is able to confer LSP-specific transcriptional capacity on RNAP. These results closely parallel observations recently reported for human mitochondria (5). Continued purification of the RNAP and TF activities will permit a critical assess-ment of the degree to which each fraction contributes to the accuracy and efficiency of initiation.

Further characterization should also reveal the extent of similarity (both biochemical and evolutionary) of the analogous mouse and human mt RNAP and TF fractions. As well, continued dissection of these two systems opens the way for heterologous "re-programming" experiments to try to define the role of the separate RNAP and TF activities in determin-ing the species specificity of mammalian mt transcription.

ACKNOWLEDGMENTS

We thank Rob Fisher for gifts of KB mt enzymes and cloned DNA template, and David Chang for a gift of LA9 mt enzymes. The advice, assistance and encouragement of these individuals in the course of this study are also gratefully acknowledged.

REFERENCES

1. Clayton, DA (1984). Transcription of the mammalian mitochondrial genome. Ann. Rev. Biochem. 53:573.
2. Walberg, MW, Clayton, DA (1983). In vitro transcription of human mitochondrial DNA: identification of specific light strand transcripts from the displacement loop region. J. Biol. Chem. 258:1268.
3. Chang, DD, Clayton, DA (1984). Precise identification of individual promoters for transcription of each strand of human mitochondrial DNA. Cell 36:635.
4. Hixson, JE, Clayton, DA (1985). Initiation of transcription from each of the two human mitochondrial promoters requires unique nucleotides at the transcriptional start sites. Proc. Natl. Acad. Sci. U.S.A. 82:2660.
5. Fisher, RP, Clayton, DA (1985). A transcription factor required for promoter recognition by human mitochondrial RNA polymerase: accurate initiation at the heavy- and light-strand promoters dissected and reconstituted in vitro. J. Biol. Chem. 260:11330.
6. Bibb, MJ, Van Etten, RA, Wright, CT, Walberg, MW, Clayton, DA (1981). Sequence and gene organization of mouse mitochondrial DNA. Cell 26:167.
7. Anderson, S, Bankier, AT, Barrell, BG, de Bruijn, MHL, Coulson, AR, Drouin, J, Eperon, IC, Nierlich, DP, Roe, BA, Sanger, F, Schreier, PH, Smith, AJH, Staden, R, Young, IG (1981). Sequence and organization of the human mitochondrial genome. Nature 290:457.
8. Anderson, S, de Bruijn, MHL, Coulson, AR, Eperon, IC, Sanger, F, Young, IG (1982). Complete sequence of bovine mitochondrial DNA. J. Mol. Biol. 156:683.
9. Chang, DD, Hauswirth, WW, Clayton, DA (1985). Replication priming and transcription initiate from precisely the same site in mouse mitochondrial DNA. EMBO J. 4:1559.

Transcriptional Control Mechanisms, pages 405–419
© **1987 Alan R. Liss, Inc.**

TRANSCRIPTION OF THE TWO δ-CRYSTALLIN GENES IN A HELA CELL EXTRACT

Gokul C. Das[1] and Joram Piatigorsky

Laboratory of Molecular and Developmental Biology
National Eye Institute
National Institutes of Health
Bethesda, Maryland 20892

ABSTRACT   δ-Crystallin is the first crystallin
synthesized in the embryonic chicken lens and is the
major structural protein in the lens before hatching.
δ-Crystallin synthesis ceases several months after
hatching. There are 2 linked δ-crystallin genes in
the chicken (5'- δ1- δ2-3'). Only the δ1 gene has
been shown definitively to be active in the lens. We
show here that both the δ1 and δ2 promoters are
active in a Hela cell extract. The δ1 promoter
appeared several-fold stronger than that of δ2 under
the present conditions. Transcription of deletion
mutants showed that the sequences necessary for the
functioning of the δ1 promoter in a Hela cell extract
are located upstream from the RNA initiation site
between nucleotide positions -121 and -38. This
region includes a number of G + C-rich motifs,
including one hexanucleotide sequence CCGCCC which is
repeated 6 times in the SV40 promoter. Competition
experiments with purified fragments from the δ1
promoter showed that binding of transcription
factor(s) from the Hela cell extract to this G +
C-rich region is required for promoter activity in
vitro. Furthermore, competition experiments using 3
different fragments from the SV40 promoter suggested
that the δ1 transcription factor(s) is similar to
Spl, which stimulates transcription by binding to the
G + C-rich 21 bp repeats of the SV40 promoter and
differs from that which interacts with the SV40
enhancer region.

1. Present address: Division of Molecular
                     Biology and Biophysics
                     School of Basic
                     Life Sciences
                     University of Missouri-
                     Kansas City
                     Kansas City, Missouri
                     64110

INTRODUCTION

Differentiation of the vertebrate eye lens is
associated with a differential expression of crystallin
gene families (1,2). There are four distinct classes of
crystallins (α, β, γ, and δ) comprising 80-90% of the
soluble protein of the lens, and each crystallin family
consists of several closely related polypeptides (3).
δ-Crystallin, unlike the other crystallins, is present
only in birds and reptiles. It is the first and principal
crystallin synthesized in the developing chicken lens.
δ-Crystallin accounts for about 70% of the embryonic lens
protein (2) and consists of two polypeptides having
molecular masses of about 50K and 48K (4).   δ-Crystallin
synthesis ceases 3 to 5 months after hatching (5).
      There are two linked, extremely similar δ-crystallin
genes in the chicken, 4 kb apart and arranged in the same
transcriptional polarity (6-8). The δ1 gene is situated
5' to the δ2 gene. Sequence studies have shown that both
genes have 17 exons and potentially can code for proteins
which are extremely homologous. All the reported cDNAs so
far have been derived from the δ1 gene (9-11), which
raises the question as to whether δ2 is a functional gene.
In the present study we have initiated an in vitro
approach to investigate the molecular basis of
δ-crystallin gene transcription.

RELATIVE PROMOTER EFFICIENCY OF THE TWO δ-CRYSTALLIN GENES IN VITRO

The two δ-crystallin genes are shown in Fig. 1a. For
transcriptional studies, a 2.9 kb Pst I fragment from the

δ1 gene and a 5.4 kb EcoRl fragment from the δ2 gene were
cloned respectively at the Pst I and EcoRl sites of
pBR322. The resulting plasmids were named pδ1.6 and pδ2.1
(Fig. 1b).

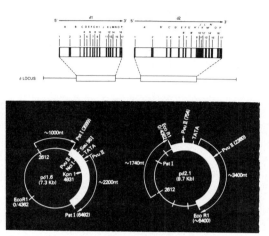

Fig. 1a (top). Schematic representation of two
δ-crystallin genes in a single chromosomal locus (8).

Fig. 1b (bottom). Recombinant plasmids containing the
promoter regions of the δ1 (left) and δ2 genes (right).

These plasmids were digested with Pst I and EcoRl,
and transcribed in a Hela cell extract, as described
previously (12). After enzymatic digestion, the δ1 gene
fragment had 662 bp and the δ2 gene fragment had 1606 bp
of 5' flanking sequences. RNA transcripts originating
from these δ-crystallin templates were 2100 (pδ1.6) and
3800 (pδ2.1) nucleotides long; transcription was inhibited
by low concentrations of α-amanitin (Fig. 2). By S1
nuclease mapping, we have demonstrated that the major site
for initiation of transcription from the δ1 promoter
occurs in vitro at the same site utilized in vivo (results
not shown). For estimation of the relative promoter
efficiency, we transcribed both genes in the same reaction
mixture. Transcription studies were performed at three
different total DNA concentrations selected from the
titration curves for the individual genes (not shown) and
examined by autoradiography. Specific transcripts obtained
from each δ-crystallin gene are marked by arrows (Fig. 1).
Inspection of lanes 1-3 in Fig. 2 indicates that more

transcripts were derived from the δ1 gene fragment than from the δ2 gene fragment. The autoradiograms were scanned at two different exposures. If the two promoters have equal efficiency, the expected ratio of the δ1 (2200 nucleotides) to the δ2 (3400 nucleotides) transcripts would be near 0.7 due to the difference in their size, since their base compositions are similar (13,14). The ratio was about 3, indicating that the δ1 promoter is several fold stronger than the δ2 promoter under this assay condition.

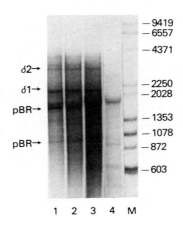

FIGURE 2   Promoter efficiency of the two δ-crystallin genes in a Hela cell extract. Pst I digest of pδ1.6 and EcoRI digest of pδ2.1 (see Fig. 1) were mixed together in equimolar concentration and transcribed at the following total DNA concentrations. Lane 1: 15 μg/ml. Lane 2:   19 μg/ml. Lane 3:   29 μg/ml. Lane 4:   38 μg/ml + 2 μg/ml α-amanitin. Lane M: DNA markers.

UPSTREAM SEQUENCE REQUIREMENT FOR TRANSCRIPTION OF THE δ1-CRYSTALLIN GENE

Using various restrictions sites, progressive deletions of the upstream sequences of the δ1 promoter were made to identify the region required for in vitro transcription. pδ1.6 was first digested with Pvu II and a 520 bp fragment was isolated. Increasing amounts of the

5' flanking sequences were next removed from the Pvu II
–Pvu II fragment by digestion with Sau 961 and Nci I.   The
resulting fragments were cloned at the Pvu II site of
pBR322 in the same orientation to create pδ1.8 and pδ1.9
respectively.   The plasmids were linearized with Pst I
(pδ1.6) or EcoRl (pδ1.7, pδ1.8 and pδ1.9) before
transcription.   Transcription was also performed with
pδ1.7 digested with the single-cut enzymes Bst EII and
EcoRl (Fig. 3).   As an internal control, pδ1.6 was
double-digested with Pst I and Kpn I and co-transcribed in
all assays. This gave rise to 706 nucleotide transcript
which was easily identified.

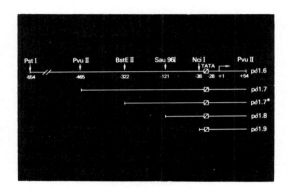

FIGURE 3   Progressive deletions of the δ1 promoter.
The name of the plasmids in which these fragments are
cloned is listed on the right.   pδ1.7* is not a
separate plasmid; it was derived by Bst EII digestion
of pδ1.7.   +1 is the transcription initiation site.
The end points of the deletion are denoted by the
nucleotide positions given for the indicated
restriction sites and are taken from Nickerson et al.
(13).

The autoradiograms in Fig. 4 show that δ1 transcripts
were obtained from pδ1.6, pδ1.7, and pδ1.8.   The ratio of
the transcripts from the fragment being tested to the 706
nucleotide internal standard was determined by
densitometric scanning.   We have compared directly pδ1.6
with pδ1.7* since these δ1 templates did not contain

pBR322 sequences at their 5' end, and pδl.7, pδl.8 and
pδl.9 with each other since each of these EcoR1 digested
templates contained the same pBR322 sequences extending
from their 5' end.   The 5' pBR322 sequence reduced the
promoter efficiency by 2-3 fold.   Inhibition of promoter
activity might be caused by the pBR322 sequence by
generating an end point distant from the normal promoter
start-site.   Comparison of pδl.6 with pδl.7* shows that
the sequences necessary for efficient δl promoter activity
reside within the 320 bp upstream of the RNA initiation
site.   A further deletion to nucleotide position -121 did
not change the transcriptional efficiency of δl promoter
(pδl.7 and pδl.8).   When an additional 83 bp were deleted
(up to nucleotide position -38 in pδl.9), no in vitro
promoter activity was observed.   These results demonstrate
that the region between nucleotide positions -121 and -38
contains sequences necessary for in vitro transcription of
the δl gene.   The striking feature of this region is that
it is highly rich in G + C content and contains one copy
of CCGCCC sequence associated with a CAT box (Fig. 5).

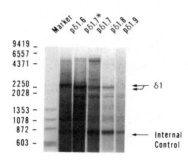

FIGURE 4   Autoradiograms of in vitro transcription
reactions. 1.0 µg of the test plasmid digested either
with Pst I (pδl.6) or EcoRI (pδl.7, pδl.8 and pδl.9)
was mixed with 1 µg of pδl.6, digested with Pst I and
Kpn I and transcribed.   pδl.7* is pδl.7 digested with
Bst EII and EcoRI.   The total DNA concentration in
the reaction mixtures was 32 µg/ml.   The 706
nucleotide transcript arising from the
double-digested pδl.6 served as an internal control.
All the transcripts originating from the δl promoter
in these tests were approximately 2120 nucleotides
long, except that from pδl.6 which was about 2200

nucleotides. The pδ1.6 transcript was longer, as
expected (see Fig. 1). The name of the plasmid which
was used for each run-off assay is indicated at the
top of each lane. The data in this figure is taken
from Das and Piatigorsky (29).

```
     -150       -140       -130       -120       -110      -100
   GGACACACAGGATAGGGGTGGGCAGCATGAGGGGGGCCAGAGGGAGAGGGGGCAG
                                   ↑
     -90        -80          CAT box  -60       -50
   AGCTGGGCTGGACGAGGGGACACCGCCCCCCAATGGGGCGTGACGAGCTGCCAGCC

     -40       -30   TATA box          -10       +1
   CAGGCTCCGGGGCACGTAAAAGCGGGGCTGTGAGACCGGAGAGCACGGAGCGAC
              ↑
```

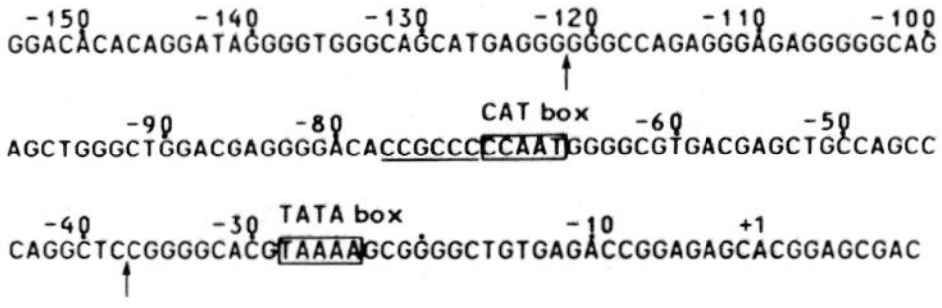

FIGURE 5   DNA sequence showing significant features
of a portion of the 5' flanking region of the
δl-crystallin gene (15). The RNA initiation site
(+1), CAAT box, TATA box and CCGCCC sequences are
marked. The region important for in vitro promoter
activity is located between nucleotide positions -121
and -38, as shown by the arrows. This figure is
taken from Das and Piatigorsky (29).

INTERACTION OF CELLULAR FACTORS WITH THE
G + C-RICH REGION FOR PROMOTER FUNCTION

Involvement of cellular factors in δl transcription
was investigated by competition analysis with different
sequences purified from the δl promoter. The 520 bp Pvu
II-Pvu II insert was purified first from pδ1.7. Further
digestion of this fragment with Msp I gave 4 fragments;
from 5' to 3' these included C-I (-466 to -154), C-II
(-154 to -38), C-III (-38 to -9) and C-IV (-9 to +54). The
first 3 fragments, which lack the RNA initiation site,
were used as competitors in the transcription assay. Each
assay contained 1.5 μg of pδ1.6 double-digested with Pst I
and Kpn I as template and 1.0 μg of the indicated
competitor DNA. A 234 bp DNA fragment lacking promoter
activity (purified from Hae-III digested replicating form
of φX174 DNA) was used as a non-competitor control.

The results of the competition analysis are shown in
Fig. 6. The transcript originating from the δ1 promoter
is shown by an arrow (lane 1). When the pδ1.6 template
was transcribed in the presence of the non-competitor DNA
(results not shown) or of the competitor fragment C-I
(lane 2), the δ1 promoter was efficiently utilized. In
both cases, we noticed an enhancement of transcription.
Possibly, the non-competitor fragment and C-I bind one or
more factors which are inhibitory in vitro. In contrast,
the C-II fragment, which contains the sequences important
for the function of the δ1 promoter, abolished
transcription (lane 3). No transcript was visible even
after overexposure of the gel. This indicates that the
C-II fragment contains sequences that compete for factors
required for the δ1 promoter to function in vitro. The
fragment C-III, which contains the TATA box, only caused a
minor inhibition of transcription when it was used at a
much higher molar ratio than presented here (lane 4). The
results of the deletion analysis and competition assays,
taken together, indicate that one or more trans-acting
factors must bind to the G + C-rich region of the δ1
promoter for transcription to occur. One or more factors
also appears to interact with the TATA box region, but
this is not critical for the overall efficiency of the δ1
promoter, since the C-III fragment was not a very
effective competitor for δ1 transcription.

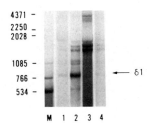

FIGURE 6 Competition analysis of the δ1 promoter
with its own promoter fragments. Competitor fragments
were made as described in the text. Each assay
contained 1.5 µg of pδ1.6 digested with Pst I and Kpn
I and 1.0 µg of the indicated competitor DNA.
Incubations were for 20 min at 30°C. Lane M: DNA
size markers; Lane 1: pδ1.6; Lane 2: pδ1.6 +
competitor C-I; Lane 3: pδ1.6 + competitor C-II;

Lane 4: pδ1.6 + competitor C-III. This data is
taken from Das and Piatigorsky (29).

COMPETITION FOR TRANSCRIPTION FACTORS BETWEEN THE δ AND SV40 PROMOTERS

The Hela cell extract contains factors that bind to
the TATA box (16,17), CCGCCC sequences (18-20), CCAAT (21)
sequences and the enhancer (22). The 21 bp repeats of the
SV40 promoter that contain 6 copies of CCGCCC bind a
protein, Sp1, which enhances in vitro transcription
several-fold (see 20). The structural similarity of the
δ1 and SV40 promoters prompted us to investigate whether
both might bind similar factor(s). Several deletion
mutants were used to make three competitor fragments from
the SV40 promoter region (Fig. 7). The first one (C1)
contains the TATA box, the 6 copies of CCGCCC and about 20
nucleotides of the first 72 bp repeat of the enhancer. The
second competitor fragment (C2) includes the 6 copies of
CCGCCC and both 72 bp repeats of the SV40 enhancer, but
lacks the TATA box. The third competitor fragment (C3)
contains the enhancer element, except for the first 20 bp
of the 72 bp repeat closest to the G + C-rich motifs. An
additional 100 bp of pBR322 extends from the 5' end of
both C2 and C3.

As a control experiment, we tested the ability of
these fragments to inhibit transcription of the SV40
promoter in the plasmid pSVK104, as described previously
(23). In run-off assays, a Bam Hl digest of the plasmid
gave rise to transcripts of 2150 and 3130 nucleotides
originating from the early and one of the late promoters,
respectively (results not shown). Addition of the
non-competitor fragment, used above, enhanced the level of
transcription. However, transcription from both promoters
was totally eliminated with both C1 and C2. By contrast,
the enhancer-containing fragment C3 did not reduce SV40
transcription. These data suggest that the G + C-rich
motifs bind transcription factor(s) from the Hela cell
which activate SV40 promoters in vitro. Binding of a Hela
cell factor (Sp1) has been reported previously (19,20).

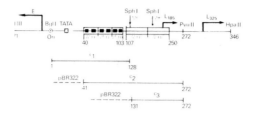

FIGURE 7   Schematic representation of the SV40
promoter region.   The competitor fragments (C_1, C_2,
and C_3) used in these experiments were made from
recombinant plasmids described previously (4).   The
dotted line is the pBR322 sequence and the solid line
represents SV40 sequence with the nucleotide
positions indicated.   The TATA box (▢), G + C-rich
motifs (▪) within the three 21 bp repeats and the 72
bp repeats are indicated. L_{185} and L_{325} are late
promoter start sites.   This figure is taken from Das
and Piatigorsky (29).

Transcription from the δ1-crystallin promoter was
next studied in the presence of the three SV40 competitor
fragments (Fig. 8).   The 700 nucleotide transcript
originating from the δ1 promoter of the plasmid pδ1.6
digested with Pst I and Kpn I is shown when transcribed
alone (lane 2) or in the presence of the non-competitor
DNA (land 3).   By contrast, δ1 transcription was not
evident in the presence of the SV40 competitor fragments
Cl (lane 4) or C2 (lane 5). However, for C2, which does
not contain the TATA box, a low level (5-10%) of
transcription was observed when the autoradiogram was
overexposed.   Competitor fragment C3, which contains the
SV40 enhancer region,did not reduce transcription directed
by the δ1 promoter (lane 6).   These experiments suggest
that similar transcription factor(s) bind to the G +
C-rich region of the δ1 and SV40 promoters and this factor
differs from that interacting with the enhancer region of
SV40 (22).

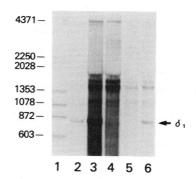

4371 —

2250 —
2028 —

1353 —
1078 —
872 —

603 —

← δ₁

1 2 3 4 5 6

FIGURE 8 In vitro competition analysis of the δ1
promoter against SV40 promoter fragments. Each assay
contained 1.5 μg of pδ1.6 digested with Pst I and Kpn
I and 0.5 μg of non-competitor DNA (Hae III fragment
of φX-174) or of competitor DNA (see Fig. 6).
Transcription reactions were incubated for 20 min at
30°C. Lane 1: DNA size marker; Lane 2: pδ1.6
alone; Lane 3: pδ1.6+ non-competitor DNA; Lane 4:
pδ1.6+ competitor C_1; Lane 5: pδ1.6+ competitor C_2;
Lane 6: pδ1.6+ competitor C_3. This data is taken
from Das and Piatigorsky (29).

DISCUSSION

Although there are two linked, extremely similar
δ-crystallin genes in the chicken (7,8,13,14), only the δ1
gene has been shown to produce mature cytoplasmic mRNA
(10,11). Our present in vitro transcription data indicate
that both δ-crystallin genes can initiate transcription in
a Hela cell extract. In addition, the data provide
evidence that the δ1 promoter is stronger than the δ2
promoter under the present assay conditions. These
results agree well with transient expression experiments
using the pSVO-CAT vector, which showed stronger δ1 than
δ2 promoter activity in transfected lens epithelia (15).
It remains to be determined whether or not the δ2 gene
functions in vivo.
    As yet, little is known about the numbers and nature
of eukaryotic transcription factors or their mode of
action on protein-coding genes. In vitro transcription

studies with partially purified cellular factors have identified the binding of a number of proteins to different promoter elements, such as the TATA box, CCGCCC sequences, CAT box and enhancers, as a prerequisite for optimum and accurate promoter function (16-22). The results of our experiments delineate the δ1 promoter region which is important for directing transcription and indicate that it must bind one or more factors for transcription to occur. This region of the δ1 promoter includes the sequence 5' CCGCCC 3', multiple copies of which in the SV40 promoter are known to interact with the Hela cell transcription factor Sp1 (see 20). We interpret the ability of the SV40 promoter fragments C1 and C2 to compete effectively for transcription directed by the δ1 promoter in the present experiments as evidence for the involvement of Sp1 or an Sp1-like factor for δ1 promoter function in the Hela cell extract, since the C1 and C2 fragments contain the six copies of CCGCCC.

The binding of Sp1 to cellular promoters is reasonable in view of its presence in uninfected Hela cells. Recent data provide evidence that Sp1 binds to CCGCCC sequences in the monkey genome as well as to this sequence in various viral and cellular promoters (24,25). Even the single copy of CCGCCC in the δ1 promoter appears to be able to bind Sp1. This is consistent with the ability of one copy of CCGCCC to bind Sp1 in other genes (20). We cannot yet rule out that the δ1 promoter binds a different factor that functions only in conjunction with Sp1, however this seems unlikely since the upstream regions of SV40 and the Ad2 major late promoter do not compete with each other in vitro and bind different transcription factors (26).

We do not know why the δ2 promoter functions less well in the Hela cell extract than the δ1 promoter. Both genes have similar TATA boxes and G + C-rich flanking sequences, including a copy of CCGCCC (15). Possibly, the presence of the CAT box associated with CCGCCC in the δ1 promoter (but not in the δ2 promoter) has a beneficial effect on transcription. There is evidence for CAT boxes having a regulatory role in transcription (27). In this connection, it is interesting to note that in the herpes simplex virus-thymidine kinase gene, the second distal promoter signal contains a CAT box in the opposite strand associated with a GGCGGG sequence which dominates over the first distal signal containing the hexanucleotide signal alone (28).

Finally, the facts that we observed a weak inhibition of transcription of the δ1 promoter when the C-III competitor fragment was used (which contains the δ1 TATA box) and a very low level of transcription of the δ1 promoter when C2 competitor fragment was used (which contains the SV40 CCGCCC sequences but not the TATA box) implicate another Hela cell factor which may bind to the δ1 TATA box region. Clearly, further experiments are needed to understand the molecular basis for the differential strength of the two δ-crystallin promoters in the Hela cell extract.

ACKNOWLEDGMENT

We thank Drs. T. Borras and J. M. Nickerson for sequence information on the δ-crystallin promoters and Dr. J. W. Hawkins for providing us with the plasmid pδ2.1. We also thank Mrs. Dawn Chicchirichi for expert secretarial assistance.

REFERENCES

1. Piatigorsky J (1981). Lens differentiation in vertebrates. A review of molecular and cellular features. Differentiation 19:134.
2. Piatigorsky J (1984). Lens crystallins and their gene families. Cell 38:620.
3. Bloemendal H (1981). The lens proteins, In Molecular and Cellular Biology of the Eye Lens, ed. Bloemendal H (Wiley, New York) pp. 1.
4. Reszelbach RT, Shinohara T, Piatigorsky J (1977). Resolution of two distinct chick δ-crystallin bands by polyacrylamide gel electrophoresis in the presence of sodium dodecyl sulfate and urea. Exp Eye Res 25:583.
5. Treton JA, Shinohara T, Piatigorsky J (1982). Degradation of δ-crystallin mRNA and the lens fiber cells of the chicken. Develop Biol 92:60.
6. Bhat SP, Jones RE, Sullivan MA, Piatigorsky J (1980). Chicken lens crystallin DNA sequences show at least two δ-crystallin genes. Nature 284:234.

7.   Yasuda K, Kondoh H, Okada TS, Nakajima N, Shimura Y (1982). Organization of δ-crystallin genes in the chicken. Nuc Acids Res 10:2879.

8.   Hawkins JW, Nickerson JM, Sullivan MA, Piatigorsky J, (1984). The chicken δ-crystallin gene family. Two genes of similar structure in close chromosomal approximation. J Biol Chem 259:9821.

9.   Jones RE, Bhat SP, Sullivan MA, Piatigorsky J (1980). Comparison of two δ-crystallin genes in the chicken. Proc Natl Acad Sci USA 77:5879.

10. Nickerson JM, Piatigorsky J (1984). Sequence of a complete chicken δ-crystallin cDNA. Proc Natl Acad Sci USA 81:2611.

11. Yasuda K, Nakajima N, Isobe T, Okada TS, Shimura Y (1984). The nucleotide sequence of a complete chicken δ-crystallin cDNA. EMBO J 3:1397.

12. Manley JL, Fire A, Cano A, Sharp PA, Gefter ML (1980). DNA-dependent transcription of adenovirus genes in a soluble whole-cell extract. Proc Natl Acad Sci USA 77:3855.

13. Nickerson JM, Wawrousek, EF, Hawkins JW, Wakil AS, Wistow GJ, Thomas G, Norman B, Piatigorsky J (1985). The complete sequence of the chicken δ1 crystallin gene and its 5' flanking region. J Biol Chem 260:9100.

14. Nickerson JM, Wawrousek EF, Hawkins JW, Norman B, Filpula DR, Nagle JW, Ally AH, Piatigorsky J (1986). Sequence of the chicken δ2 crystallin gene and its intergenic spacer. Extreme homology with the δ1 crystallin gene. J Biol Chem 261:552.

15. Borras T, Nickerson JM, Chepelinsky AB, Piatigorsky J (1984). Structural and functional evidence for differential promoter activity of the two linked δ-crystallin genes in the chicken. EMBO J 4:445.

16. Davison BL, Egly JM, Mulihill ER, Chambon P (1983). Formation of stable preinitiation complexes between eukaryotic class B transcription factors and promoter sequences. Nature 301:680.

17. Carthew RW, Chodosh A, Sharp PA (1985). An RNA polymerase II transcription factor binds to an upstream element in the adenovirus major late promoter. Cell 43:439.

18. Dynan WS, Tjian R (1983). Isolation of transcription factors that discriminate between different promoters recognized by RNA polymerase II. Cell 32:669.

19. Dynan WS, Tjian R (1983). The promoter-specific transcription factor Spl binds to upstream sequences in the SV40 early promoter. Cell 35:79.

20. Gidoni D, Kadonaga JT, Barrera-Saldana H, Takahashi K, Chambon P, Tjian R (1985). Bidirectional SV40 transcription mediated by tandem Spl binding interactions. Science 230:511.

21. Jones KA, Yamamoto KR, Tjian R (1985). Two distinct transcription factors bind to the HSV thymidine kinase promoter in vitro. Cell 42:559.

22. Sassone-Corsi P, Corden J, Kedinger C, Chambon P (1981). Promotion of specific in vitro transcription by excised TATA:box sequences inserted in a foreign nucleotide environment. Nuc Acids Res 9:3941.

23. Das GC, Salzman NP (1985). Simian virus 40 early promoter mutations that affect promoter function and autoregulation by large T-antigen. J Mol Biol 182:229.

24. Gidoni D, Dynan WS, Tjian R (1984). Multiple specific contacts between a mammalian transcription factor and its cognote promoters. Nature 312:409.

25. Dynan WS, Saffer JD, Lee WS, Tjian R (1985). Transcription factor Spl recognizes promoter sequence from themonkey genome that are similar to the simian virus 40 promoter. Proc Natl Acad Sci USA 82:4915.

26. Miyamoto NG, Moncollin V, Wintzerith M, Hen R, Egly JM, Chambon P (1984). Stimulation of in vitro transcription by the upstream element of the adenovirus-2 major late promoter involves a specific factor. Nuc Acids Res 12:8779.

27. Grosveld GC, de Boer E, Shewmaker CK, Flavell RA (1982). DNA sequences necessary for transcription of the rabbit beta-globin gene in vivo. Nature (Lond) 295:120.

28. McKnight SL, Kingsbury RC, Spence A, Smith M (1984). The distal transcription signals of the herpesvirus tK gene share a common hexanucleotide control sequence. Cell 37:253.

29. Das GC, Piatigorsky J (1986). The chicken δl-crystallin gene promoter: Binding of transcription factor(s) to the upstream G + C-rich region is necessary for promoter function in vitro. Proc Natl Acad Sci USA (in press).

Transcriptional Control Mechanisms, pages 421–435
© 1987 Alan R. Liss, Inc.

Adenoviral Vectors and Liver Specific Gene Control

Jeffrey M. Friedman[1] and Lee E. Babiss[2]

[1]Howard Hughes Medical Institute and

[2]Laboratory of Molecular Cell Biology
Rockefeller University
New York, New York

Abstract

We have employed an adenovirus vector in order to
test the transcripitonal activity of tissue
specific promoters in a wide range of host cells.
In this vector, the adenoviral E1A gene and E1B
promoters, which are situated at the extreme left
end of the viral genome have been deleted and
replaced by subgenomic fragments of the rat albumin, mouse B globin or mouse immunoglobulin heavy
chain gene. In all cases the tissue specific promoter directs the synthesis of the 3' exon of the
E1B gene. After infection of hepatoma cells, E1B
transcription could be detected only from the
virus containing the albumin promoter. Conversely
in myeloma cells only the immunoglobulin promoter
transcribed the E1B sequences. We also infected
primary liver cells with these viruses and found
that only the albumin promoter was functional but
that the activity of the viral promoter declined
in paralell with the endogenous gene with time in
culture. In spite of the fact that tissue specificity was maintained, the transcription rate of
the virally encoded albumin promoter was considerably lower than that of the single endogenous
gene.

Introduction

The most prominent level of gene control in

vertebrates is at the level of transcriptional
initiation (1,2,3,4,5,6). In previous work from
this laboratory on gene control in liver, two
aspects of transcriptional control have been
noted: specificity among cell types (5,6) and a
higher   differential rate of transcription for
liver-specific genes by adult hepatocytes on the
one hand and fetal hepatocytes, hepatoma cells,
and cultured adult cells on the other (6,7,8,9,).
It has been our objective to define those regulatory elements involved in the quantitative and
qualitative control of liver specific transcripi-
tion. Toward that end we have used an adenoviral
vector that is capable of introducing recombinant
DNA into a wide range of mammalian celss (including all the cell types listed above) and which
allows the direct measurement of the transcripiton
rate of newly introduced genes (10,11).

The role of tissue specific trans-acting factors in gene control has been most frequently studied by scoring the level of expression of recombinant DNAs introduced into cultured cells by
transfection.   Thus, tissue specific expression
has been observed for a number of genes introduced
by transfection of recombinant plasmids into cultured cells that exhibit phenotypic characteristics of differentiated cells for example, pancreas
(12), lens (13) β-lymphocytes (14,15) and, liver
(16,17,18). These results have led to the hope
that factors responsible for cell specific gene
expression can be identified.

However, there are several potential disad-
vantages with this experimental approach.   In the
reported transfection experiments of differentiated cells only a small fraction of the exposed
cells actually take up and express the plasmid DNA
and they often do so at a relatively low rate.
Since only a subpopulation of cells express these
transfected plasmids, it is impossible to directly
compare the level of expression of the newly
introduced DNA to the endogenous gene.   An additional problem with this approach is that all cell
types are not equally receptive to uptake and/or
the expression of newly introduced DNA making comparisons between different cell types difficult.

This is particularly problematic in circumstances
where the molecular basis of differences in transcription rates among different cell types is to
be considered.

       As an alternative to transfection, we have
utilized recombinant adenoviruses to introduce
cellular promoters into cultured cells. This
approach depends on technology that permits the
introduction of DNA sequences into the left end of
human adenovirus 5. (19,20,21). Adenovirus will
adsorb to and promptly initiate the early stages
of infection in the great majority of rodent and
human cells, including primary cells (22,23).
Moreover, since an assay of the activity of
virally encoded test genes can be made within a
few hours and because many genomes are introduced
into all cells transcription can be measured
directly by performing in vitro nuclear chain
elongation assays on infected nuclei.
(10,11,24,25) This is especially important since
the gene expression assay in most reported
transfection experiment relies on the steady-state
amount of an mRNA or protein product, often the
levels of bacterial chloramphenicol acetylase
(CAT) activity produced from a fusion of the regulatory region of the tissue specific gene with the
gene encoding the CAT protein. Such steady state
measurements are usually made only after prolonged
incubation time (24-72 hours). Use of this assay
to study transcriptional control requires the
assumption that both an unusual mRNA and protein
product are equally stable in different cells
under differing conditions. Thus in these experiments we describe initial experiments in which the
activity of tissue specific promoters incorporated
into the adenoviral chromosome is dtermined after
infection of differentiated cultured cells.

Methods

Materials and Methods

Cells and viruses. Monolayer cultures of HepG2
and 293 cells, rat H4II, mouse MPCll and monolayer
cultures of mouse hepatocytes were all grown or
maintained in Eagle's medium supplemented with 10%

fetal calf serum. The dispersal of primary mouse
hepatocytes has been described (7).

H5in340, a type 5 adenovirus with wild-type
properties (20) was provided by P. Hearing and all
the recombinant virus as described in this paper
were propagated in 293 cell monlayers. Virus
stocks were titrated by plaque titration on 293
cells, and all of the virus-infections described
were at a multiplicity of 20PFU/cell.

Plasmid and Recombinant Virus Construction. The
plasmid pMLP6 (21) which contains the 0-15.5 map
unit segment of type 5 adenovirus was treated with
the endonculeases Bgl II and Rsa to remove nucleotides 194 to 3320. The site at nucleotide 191 was
filled in using DNA polymerase and a Bg1 II linker
ligated to the blunt end. Segments of various
cell genes (rat albumin, mouse β-globin, and mouse
immunoglobulin) were isolated and modified to contain Bg1 II linkers at their termini and ligated
into the newly created BgII site and the 194/3320
viral junction to produce the desired plasmids.
Plasmid DNAs were prepared, linerarised with an
appropriate restriction endonuclease and mixed
with the sub-genomic adenovirus DNA representing
3.85 to 100 map units. This adenovirus segment
was prepared by digesting the In340 genome once
with ClaI, that digests the viral DNA at 2.0 m. u.
and once with xbaI that digests the viral DNA at
3.85 map units (21). These two digestions
resulted in DNA that was almost completely free of
whole infectious DNA molecules. Intracellular
recombination between the 3.85 to 100 and each of
the plasmids (containing cellular promoter
sequences) resulted in three recombinant viruses.
Each was plaque purified and used in this study:
Alb194 (adenovirus strain name) that contained rat
DNA from -441 to +957 of the rat albumin gene
(26); Igl94 (adenovirus strain name) that contained -2000 to +1700 of the mouse immunoglobulin
heavy chain (14,27); and Glol94 (adenovirus strain
name) that contained -1225 to +1075 of the mouse β
globin gene (28).

Transfection. HepG2 cells were transfected with

DNA by the calcium phosphate precipitation (29) method and RNA was extracted after 48 hours using guanidinum isothicyanate (30) to assay for specific expression by the Sp6 assay described below.

Sp6 Assays.   The type 5 adenovirus E1B sequences between nucleotides 3320 (BglII) and 3780 (PstI) were cloned into pGEM2 (31).   The plasmid was linearized with EcoR1 prior to transcription.   The transscription reaction using the sp6 RNA polymerase, hybridization at 65°C and trimming with RNaseT2 were as described (31).

Preparation and hybridization of ^{32}P-labeled nuclear RNA.   Nuclei from infected or uninfected cells were isolated and allowed to chain elongate nascent RNA by established procedures (5).   Following RNA extraction, labeled RNA was hybridized to excess DNA affixed as "dots" (7) to nitrocellulose and after RNAse digestion, autoradiography was employed to detect hybridized RNA.

Results

It has been known for several years that the first gene that is transcribed in cells infected with type 5 adenovirus is the E1A gene, which is situated at the extreme left end of the virus (32,33) We reasoned that this property of the E1A gene was a consequence of its promoter containing regulatory elements that are recognized in a wide range of cells.   In order to test this possiblity, we sought to replace the normal regulatory elements found at the left end of the virus genome with promoter elements from a series of tissue specific genes.   Thus, it was our objective in these studies to prepare recombinant adenoviruses in which the entire E1A gene (including the E1A enhancer) and the E1B promoter were replaced by tissue specific promoters. Each of three separate viruses were constructed: alb 194, in which 1400 bp of the rat albumin gene including -441 bp upstream of the cap site, glo 194 which included 2.2 kb of the mouse β globin gene including -1.2 kb upstream of the cap site, and Ig194 in which

3.7 kb were inserted including -2 kb of upstream
sequence plus the first intron (which contains the
enhancer) upstream of the E1B coding sequences.

    The DNA of each recombinant virus prepared in
this way (Fig. 1) contained at its left end the
first 194 nucleotides of type 5 adenovirus, followed by a transcription unit constituted of these
sequences: the promoter region, a cap site, one
or two exons, and the 5' portion of an intron of a
cellular gene linked to viral sequences, namely,
the E1B intron, the E1B 3' splice site, the E1B
exon and poly(A) site. The viral sequences downstream of the E1B coding sequence are idenical
between the recombinants and wild type adenovirus.

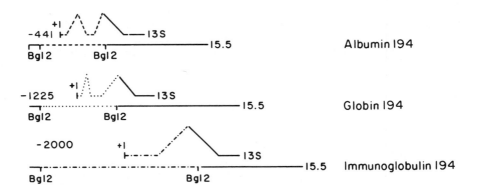

Fig.1 Recombinant adenoviruses were prepared,
using standard methodologies, in which the E1A
gene and the E1B promoter (normally situated at
the left end of the virus) were deleted and
replaced with each of three tissue specific promoters; alb 194 which contains the rat albumin
gene gene from -441 to +957, glo 194 which contains the mose golibin gene between -1225 and
+1075 and Ig 194 which contains a mouse heavy
chain immunoglobulin fragment between -2000 and
+1700. In all cases, the tissue specific promoter
directs the synthesis of E1B RNAs, designated 13S
in the figure. These viruse are defective (they
will not replicate) because they are missing the
E1A gene, and can only be propogated on 293 cells

which express E1A proteins constituitively. The
lack of the E1A and E1B proteins does not alter
the ability of the virus to introduce DNA and
establish the early phase of the viral infection.

The activity of each tissue promoter was
first tested by infecting HepG2 (human hepatoma)
cells and H411 cells (rat hepatoma), which transcribe albumin, and mpc11 cells, which transcribe
immunoglobulin, with 20 PFU/cell of the alb194,
glo194 and Ig194 viruses. At 24 hours RNA was
extracted and scored for the presence of E1B RNAs
using an RNAse protection assay. After infection
of hepatoma cells, E1B containing RNAs accumulated
only after infection with the alb194 virus. No
E1B RNA could be detected after infection with the
Ig194 or glo194 viruses, even though the level of
expression of the early viral E2a and E4 genes was
identical among the three viruses.. Similar
amounts of E1B RNAs were produced by the alb194
virus in human hepatoma cells when compared to rat
hepatoma cells suggesting that the interaction
between the albumin specific transacting factors
have been conserved between rodents and humans.

When the same viruses were used to infect
mpcll cells (mouse myeloma) E1B mRNAs could be
shown to accumulate only for the Ig194 virus.
Thus for these promoters, the upstream sequences
present in the adenovirus chromosome function in a
qualitatively similar fashion to the endogenous
gene. In separate experiments, it was also demonstrated ,using RNAse protection assays specific
for the 5' end of these promoters, that the RNA
molecules synthesized on the virus were correctly
initiated.

We next wished to consider the activity of
the albumin promoter in primary liver cells, a
circumstance where the rate of transcription of
the endogenous gene can be shown decline dramatically during the first 24 hours in culture. In
fact the highest rate of transcription of the
albumin gene is in hepatocytes fresh from the
animal (7,8,9) It has been shown previously that

the rate of cell-specific gene transcription but
not of "common" gene transcription (e.g. actin,
tubulin and others) declines when isolated hepatocytes are tested after 24 hours or more in culture. We next infected freshly plated primary
mouse hepatocytes as well as hepatocytes plated
for 24 hours with the alb194 and Ig194 viruses.
After five hours of infection total cell RNA was
prepared and assayed for the presence of E1B containing RNA The fresh mouse hepatocytes produced
E1B containing mRNA when infected with alb194
viruses but not when infected with Ig194 virus.
After the five hours of mRNA accumulation in mouse
hepatocytes the E1B signal was considerably
stronger (i.e. the E1B RNA concentration was
higher) than in HepG2 cells or rat hepatoma cells
infected for an equivalent length of time. After
24 hours in culture disaggregated primary liver
cells in culture transcribe the albumin gene at a
50 fold lower rate than after one hour in culture.
Likewise, cells in culture for 24 hours infected
with the alb194 virus for four hours accumulated
approximately fifty fold less E1B mRNA than
infected cells in culture for one hour. Thus the
activity of those trans acting factor(s) controlling albumin transcription declines considerably
during the first 24 hours in primary culture, and
suggests that the activity of this factor is
dependent on the presence of an external signal,
perhaps a cell-cell contact (8).

Finally, in order to demonstrate that gene
regulation in these experiments was the result of
transcriptional control, HepG2 cells were infected
with 20 PFU/cell of the alb194. glo194 and Ig194
viruses. After 24 hours, nuclei were prepared and
nascent RNA chains were elongated in each nuclear
sample in the presence of α-^{32}P-UTP and the
labeled RNA was hybridized to DNA affixed to
nitrocellulose filters. RNAse resistant hybridized RNA was scored. The transcription of the
early viral transcription units E4 and E2a was
equal after infection of HepG2 cells for each of
the viruses and similar to transcription in HeLa
cell nuclei, indicating that adsorption and initiation of transcription of viral transcription
units was normal for each of the viruses, and

similar between HepG2 and HeLa cells. [It should be noted that in each of these infections no E1A protein was present and the viral transcription observed was at the low levels established after 24 hours for early transcription units in such cases (34).] Transcription could only be scored from the alb194 virus in HepG2 cells. No transcription was detected for the other two viruses in these cells.

We also noted that in HepG2 cells the transcription of the endogenous human albumin gene (which can be scored independently from the endogenous gene) is easily detected and the signal from the endogenous albumin gene is fifty-fold higher than the viral gene. Thus the albumin promoter in the adenovirus functions in a tissue specific fashion, but at a reduced rate relative to the endogenous gene.

Discussion

In this report, we describe initial experiments charactarizing an adenoviral vector that may be useful in the qualitatative and quantitative ananlysis of tissue specific transcription. This approach may be of general utility since adenoviruses will efficiently deliver DNA into almost all mammalian cell types (including many primary cells).

The experiments in this report demonstrate, that at least in differentiated cells, appropriate patterns of tiisue specific transcription can be demonstrated for tissue specific promoters incorporated into the adenovirus genome. Thus, the albumin promoter situated on the viral genome is transcribed only in hepatoam cells and in primary liver cells. Conversely, the immunobglobulin promoter is functional only in myeloma cells. While tissue specificity is maintained for these promoters, the rate of transcription for the albumin promoter on the newly introduced viral DNAs is considerably lower than that of the single enogenous gene. This difference in rate is meaningful since each cell in the population can be shown to

have received at least one active viral genome.
This discrepancy between the quantitative perfor-
mance of the endogenous and exogenous albumin
genes could not have been discerned using a transient transfection assay since transfection introduces DNA into only a small subpopulation of
cells. This makes comparisons between the
activity of the newly introduced and chromosomal
genes impossible.

There are several possible ways to explain
the low rate of transcription of the albumin promoter in the virus genome. It is possible that a
maximal rate of transcription depends on the
activity of more than one regulatory sequence and
that a maximal rate is not achieved unless all of
these elements are present. It is unclear at this
point whether all the control elements involved in
transcription of the albumin gene were included in
the virus. It has been shown for example, that
several different tissue specific regulatory
sequences may contribute to the activity of the
immunoglobulin heavy chain gene (27) during
transfection experiments. Likewise, multiple regulatory elements are known in the vicinity of viral
promoters (20,35,36). To test this possibility
for the rat albumin gene we are constructing
viruses in which about 5 kb of albumin sequence
upstream of the cap site have been included.

An additional possibility is that the factors
controlling albumin transcription are unavailable
(i.e.; stably bound to genomic copy) to the newly
introduced DNA. Stable complexes for polymerase
II transcription units have been described (33)
and it is possible that the enogenous gene has
sequestered a large proportion of the requisite
transcription factors . We do not believe that
the low rate of transcription of the viral albumin
promoter reflects some interfering effect of the
viral chromosome because the albumin promoter can
be made to function at ahigh rate if the viral E1A
enhancer is insereted upstream of the albumin promoter (11).

It is of great long term interest to understand, not only how these factor(s) regulate albu-

min transcription, but to understand how these factor(s) are themselves regulated. One clear example of regulation occurrs in cultured primary liver cells where the rate of albumin transcription can be shown to decreas 50 fold in the first 24 hours after disaggregation and plating in culture (7). On the basis of the findings reported here, we can account for at least part of this decrease in albumin transcription as being a result of a loss of activity of a trans acting factor(s) that interacts with DNA sequences between -441 and +1075 of the albumin gene. The relationship between the external signal that is lost during liver cell culture and the trans acting factor that it controls is unknown but this interaction between the cell surface and the transcriptional machinery may be of great importance. The ability of the virus to probe transcriptional changes during cell culture may be of great advantage in studying developmental processes that are not amenable to analysis in cultured tumor cells.

In summary, these experiments suggest that cell specific factors are present in differentiated cells and that they recognize the regulatory sequences in the vicinity of their target genes. At least some of these factors are available for interaction with the extrachromosomal adenovirus template. Even though the albumin promoter in the virus functions in a cell-specific manner, expression occurs at a greatly reduced rate compared to the endogenous albumin gene and to the other viral genes. Finally, the activity of the factor(s) controlling the albumin gene declines significantly after disaggragated primary liver cells are placed in culture suggesting that this transcription factor(s) is controlled by an external signal. The nature of this signal remains to be elucidated.

Acknowledgements

J.M.F. is an assistant investigator of the Howard Hughes Medical Institute. L.E.B. is the recipient of a Damon Runyon Fellowship (DRG-794). This work was supported by grants from the National Institutes of Health (CA16006-13 and CA 18213-10 and the American Cancer Society Grant MV2710.

References

1. Darnell, J. E., Jr. (1982). Variety in the level of gene control in eukaryotic cells. Nature (London) 297, 365-371.

2. McKnight, G. S. and Palmiter, R. D. (1978). Transcriptional regulation of the ovalbumin and conalbumin genes by steriod hormones in chick oviduct. J. Biol. Chem. 254, 9050-9058.

3. Swaneck, G. E., Nordstrom, J. L., Kreuzaler, F., Tsai, M.-J. and O'Malley, B. (1979). Effect of estrogen on gene expression in chicken oviduct: evidence for transcriptional control of ovalbumin gene. Proc. Nat. Acad. Sci. USA 76, 1049-1053.

4. Weintraub, H., Lasen, A. and Groudine, M. (1981). ∝-globin-gene switching during the development of chicken embryos: Expression and chromosome structure. Cell 24, 333-344.

5. Derman, E., Krauter, K., Walling, L., Weinberger, C., Ray, M., and Darnell, J. E., Jr. (1981). Transcriptional control in the production of liver-specific mRNAs. Cell 23, 731-739.

6. Powell, D. J., Friedman, J. M., Oulette, A. J., Krauter, K. S., and Darnell, J. E., Jr. (1984). Transcriptional and posttranscriptional control of specific messenger RNAs in adult and embryonic liver. J. Mol. Biol. 179, 21-36.

7. Clayton, D. F., and Darnell, J. E., Jr. (1983). Changes in liver-specific compared to common gene transcription during primary culture of mouse hepatocytes. Mol. Cell Biol. 3, 1552-1561.

8. Clayton, D. F., Harrelson, A. L. and Darnell, J. E., Jr. (1985). Dependence of liver-specific transcription on tissue organization. Mol. Cell. Biol. 5, 2623-2632.

9. Clayton, D. F., Weiss, M., and Darnell, J. E., Jr. (1985). Liver specific-compared to common gene transcription during primary culture of mouse hepatocytes. Mol. Cell Biol. 3, 1552-1561.

10. Friedman, J.M., Babiss, L.E., Clayton, D.C., and Darnell, J.E, Cellular promoters incorporated into the adenovirus genome: regulation of transcriptional specificty and rate, Molecular Cell Biology, submitted

11. Babiss, L. E., Friedman, J. M. and Darnell, J.

E., Jr. (1986). Cellular promoters incorporated
into the adenovirus genome: Effects of viral
regulatory elements on transcription rates and
cell specificity of albumin and β-globin promoters. Molecular Cell Biology, submitted
12. Walker, M. D., Edlund, T., Boulet, A. M., and
Rutter, W. J. (1983). Cell-specific expression
controlled by the 5'-flanking region of insulin
and chymotrypsin genes. Nature 306, 551-561.
13. Hayashi, S., Kondoh, H., Yasada, K., Soma, G.,
Ikawa, Y. and Okada, T. S. (1985). Tissue-
specific regulation of a chicken 6 (delta) crystallin gene in mouse cells: involvement of the 5'
end region. EMBO J. 4, 2201-2208.
14. Queen, C., and Baltimore, D. (1983). Immunoglobulin transcription is activated by downstream
sequence elements. Cell 33, 741-748.
15. Gilles, C. D., Morrison, S. L., Oi, V. T., and
Tonegawa, S. (1983). A tissue-specific transcription enhancer element is located in the major
intron of a rearranged immunoglobulin heavy chain
gene. Cell 33, 717-728.
16. Ott, M., Sperling, L., Herbomel, P., Yaniv,
M., and Weiss, M. C. (1984). Tissue-specific
expression is conferred by a sequence from the 5'
end of the rat albumin gene. EMBO J. 3, 2505-
2510.
17. Ciliberto, G., Dente, L., and Cortese, R.
(1985). Cell-specific expression of a transfected
human d_1-antitrypsin gene. Cell 41, 531-540.
18. D'Onofrio, C., Colantuoni, V., and Cortese, R.
(1985). Structure and cell-specific expression of
a cloned human retinol binding protein gene: the
5' flanking region contains hepatoma-specific
transcriptional signals. EMBO J. 4, 1981-1989.
19. Stowe, N. D. (1981). Cloning of a DNA
fragment from the left-hand terminus of the
adenovirus type 2 genome and its use in site
directed mutagenesis. J. Virol. 37, 171-180.
20. Hearing, P., and Shenk, T. (1983). The adenovirus type 5 E1A transcriptional control region
contains a duplicated enhancer element. Cell
33, 695-703.
21. Babiss, L. E. and Ginsberg, H. S. (1984).
Adenovirus type5 early region 1B gene products is
required for efficient shut-off of host protein

synthesis. J. Virol. 50, 202-212.
22. Philipson, L., and Lindberg, U. (1974).
"Reproduction of adenovirus" pp 143-207 in
Comprehensive Virology, Vol. 3, ed. H. Fraenkel-
Conrat and R. R. Wagner. Plenum.
23. Klessig, D. F. (1984). "Adenovirus-Simian
virus 40 interactions" pg. 399-439, in The Adenoviruses, ed. H. S. Ginsberg. Plenum.
24. Nevins, J. R., Ginsberg, H. S., Blanchard, J.
M., Wilson, M. C. and Darnell, J. E., Jr. (1979).
Regulation of the primary expression of the early
adenovirus transcription units. J. Virol. 32,
727-733.
25. Cross, F., and Darnell, J. E. (1983).
Cycloheximide stimulates early adenovirus transcription if early gene expression is allowed
before treatment. J. Virol. 45, 683-692.
26. Sargent, T. D. (1981). The rat serum albumin
gene. Ph.D. Thesis, California Institute Technology.
27. Grosschedl, R. and Baltimore, D. (1985).
Cell-type specificty of immunoglobulin gene
expression is regulated by at least three DNA
sequence elements. Cell 41, 885-897.
28. Konkel, D. A., Tilghman, S. M., and Leder, P.
(1978). The sequence of the chromosomal mouse β-
globin major gene: homologies in capping splicing
and poly(A) sites. Cell 15, 1125-1132.
29. Frost, E., and Williams, J. (1978). Mapping
of temperature-sensitive mutants of adenovirus
type 5 by marker rescue. Virol. 91, 39-50.
30. Chirgwin, J. M., Przybila, A. E., MacDonald,
R. and Rutter, W. J. (1979). Isolation of biologically active ribonucleic acid from sources
enriched in ribonuclease. Biochem. 18, 5294-5299.
31. Melton, D. A., Krug, P. A., Rebaglicti, M.R.,
Maniotis, T., Zinn, K., and Green, M. R. (1984).
Efficient in vitro synthesis of biologically
active RNA andRNA hybridization probes for
plasmids containing bacteriophage sp6 promoter.
Nuc. Acid Res. 12, 7035-7056.
32. Jones,N. and Shenk, T. (1979) An adenovirus
type 5 early gene function regulates the expression of other early viral genes. Proc. Nat. Acad.
of Sci76, 3665-3669

33. Nevins, J. R. (1981) Mechanism of activation
of early viral transcription by the adenovirus E1A
gene product. Cell 26, 213-220.
35. Imperiale, M. J., Feldman, L. J., and Nevins,
J. R. (1983). Activation of gene expression by
adenovirus and herpes virus regulatory genes acting in trans and by a cis-acting adenovirus
enhancer element. Cell 35, 127-136.
36. Moreau, P., Hen, B., Wasylyk, B., Everett, R.,
Gaub, M. P. and Chambon, P. (1981). The SV40 72
base pair repeat has a striking effect on gene
expression both in SV40 and other chimeric recombinants. Nuc. Acids Res. 9, 6047-6068.

Transcriptional Control Mechanisms, pages 437–450
© 1987 Alan R. Liss, Inc.

ANALYSIS OF THE PROMOTER REGION OF THE HUMAN PROTO-ONCOGENES ENCODING THE EGF RECEPTOR AND c-HARVEY ras1[1]

Glenn T. Merlino, Shunsuke Ishii, Alfred Johnson, James Kadonaga[2], Robert Tjian[2], and Ira Pastan

Laboratory of Molecular Biology, National Cancer Institute, National Institutes of Health, Bethesda, Maryland 20892

ABSTRACT   The hypothesis that the control of cellular growth can be intimately associated with the quality and quantity of proteins encoded by the cellular homologs of viral oncogenes (proto-oncogenes) is gaining considerable support. We decided to examine control of cellular oncogene expression at the level of transcription. The promoter regions of two proto-oncogenes, the epidermal growth factor (EGF) receptor gene and the cellular Harvey ras1 (c-Ha-ras1) gene, have been identified and compared in the hope that such an analysis will clarify our understanding of the mechanism by which these genes are regulated. Both promoters were found to be highly G+C rich, and not to contain a TATA box or a CAAT box in their characteristic upstream locations. Both EGF receptor and c-Ha-ras1 RNAs are intiated at multiple sites. Furthermore, both promoters contain multiple copies of GC boxes (GGGCGG or CCGCCC), which can bind Sp1 factor as determined by an in vitro DNase I footprinting assay.

[1]This work was supported in part by an NIH grant to R.T.
[2]Department of Biochemistry, University of California, Berkeley, California 94720

INTRODUCTION

Epidermal growth factor (EGF) stimulates cellular growth by binding to a 170 kDA membrane-spanning glycoprotein receptor (1,2). The EGF receptor contains an external amino terminal domain capable of binding EGF, and an internal carboxy terminal domain possessing tyrosine kinase activity (3-5). EGF binding induces internalization, or downregulation, of the receptor, and stimulates the tyrosine kinase to phosphorylate itself and other substrates (6-8). The oncogene erb B, carried by the avian erythroblastosis retrovirus, is homologous to the kinase and transmembrane domains of the EGF receptor gene, suggesting that the latter is the cellular homology of v-erb (9-12).

The ras proto-oncogenes constitute a family of conserved mammalian genes that encode related proteins of about 21 kDA (p21) (13-15). The ras genes were initially characterized as the transforming genes of the Harvey (Ha-ras) and Kirsten (Ki-ras) murine sarcoma viruses (13). Mutant cellular ras genes encoding altered p21 proteins are found in numerous human and rodent tumor cells and are capable of transforming the murine cell line NIH 3T3 (14,16-21).

Regulation of normal transcription of these proto-oncogenes undoubtably plays a major role in controlling cellular growth. We have identified transcriptional promoter regions of these two human proto-oncogenes (EGF receptor and c-Ha-ras1) and discovered several notable similarities.

MATERIALS AND METHODS

Genomic clones were isolated and analyzed as detailed elsewhere (22). DNA sequencing was performed according to Maxam and Gilbert (23). Primer extension of mRNA by reverse transcriptase was described previously (22,24). DNA transfections into African green monkey CV-1 cells was performed as detailed elsewhere (22,24). Chloramphenicol acetyltransferase (CAT) assays were as described (25,26). DNase I footprinting of the transcription factor Sp1 was performed according to Tjian and coworkers (27,28).

RESULTS AND DISCUSSION

The Promoter Regions of EGF Receptor and c-Ha-ras1

Genomic clones containing the 5' end of the EGF receptor gene were isolated from a human fetal liver genomic DNA library (22). The sequence of the 5' flanking region is shown in Fig. 1. Several techniques were utilized to identify this DNA as containing the active promoter and RNA start sites (22).

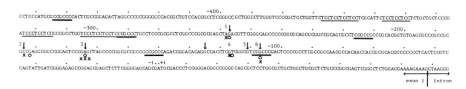

FIGURE 1. Nucleotide sequence of the 5' end of the human C-Ha-ras1 gene. +1 corresponds to the "A" of the initiator methionine codon, and residues preceding it are represented by negative numbers. The first exon-intron border is shown at lower right. The multiple initiation sites (numbered 1-6) demonstrated by primer extension, S1 nuclease mapping, and in vitro transcription are indicated by x, o, and ↓, respectively. The GC box sequences are underlined by a thick solid line, while another repeated sequence, TCCTCCTCC is underlined by a thin solid line. Taken from ref. 22.

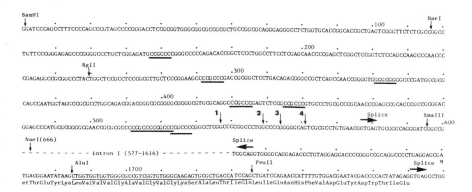

FIGURE 2. Nucleotide sequence of the 5' end of the human C-Ha-ras1 gene. Sequence and numbering is from

elsewhere (37,38). The transcriptional initation sites as
determined by primer extension are indicated by arrows 1 to
4. The DNA fragment used as a primer (1735 to 1684) is
underlined by a thin solid line. The GC box sequences are
underlined by thick solid lines. The donor and acceptor
splice signals of what is thought to be the first intron
are indicated. Taken from ref. 24 (Copyright 1985 by the
AAAS).

     The promoter region of the c-Ha-ras1 gene was
identified as described by Ishii et al. (24). The sequence
of the 5' flanking region is shown in Fig. 2. A comparison
of the two promoter regions reveals that both are highly
enriched in G+C residues: 88% for the EGF receptor gene
and 80% for the c-Ha-ras1 gene. In addition, neither
promoter contains a TATA box nor a CAAT box in their
characteristic upstream positions. Both promoters also
possess numerous CCGCCC or GGGCGG sequences (GC boxes,
29), similar to those shown to bind the SV40 transcription
factor Sp1 (see below). The GC boxes of the c-Ha-ras1
promoter more closely conform to the consensus sequence
proposed by Tjian ($\frac{G}{T}$GGGCGG$\frac{GGC}{AAT}$) (30).

Multiple RNA Transcriptional Start Sites

     To determine the location of the RNA start sites in
these proto-oncogene promoters, primer extension assays
were utilized. Fig. 3A shows that when a 76 bp DNA primer
complementary to a downstream site of EGF receptor mRNA
isolated from A431 epidermoid carcinoma cells was extended
by reverse transcriptase, six RNA start sites were
detected. These sites were confirmed by S1 nuclease protection mapping and in vitro transcription (22).

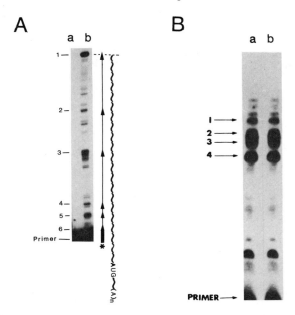

FIGURE 3. Determination of transcriptional start sites for EGF receptor gene (A) and c-Ha-ras1 gene (B) by primer extension. (A) The primer used for the EGF receptor, a 5' end labeled fragment (-101 to -25), was hybridized to either 5 µg of A431 poly (A)+ RNA (b) or 5 µg of calf liver tRNA (a) and extended by reverse transcriptase; the resulting products were analyzed by denaturing polyacrylamide gel electrophoresis (PAGE). RNA is represented by a wavy line. Stop sites are numbered. (B) The primer for c-Ha-ras1, a 5' end labeled fragment (1735 to 1684), was hybridized to either 5 µg A431 poly (A)+ RNA (a) or 5 µg of pEJ6.6-transformed NIH3T3 cells (b), and extended and analyzed as above. Locations of primers is indicated. Taken from refs. 22, and 24 (Copyright 1985 by the AAAS).

Extention of a 50 bp 5'-specific c-Ha-ras1 primer after hybridization to A431 mRNA revealed that Ha-ras RNAs also start at multiple sites (Fig. 3B). Four major sites were detected in this experiment, although other sites could exist further upstream (24,31). We have therefore determined that both proto-oncogene promoters were capable of initiating RNA synthesis at multiple sites.

Activity of the Two Proto-oncogene Promoters

The activity of these promoter regions was tested by creating chimeric plasmids containing the proto-oncogene promoters upstream of the bacterial gene encoding the CAT marker enzyme. The resulting plasmid DNAs (pERCAT1 and prasCAT1, Fig. 4) were $CaPO_4$-precipitated and transfected into African green monkey kidney CV-1 cells. Fig. 4 shows that both pERCAT1 and prasCAT1 were capable of stimulating CAT RNA and enzyme synthesis. CAT activity was measured after 48 hrs (transient transfection assay) and found to be almost as high as that determined for RSV-CAT, a very active Rous sarcoma virus LTR promoter (26).

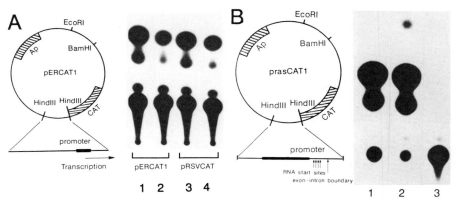

FIGURE 4. Structure and expression of DNA plasmids containing the promoter regions of EGF receptor gene (A) and c-Ha-ras1 gene (B). The promoters were placed upstream of the bacterial CAT gene. RNA start sites and exon-intron boundaries are indicated where appropriate. Ap = ampicillin resistance gene. Activity was determined in a transient (48 hrs.) transfection assay in CV-1 cells as described elsewhere (22,24-26). (A) DNA used was either pERCAT1 (lanes 1,2) or pRSV-CAT (lanes 3,4), at either 5 µg (lanes 2,4) or 10 µg (1,3). (B) DNA (10 µg) used was either prasCAT1 (1), pRSV-CAT (2), or none (3). Taken from refs. 22, and 24 (Copyright 1985 by the AAAS).

The Two Proto-oncogene Promoters Bind Sp1

To determine if the GC boxes within the proto-oncogene promoter DNA bind the transcription fractor Sp1, DNase I footprint analysis was used. A purified preparation of Sp1 (about 90%) was mixed with end-labeled c-Ha-ras 1 promoter DNA; the resulting complexes were treated with DNase I and analyzed by denaturing PAGE. Fig. 5 shows that DNase I-resistant regions appear in the Ha-ras DNA where the GC boxes are localized, suggesting that Sp1 recognizes and binds to these sites.

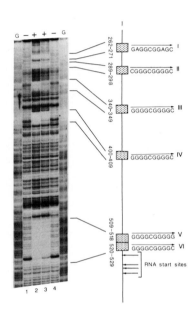

FIGURE 5. Determination of Sp1 binding to c-Ha-ras1 gene promoter region by in vitro DNase I footprint analysis. 5' end labeled c-Ha-ras1 promoter fragment was incubated with purified Sp1 (about 70% pure), and then with DNase I; the resulting fragments were fractionated on 6% denaturing PAGE, as detailed elsewhere (24). Lanes 1 and 4 (-) show autoradiographic DNase I pattern in absence of Sp1. Lanes 2 and 3 (+) show pattern in presence of 80 ng or 40 ng of Sp1, respectively. Protected regions are bracketed. The arrows over the GC boxes indicate the 5'→3' direction of the G-rich strand of DNA. Taken from ref. 31 (Copyright 1986 by AAAS).

EGF receptor promoter GC boxes were also found to bind Sp1, but much more weakly[3]. The presence of GC boxes is shared by several other promoters, including c-myc (29,30, 32,33). Sp1 binds to SV40 early promoter DNA where it acts as a positive transcription factor (34-36). The role of Sp1 in the expression of these proto-oncogenes cannot be determined by in vitro binding alone. We therefore tested the requirement for Sp1 by using an in vivo competition assay.

Competition in an in Vivo Transfection Assay

CV-1 cells were transfected with prasCAT1 DNA alone or in addition to increasing amounts of pSVneo DNA, containing the 6 GC box repeats of the SV40 promoter (31). Fig. 6 shows that as more pSV2neo plasmid is added (up to a 10-fold molar excess), levels of c-Ha-ras 1 RNA as measured by CAT activity, are reduced over 10-fold, suggesting that Sp1 or a related factor is required for Ha-ras transcription. Further experiments have localized the competitive region to the 21 bp GC-box containing repeats (31). Similar experiments with the EGF receptor promoter pERCAT1 plasmid are in progress.

[3]Ishii, Johnson, Kadonaga, Tjian, Merlino, and Pastan, manuscript in preparation.

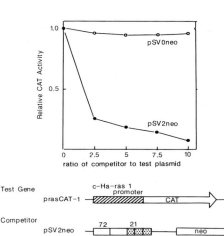

FIGURE 6. Competition between the SV40 promoter and the c-Ha-ras1 gene promoter in a transient transfection assay. CV-1 cells were transfected with 2 μg of prasCAT1 test plasmid, 5-20 μg of competitor DNA and an appropriate amount of carrier DNA to a total of 22 μg (31). Transfection and CAT enzyme assays are described previously (24-26). CAT activity was quantitated and the resulting data were normalized relative to the value obtained for cells with no competitor added (1.0). The structure of the test gene and the competitor DNAs are shown at the bottom, containing CAT gene, neo resistance gene, 72 bp repeat region, or 21 bp repeat region (containing GC boxes) of the SV40 promoter. Taken from ref. 31 (Copyright 1986 by the AAAS).

These results describe two promoters controlling proto-oncogene protein production which share some similar properties, as listed in Table 1. However, the similarities could be quite superficial; the promoters are obviously very different in a variety of other ways. Although conclusions based on these preliminary studies are intriguing, they can lead to oversimplified generalizations. Further analysis and comparison of the EGF receptor and the Ha-ras gene promoter, as well as other proto-oncogene promoters is warranted and is currently under

way; results from these studies may provide clues as to
how the expression of certain proto-oncogenes is regulated.

TABLE 1
COMPARISON BETWEEN EGF RECEPTOR AND
c-Ha-ras1 GENE PROMOTERS

	EGF Receptor	c-Ha-ras1
Number of RNA start sites[a]	6	4
TATA box	No	No
CAAT box	No	No[b]
G+C content[c]	88%	80%
GC boxes[d]	5	10
In vitro Sp1 binding sites[e]	4	5

[a]This is the number of start sites detected to date,
using techniques described elsewhere (22,24).
[b]In the c-Ha-ras1 promoter region a single CAAT box
consensus sequence exists, but not in the expected upstream
location (31).
[c]The percentages of the G+C content are for the
following regions: EGF receptor, -540 to -1 (22), c-Ha-
ras1, 117 to 536 (24).
[d]GC boxes here are defined as the sequences CCGCCC or
GGGCGG. If this 6-bp sequence is redefined as a 10-bp consensus sequence (see 30), then these boxes conform to the
consensus by differing extents (31).
[e]This number represents those sites which were clearly
defined by in vitro DNase I footprinting (31); more sites
may exist further upstream.

ACKNOWLEDGMENTS

    The authors wish to thank Dr. John Brady for useful
suggestions, Ray Steinberg for photography, and Althea
Gaddis for editorial assistance.

REFERENCES

1.   Savage CR, Cohen SJ (1972). Epidermal growth factor and a new derivative: rapid isolation procedures and biolgical and chemical characterization. J Biol Chem 247:7609.

2.   Carpenter G, Cohen S (1979). Epidermal growth factor. Annu Rev Biochem 48:193.

3.   Ushiro H, Cohen S (1980). Identification of phospho-tyrosine as a product of epidermal growht factor activated protein kinase in A431 cell membranes. J Biol Chem 255:8363.

4.   Das M, Miyakawa T, Fox CF, Pruss RM, Aharonov A, Herschman HR (1977). Specific radiolabeling of a cell surface receptor for epidermal growth factor. Proc Natl Acad Sci USA 74:2790.

5.   Sahyoun N, Hock RA, Hollenberg MD (1978). Insulin and epidermal growth factor urogastrone: affinity cross-linking to separate binding sites in rat liver membranes. Proc Natl Acad Sci USA 75:1675.

6.   Beguinot L, Lyall RM, Willingham MC, Pastan I (1984). Down-regulation of the EGF receptor in KB cells is due to receptor internalization and subsequent degradation in lysosomes. Proc Natl Acad Sci USA 81:2384.

7.   Hunter T, Cooper JA (1981). Epideral growth factor induces rapid tyrosine phosphorylation of proteins in A431 human tumor cells. Cell 24:741.

8.   Cohen S, Ushiro H, Stoscheck C, Chinkers M (1982). A native 170,000 epidermal growth factor receptor-kinase complex from shed plasma membrane vesicles. J Biol Chem 257:1523.

9.   Downward J, Yarden Y, Mayes E, Scarce G, Totty N, Stockwell P, Ullrich A, Schlessinger J, Waterfield MD (1984). Close similarity of epidermal growth factor receptor and v-erb B oncogene protein sequence. Nature (London) 307:521.

10. Xu Y-H, Ishii S, Clark AJL, Sullivan M, Wilson RK, Ma DP, Roe BA, Merlino GT, Pastan I (1984). Human epi-dermal growth factor receptor cDNA is homologous to a variety of RNAs overproduced in A431 carcinoma cells. Nature (London) 304:806.

11. Ullrich A, Coussens L, Hayflick JS, Dull TJ, Gray A, Tam AW, Lee J, Yarden Y, Libermann TA, Schlessinger J, Downward J, Mayes ELV, Whittle N, Waterfield MD, Seeburg PH (1984). Human epidermal growth factor receptor cDNA sequence and aberrant expression of the

amplified gene in A431 epidermal carcinoma cells. Nature (London) 304: 418.

12. Lin CR, Chen WS, Kruiger W, Stolarsky LS, Weber W, Evans RM, Verma IM, Gill GN, Rosenfeld MG (1984). Expression cloning of human EGF receptor complementary DNA: gene amplification and three related messenger RNA products in A431 cells. Science 224:843.

13. Ellis RW, Defeo D, Shih TY, Gonda MA, Young HA, Tsuchida N, Lowy DR, Scolnick EM (1981). The p21 src gene of Harvey and Kirsten sarcoma viruses originate from divergent members of a family of normal verte-brate genes. Nature (London) 292:506.

14. Taparowsky E, Shimizu K, Goldfarb M, Wigler M (1983). Structure and activation of the human N-ras gene. Cell 34:581.

15. Madaule P, Axel R (1985). A novel ras-related gene family. Cell 41:31.

16. Tabin CJ, Bradley SM, Bargmann CI, Weinberg RA, Papageorge AG, Scolnick EM, Dhar R, Lowy DR, Chang EH (1982). Mechanism of activation of a human oncogene. Nature (London) 300:143.

17. Reddy E, Reynolds R, Santos E, Barbacid M (1982). A point mutation is responsible for the acquisition of transforming properties by the T24 human bladder car-cinoma oncogene. Nature (London) 300:149.

18. Taparowsky E, Suard Y, Fasano O, Shimizu M, Goldfarb M, Wigler M (1982). Activation of the T24 bladder carcinoma transforming gene is linked to a single amino acid change. Nature (London) 300:762.

19. Shimizu K, Birnbaum D, Ruley MA, Fasano O, Suard Y, Edlund L, Taparowsky E, Goldfarb M, Wigler M (1983). Structure of the Ki-ras gene of the human lung carci-noma cell line Calu-1. Nature (London) 304:497.

20. Yuasa Y, Srivastara S, Dunn C, Rhim J, Reddy E, Aaronson S (1983). Acquisition of transforming prop-erties by alternative point mutations within c-bas/has human proto-oncogene. Nature (London) 303:775.

21. Capon D, Seeburg PH, McGrath JP, Hayflick JS, Edman U, Levinson AD, Goeddel DV (1983). Activation of Ki-ras2 gene in human colon and lung carcinomas by two different point mutations. Nature (London) 304:507.

22. Ishii S, Yu Y-H, Stratton RM, Roe BA, Merlino GT, Pastan I (1985). Characterization and sequence of the promoter region of the human epidermal growth factor receptor gene. Proc Natl Acad Sci USA 82:4920.

23. Maxam AM, Gilbert W (1980). Sequencing end-labeled DNA with base-specific chemical cleavages. Methods Enzymol 65:499.

24. Ishii S, Merlino GT, Pastan I (1985). Promoter region of the human Harvey ras proto-oncogene: similarity to the EGF receptor proto-oncogene promoter. Science 230:1378.

25. Gorman C, Moffat L, Howard B (1982). Recombinant genomes which express chloramphenicol acetyltransferase in mammalian cells. Molec Cell Biol 2:1044.

26. Gorman C, Merlino GT, Willingham M, Pastan I, Howard B (1982). The Rous Sarcoma virus long terminal repeat is a strong promoter when introduced into a variety of eucaryotic cells by DNA mediated transfection. Proc Natl Acad Sci USA 79:6777.

27. Dynan WS, Tjian R (1983). The promoter-specific transcription factor Sp1 binds to upstream sequences of the SV40 early promoter. Cell 35:79.

28. Jones KA, Yamamoto KR, Tjian R (1985). Two distinct transcription factors bind to the HSV thymidine kinase promoter in vitro. Cell 42:559.

29. Dynan WS, Tjian R (1985). Control of eukaryotic messenger RNA synthesis by sequence-specific DNA-binding proteins. Nature (London) 316:774.

30. Kadonaga JT, Jones KA, Tjian R (1986). Promoter-specific activation of RNA polymerase II transcription by Sp1. Trends Biochem Sci 11:20.

31. Ishii S, Kadonaga JT, Tjian R, Brady JN, Merlino GT, Pastan I (1986). The human Harvey ras 1 proto-oncogene promoter binds the Sp1 transcription factor. Science 232:1410.

32. Watt R, Nishikura K, Sorrentino J, ar-Rushdi A, Croce CM, Rovera G (1983). The structure and nucleotide sequence of the 5' end of the human c-myc oncogene. Proc Natl Acad Sci USA 80:6307.

33. Stanton LW, Yang J-Q, Eckhardt LA, Harris LJ, Birschtein BK, Marcu KB (1984). Product of a reciprocal chromosome translocation involving the c-myc gene in a murine plasmacytoma. Proc Natl Acad Sci USA 81:829.

34. Benoist C, Chambon P (1981). In vivo sequence requirements of the SV40 early promoter region. Nature (London) 290:304.

35. Hansen U, Sharp PA (1983). Sequences controlling in vitro transcription of SV40 promoters. EMBO J 2:2293.

36.   Tjian R (1981). T antigen binding and the control of SV40 gene expression. Cell 26:1.
37.   Capon DJ, Chen EY, Levinson AD, Seeburg PH, Goeddel DV, (1983). Complete nucleotide sequences of the T24 human bladder carcinoma oncogene and its normal homologue. Nature (London) 302:33.
38.   Reddy EP (1983). Nucleotide sequence analysis of the T24 human bladder carcinoma oncogene. Science 220: 1061.

Transcriptional Control Mechanisms, pages 451–462
© **1987 Alan R. Liss, Inc.**

TRANSCRIPTIONAL QUIESCENCE AND METHYLATION OF
MLV LTR FUSION GENES IN TRANSGENIC MICE

Timothy A. Stewart and Sharon L. Pitts

Department of Developmental Biology
Genentech, Inc., 460 Point San Bruno Boulevard
South San Francisco, CA   94080

ABSTRACT   Fusion genes, made by linking the Moloney
leukemia virus long terminal repeat to either the
mouse c-myc gene or to the human interleukin 2 gene,
were introduced into the mouse germline. These genes
were in general transcriptionally inactive although
there was a low level of expression of the c-myc
containing fusion gene in the thymus of one transgenic
mouse. This lack of expression correlated with
methylation of the fusion genes.

INTRODUCTION

Progress in the analysis of the hematopoietic pathway
has been made possible, at least in part, by the advance
of two complementary technologies, that of cell culture
which has enabled the development and study of either
permanent hematopoietic cell lines or partially purified
cell populations and that of gene transfection which has
permitted the analysis of the control regions of genes
expressed in hematopoietic cells. More recently the use
of transgenic mice has made feasible the study of gene
regions that participate in more complex phenomena (for
example, the switching from embryonic to adult hemoglobins
(1)). We have also used transgenic mice to explore the
consequences of inappropriately regulated gene expresssion
in hematopoietic cells as a means of understanding the
early events in hematopoietic pathology. These
experiments require that particular genes be expressed
under the control of heterologous promoters that are
active in the targeted hematopoietic cells. The murine
leukemia viruses have been reported to have a T-cell type

tropism (2), thus we fused the Moloney leukemia virus
(MLV) long terminal repeat (LTR) to the mouse c-myc gene
and also to the human interleukin 2 gene. These fusion
genes then allowed us to ask whether the reported MLV
tropism for T cells is at the level of expression or
infection. An unusual feature of the murine leukemia
viruses is that proviruses transmitted through the germ
line are, to a large extent, transcriptionally silent (3)
and this lack of expression is accompanied by extensive
methylation of the proviral sequences (4). Furthermore,
MLV-LTR fusion genes transfected into teratocarcinoma
cells (5,6) are also expressed at levels substantially
below that seen using fibroblast cells as recipients. The
transgenic mice carrying MLV-LTR fusion genes described
here have allowed us to examine whether the lack of
proviral expression is due to the process of retroviral
integration in embryonic cells (for example a preference
for transcriptionally inert regions for integration in
germ cells or early embryonic cells) or is inherent in the
viral sequences and their interaction with transacting
factors in the early embryo.

RESULTS

Two recombinant DNA constructions are detailed in
Figure 1. The MLV-myc gene (Fig. 1) has the second and
third exons of the mouse myc gene (which includes the
entire myc protein coding region) placed under the control
of the Moloney leukemia virus long terminal repeat (MLV
LTR). The MLV-IL-2 (Fig. 1) has the cDNA encoding human
interleukin 2 followed by the poly A addition signal from
the hepatitis B virus and similarly placed under the
control of the MLV LTR. These plasmids were separately
injected into the pronucleus of one-cell mouse eggs using
established protocols (7). One mouse carrying the MLV-myc
construction and three carrying the MLV-IL-2 construction
were identified by analysis of tail DNA; all four were
mated to animals of the same genotype (C57BL/6J) and
except for one of the MLV-IL-2 mice, the sequences were
transmitted through the germline as expected for an
integrated gene giving rise to three MLV fusion gene
transgenic families.

pMLV-myc

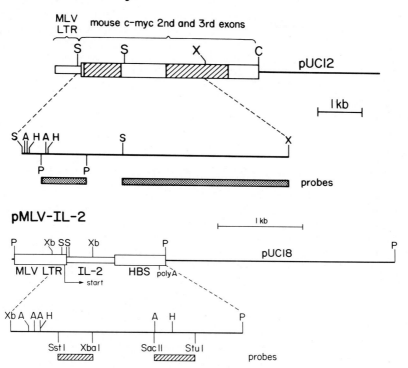

FIGURE 1. MLV-myc and MLV-IL-2 fusion genes and
probes used in the Southern analysis. MLV-myc was
constructed by a three way ligation between 1) the 618 bp
ClaI-SmaI MLV fragment (includes the MLV promoter and RNA
transcription start sites and approximately 500 bp 5' of
the start, and was taken from the pZAP subclone (18); 2)
the 3.5 kb XbaI to EcoRI fragment of the mouse c-myc gene
(the XbaI site is immediately 5' to the second exon and
was made blunt by Klenow polymerase); the EcoR1 site was
added 3' to the poly A addition site in the original
subcloning (19); and 3) PUC12 digested with AccI and
EcoRI. MLV-IL-2 was constructed in three steps. First
the 2.0 kb MLV Sma-SmaI fragment that includes the 3' LTR
was subcloned into the SmaI site of PUC12 and the 625 bp
ClaI to EcoRI LTR fragment prepared. Second, the 1150 bp
IL-2-HbS poly A EcoRI to EcoRI fragment was prepared
following subcloning of the IL-2-HBS fragment (taken from

an expression plasmid (20)) into a PUC 12 vector. Third, these two fragments were then ligated into a Puc12 vector that had been cut with Accl and EcoRI. All constructions were tested by multiple restriction enzyme analysis and standard recombinant DNA techniques were used (21). The MLV-myc plasmid was linearized 5' to the poly A addition site (at ClaI) and injected as a linear molecule. The MLV-IL-2 plasmid was cut twice (at the PstI sites and injected as a mixture of the two fragments. Restriction enzymes are: A, AvaI; C, ClaI, H, HpaII, P, PstI; S, SstI; X, XhoI; Xb, XbaI.

One positive F1 animal from each of the three families (one MLV-myc and two MLV-IL-2) was sacrificed and RNA, extracted from the indicated tissues and organs using the guanidine HCl/ethanol procedure (8), analyzed for transcription originating from the newly introduced gene. In the case of the MLV-IL-2 mice no transcription was detected in any of the tissues or organs from either of the two families (shown for one of the MLV-IL-2 mice in Fig. 2). The band at approximately 620 bases is due to incomplete digestion of the probe.

Analysis of expression of the MLV-myc gene requires an assay that can detect transcription from the newly introduced gene over the background of the endogenous mouse myc gene which is expressed in almost all dividing cells (9). This was achieved in an S1 nuclease protection assay by using a probe that spanned the junction between the MLV LTR and the myc gene. The results of this assay are shown in Fig. 3. In all tissues a protected fragment is seen at 486 bases representing the endogenous message spliced correctly to the second myc exon; there is also a longer protected fragment of 514 bases seen in both the control spleen RNA and the MLV-myc spleen RNA. As there are 28 bases of the first myc exon remaining in the MLV myc construction, this 514 base band could be due to either unspliced message or to initiation within the first myc intron. In addition there is in the thymus of the MLV-myc mice (but not control mice) an RNA species that protects 544 bases of the probe as would be expected for a transcript that initiates at the MLV start site. This protected fragment is not reliably seen in any other tissues from the MLV-myc mice nor is it seen in any tissues in control mice (shown for thymus in Fig. 3).

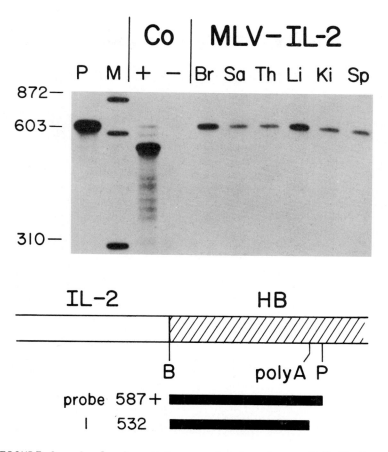

FIGURE 2. Analysis of transcription in an MLV-IL-2 transgenic mouse. RNA was prepared (8) from the indicated tissues (Br, brain; Sa, salivary glands; Th, thymus; Li, liver; Ki, kidney; Sp, spleen) and analyzed by an S1 nuclease protection assay (7) using the BamHI to PstI M13 synthesized probe diagrammed. P, undigested probe; M, size marker (QX174 DNA digested with HaeIII and 5' end labelled). The control lanes are: a) positive control (+), RNA taken from a cell line expressing an unrelated gene (the herpes gD gene) using the same hepatitis B poly A addition signal and b) negative control (−), spleen RNA taken from a wild type C57B1/6 animal. Abbreviations: B, BamHI; P, PstI.

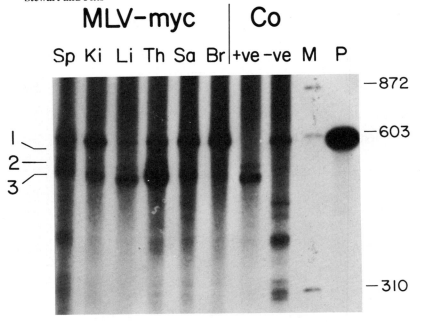

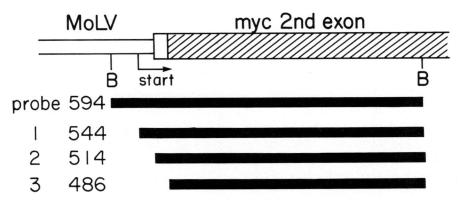

FIGURE 3. Analysis of transcription in an MLV–myc transgenic mouse. RNA was prepared from the indicated organs (same abbreviations as Fig. 2) and analyzed in an S1 nuclease protection assay using the BSSHII to BSSHII 5' end labelled fragment indicated. P, undigested probe; M–size markers (as in Figure 2). The controls are thymus RNA from a wild type C57BL/6 mouse (+ve) and yeast tRNA (–ve). (B, BssHII)

   The lack of expression in either of the two MLV-IL-2
mice and the relatively low level of expression in the
MLV-myc mouse led us to examine the methylation pattern of
the newly introduced genes. DNA was extracted from the
spleens and brains of one of the MLV-myc mice and digested
first with either of AvaI, MspI, or HpaII and then further
digested with SstI and HindIII. The results obtained from
the MLV-myc samples, shown in Fig. 4, can be summarized as
follows: there was no substantial difference between the
spleen and brain with respect to the methylation pattern
of either the endogenous myc sequences or the MLV-myc
sequences. There are, however, major differences between
the endogenous myc gene and the newly introduced MLV-myc
genes. The endogenous myc gene (1400 bp band, lanes 5 and
13) was completely sensitive to digestion with both HpaII
and with AvaI (obvious for brain only as the spleen DNA
was not digested with AvaI) when the digested DNA was
probed with a second myc exon probe. In contrast the
newly introduced MLV-myc sequences are completely
resistant to the methylation sensitive enzymes AvaI and
HpaII but are digested with the methylation insensitive
MspI. When the same samples are analyzed using a probe
that spans the 3' end of the second intron and most of the
third exon (Fig. 4B) the results were similar although the
endogenous myc sequences appear to be at least partially
methylated (for example compare lanes 5, 7, and 8 in Fig.
4A and B). Hybridization with this same probe indicated
that the MLV-myc sequences are again substantially
resistant to AvaI and HpaII digestion although the major
band in lanes 2 and 10 (Fig. 4B) at approximately 1400 bp
can be seen at shorter exposures to be a doublet
indicating partial sensitivity to AvaI digestion.
Similarly, analysis of one of the MLV-IL-2 mice with a
probe that spanned the junction between the MLV LTR and
the human IL-2 indicated that the MLV-IL-2 genes were also
resistant to AvaI and HpaII digestion (data not shown).

                         DISCUSSION

   Infection of mice by Moloney leukemia virus (MLV)
reveals an apparent tropism for hematopoietic cells,
primarily thymocytes, although the virus is obviously
capable of infecting other cell types in vitro (2). By
directly injecting MLV-LTR fusion genes into one cell
mouse eggs we were able to ask whether the apparent

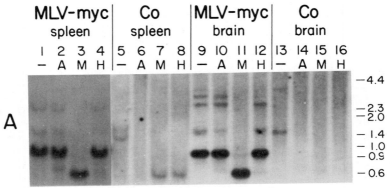

A

probe: Pst I–Pst I   2nd myc exon

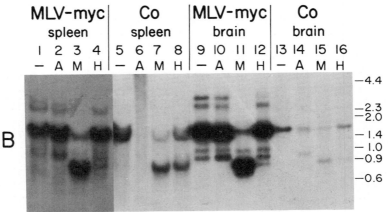

B

probe: Sst I–Xho I   2nd intron/3rd exon

FIGURE 4. Analysis of the methylation pattern of the myc genes in an MLV–myc mouse. DNA was extracted from the spleen and brain of an MLV–myc and a control mouse, digested with AvaI (lane 2), MspI (lane 3) or HpaII (lane 4) and then digested with SstI and HindIII. The digested DNA was analyzed following electrophoresis and transfer to nitrocellulose using a 400 bp PstI–PstI second myc exon probe (panel A) or a 1400 bp SstI–XhoI second myc intron/third myc exon probe. The size markers on the left are obtained using a combination of λ phage DNA digested with HindIII and ΦX174 phage DNA digested with HaeIII.

in vivo cell type selectivity is the result of an enhanced
level of MLV transcription in hematopoietic cells or
whether there are other contributing factors, for example
at the level of infection. Although we were not able to
detect transcription of the MLV-IL-2 genes in any of the
six tissues and organs examined in the two MLV-IL-2
families, in the one MLV-myc transgenic mouse there was a
low level of expression initiated within the MLV LTR in
the thymus. These results then are consistent with there
being a specific enhanced level of MLV LTR initiated
transcription within thymic cells although we have not
excluded the possibility that there may also be
differences in RNA stability.

We have also examined the newly integrated MLV LTR
fusion genes with regard to the methylation pattern of at
least some of their CpG sites. In two mice carrying two
different fusion genes the sites examined (AvaI and
HpaII/MspI) are heavily methylated both in absolute terms
and, where we were able to make comparisons with
endogenous sequences, also in relative terms.

The absence of detectable expression in the two
MLV-IL-2 mice and the low level in the MLV-myc mouse and
the extensive methylation of the fusion genes are similar
to the results obtained by introducing proviruses into the
germ line by infection of early embryos. Jaenish and
colleagues demonstrated that MLV proviruses introduced
into the germ line in this manner are largely inactive and
that this inactivity is correlated with extensive
methylation of the proviral sequences (3,4). Although it
has been suggested that this correlation is causal (3),
there is some evidence that the block to transcription
precedes methylation (10,11) and that this methylation
maintains the quiescence imposed by other mechanisms.
Transfection of MLV-CAT (CAT:chloramphenicol acetyl
transferase) (5) or MLV-neo (6) constructions into
embryonal carcinoma cells (which have been used as model
early embryonic cells) suggest that there are transacting
factors that can repress transcription from the MLV LTR.
It may be that this initial factor-mediated repression is
perpetuated by methylation. The results reported here
indicate that the inactivity and methylation of the
MLV-LTR in transgenic mice is not dependent on viral
infection but is also seen with microinjected MLV LTR
fusion genes.

One goal of these experiments was to examine whether a
pathological change would result from the uncontrolled

expression of either the c-myc or the human IL-2 genes in hematopoietic cells. Many lymphomas originating following murine leukemia virus infection have the provirus integrated near the endogenous c-myc gene (12-15). Although transcription of the myc gene is high in these cells, in most cases RNA initiation does not occur within the provirus (12,13) but at the normal myc promoters and the level is not consistently higher than in tumors in which there is no obvious c-myc rearrangement. A more direct link between unregulated myc expression and hematopoietic pathology has been obtained from transgenic mice carrying mouse mammary tumor virus (MTV)-myc fusion genes (7). These mice were originally described in terms of the contribution these MTV-myc fusion genes made to the development of mammary adenocarcinomas but recent analysis has revealed that these mice also develop a variety of hematopoietic neoplasms (16). Recently Adams et al. (17) have also demonstrated that transgenic mice carrying a mouse myc gene linked to an immunoglobulin enhancer will rapidly develop B cell lymphomas. Although we did detect transcription of the MLV-myc gene in the thymus of the MLV-myc mice there was no obvious thymic hyperplasia. Two possibilities exist for this lack of phenotypic change. Firstly, many tissues can maintain normal growth characteristics in the presence of deregulated myc expression (7) and secondly the level of transcription of the MLV-myc gene in the thymus is many fold lower than the endogenous levels and so may be insignificant in terms of altering the cellular growth characteristics. With respect to this last point we were not able to compare MLV-myc and endogenous c-myc RNA levels in individual cells.

As we did not detect transcription of the MLV-IL-2 genes these experiments have not been able to address the involvement of IL-2 in hematopoietic abnormalities. More recent experiments in which IL-2 has been expressed in transgenic mice using the immunoglobulin heavy chain promoter and enhancer should allow us to examine this possibility.

ACKNOWLEDGEMENTS

We thank Mike Cagle and Philip Hollingshead for technical assistance and Drs. Cori Gorman, JoAnne Leung and William Wood for comments on the manuscript.

REFERENCES

1. Chada, K., Magram, J., Costantini, F. (1986) An embryonic pattern of expression of a human fetal globin gene in transgenic mice. Nature 319:685.
2. Goff, S.P. (1984) The genetics of murine leukemia viruses. Current Topics in Microbiology and Immunology 112:45.
3. Stuhlmann, H., Jahner, D., Jaenish, R. (1981) Infectivity and methylation of retroviral genomes is correlated with expression in the animal. Cell 26:221.
4. Jahner, D., Jaenisch, R. (1985) Chromosomal position and specific demethylation in enhancer sequences of germ line transmitted retroviral genomes during mouse development. Mol. Cell. Biol. 5:2212.
5. Gorman, C.M., Rigby, P.W.J., Lane, D.P. (1985) Negative regulation of viral enhancers in undifferentiated embryonic stem cells. Cell 42:519.
6. Niwa, O. (1985). Suppression of the hypomethylated Moloney leukemia virus genome in undifferentiated teratocarcinoma cells and inefficiency of transformation by a bacterial gene under the control of the long terminal repeat. Mol. Cell. Biol. 5:2325.
7. Stewart, T.A., Pattengale, P.K., Leder, P. (1984) Spontaneous mammary adenocarcinomas in transgenic mice that carry and express MTV/myc fusion genes. Cell 38:627.
8. Chirgwin, J., Aeybyle, A., McDonald, R., Rutter, W. (1979). Isolation of biologically active ribonucleic acid from sources enriched in ribonuclease. Biochemistry 18:5294.
9. Stewart, T.A., Bellve, A.R., Leder, P. (1984) Transcription and promoter usage of the myc gene in normal somatic and spermatogenic cells. Science 226:707.
10. Gantsch, J.W., Wilson, M.C. (1983). Delayed de novo methylation in teratocarcinoma suggests additional tissue-specific mechanisms for controlling gene expression. Nature 301:32.
11. Niwa, O., Yokota, Y., Ishida, H., Sugahara, T. (1983). Independent mechanisms involved in suppression of the Moloney leukemia virus genome during differentiation of murine teratocarcinoma cells. Cell 32:1105.

12. Corcoran, L.J., Adams, J.M, Dunn, A.R., Cory, S. (1984) Murine T lymphomas in which the cellular myc oncogene has been activated by retroviral insertion. Cell 37:113.

13. Li, Y., Holland, C.A., Hartley, J.W., Hopkins, N. (1984). Viral integration near c-myc in 10-20 percent of MCF247-induced AKR lymphomas. Proc. Natl. Acad. Sci. USA 81:6808.

14. Selten, G., Cuypers, HT, Zijlstra, M., Melief, C., Berns, A. (1984) Involvement of c-myc in MuLV-induced T cell lymphomas in mice: frequency and mechanisms of activation. EMBO J. 3:3215.

15. O'Donnell, P.V., Fleissner, E., Lonial, H., Koehne, C.F., Reicin, A. (1985). Early clonality and high frequency integration into the c-myc locus in AKR leukemias. J. Virol. 55:500.

16. Leder, A., Pattengale, P.K., Kuo, A., Stewart, T.A., Leder, P. (1986) Consequences of widespread deregulation of the c-myc gene in transgeneic mice: multiple neoplasms and normal development. Cell, in press.

17. Adams, J.M., Harris, A.W., Pinkert, C.A., Corcorar, L.M., Alexander, W.S., Cory, S., Palmiter, R.D., Brinster, R.L. (1985). The c-myc oncogene driven by immunoglobulin enhancers induces lymphoid malignancy in transgenic mice. Nature 318:533.

18. Hoffman, J.W., Steffen, D., Gusella, J., Tabin, C., Bird, S., Cowing, D., Weinberg, R. (1982) DNA methylation affecting the expression of murine leukemia proviruses. J. Virol. 44:144.

19. Battey, J., Moulding, C., Taub, R., Murphy, W., Stewart, T., Potter, H., Lenoir, G., Leder, P. (1983). The human c-myc oncogene: structural consequences of translocation into the IgH locus in Burkitt lymphoma. Cell 34:779.

20. Svedersky, L.P., Shepard, H.M., Spencer, S.A., Shalaby, M.R., Palladino, M.A. (1984). Augmentation of human natural cell mediated cytotoxicity by recombinant human interleukin-2. J. Immunl. 133:714.

21. Maniatis, T., Fritsch, E.F., Sambrook, J. (1982) Molecular Cloning: A laboratory manual (Cold Spring Harbor, New York) Cold Spring Harbor laboratory.

Transcriptional Control Mechanisms, pages 463–477
© 1987 Alan R. Liss, Inc.

The HTLV *x* Gene: Function and Relation
to the Adenovirus E1A Gene

Janice L. Williams, Alan J. Cann, Joseph D. Rosenblatt,
Neil P. Shah, William Wachsman, and Irvin S.Y. Chen

Division of Hematology-Oncology, Department of Medicine,
UCLA School of Medicine, Los Angeles, CA   90024

ABSTRACT   The human T-cell leukemia viruses (HTLV-I
and -II) are associated with certain T-cell
malignancies. HTLV-I is the etiologic agent of adult
T-cell leukemia, while HTLV-II is associated with a
less severe T-cell variant of hairy-cell leukemia.
Despite the differences in the diseases associated
with these viruses, both HTLV-I and -II transform the
same T helper/inducer subclass of T cells *in vitro*
(9). A unique gene, termed *x*, is present in HTLV-I,
HTLV-II, and bovine leukemia virus (BLV), but not in
other retroviruses. The *x* protein activates
transcription from the viral LTR and is necessary for
viral replication (20). We have shown that the EIII
promoter of adenovirus is activated by the HTLV-II *x*
protein (20). Comparisons between the transcriptional
activating E1A protein of adenovirus and the *x*
protein of HTLV-II reveal that *x* and E1A activate
the EIII promoter in common, while other promoters are
activated either by E1A alone or by *x* alone. These
results suggest that E1A and *x* activate
transcription by different mechanisms. Additional
studies of *x* protein function should clarify the
sequences within promoters necessary for
transcriptional activation by *x*, and may provide

This work was supported by NCI grants CA30388, CA32737,
CA38597, CAO9297 and CA16042; grants PF-2182 and JFRA-99
from the American Cancer Society; grants from the
California Institute for Cancer Research and California
University-wide Task Force on AIDS.

insight into the mechanism of cellular transformation by HTLV.

INTRODUCTION

The HTLV *x* Gene.

The human T-cell leukemia viruses, HTLV-I and HTLV-II, are associated with some T-cell leukemias in man (1,2). HTLV-I is the etiologic agent of an aggressive T-cell leukemia/lymphoma known as adult T-cell leukemia (ATL), and is endemic to regions of Southern Japan, the Caribbean, and Africa. Numerous isolates of HTLV-I have been obtained from tumor cells of ATL patients (3). HTLV-II has been isolated from two cases of T-cell variant hairy-cell leukemia, which is a less severe malignancy in comparison to ATL (4-6). HTLV-I and -II share approximately 65% nucleic acid sequence homology, and both viruses transform cells of primarily T-helper/inducer phenotype *in vitro*, as defined by their continuous proliferation in the absence of exogenous interleukin 2 (7-9). HTLV-transformed cells isolated from ATL patients contain provirus which is integrated in a monoclonal or oligoclonal fashion, indicating that the virus plays a role in the transformation event. Comparisons between HTLV-transformed cell lines isolated from different ATL patients show that there is no specific site of provirus integration within the human genome, thereby ruling out a possible mechanism of transformation by viral integration next to a cellular proto-oncogene, leading to aberrant expression of the cellular gene. Furthermore, the HTLV-I and -II genomes share no sequence homology to any known oncogene, and no homology to cellular DNA. However, in addition to the *gag*, *pol* and *env* genes found in all replication-competent retroviruses, HTLV-I, HTLV-II, and the related oncogenic bovine leukemia virus (BLV) contain a novel gene, *x* [also known as *x-lor* or *tat* (10)] which is not found in other retroviruses. Though no direct proof exists, current evidence suggests that the *x* gene may be the transforming gene of the virus.

The *x* gene is located between the *env* gene and 3' LTR and encodes a protein of 40 Kd and 37 Kd in HTLV-I- and HTLV-II-infected cells, respectively (11,12). We identified these proteins by generating specific antibodies against synthetic peptides deduced from the major X open reading frame. A similar approach has recently been used to identify the *x* gene product of BLV as a 38 Kd nuclear protein (13). The HTLV-I and -II *x* proteins share 82% amino acid homology, are localized predominantly in the nucleus of infected cells, and have a short half-life of approximately 120 minutes (14,15). The *x* gene mRNA is composed of three exons. The first exon comprises a region of the 5' LTR, while the second exon contains the methionine initiation codon and one additional nucleotide of the second codon. The third exon contains the main body of the *x* protein. The methionine codon present in the second exon is the same methionine which serves as the initiation codon for the *env* protein (16-18). BLV also utilizes a similar double splicing mechanism for expression of the *x* protein (19).

Transcriptional Activation by the HTLV *x* Protein.

We have shown that the *x* gene is necessary for viral replication (20). Mutants were constructed within the X region of an HTLV-II infectious clone and transfected into B cells to generate stable transformants. The results were that proviruses with mutations within *x* failed to replicate. S$_1$ nuclease analysis revealed that the defect resulted in an inability of the mutant provirus to transcribe high levels of mRNA. Furthermore, by superinfecting the stably transformed clones with wild-type virus, mutant viral transcription was rescued *in trans* by the wild-type *x* gene product. These results show that the *x* protein acts *in trans* to increase viral transcription and thus, replication.

Low levels of transcription from the mutant provirus were observed in the absence of the *x* gene product. These low levels, less than 1/100th the amount found in wild-type infected cells, could explain how viral replication initiates after HTLV infection. Following integration of the provirus, low levels of viral RNA would

be produced, some of which would encode the *x* protein.
x would then act in a positive fashion to increase
transcription of the proviral DNA, ultimately leading to
production of large amounts of viral mRNA.

To further investigate the transcriptional
trans-activating properties of the *x* protein, we
utilized a transient transfection system with recombinant
constructs containing the HTLV LTR linked to the
chloramphenicol acetyl transferase gene (CAT). CAT gene
expression is a rapid and quantitative assay which
measures the level of transcription from a given promoter
(21). The LTR CAT constructs were co-transfected with
vectors which express the HTLV genome from the SV40 early
promoter. By introducing deletions throughout the HTLV
genome of this expression vector, we demonstrated that
trans-activation of the HTLV LTR is due to the *x* gene
alone (22). Other investigators have used similar
approaches to show that the *x* gene is responsible for
trans-activation of the HTLV LTR (23,24).

An interesting result which we observed from our
co-transfection experiments was that the HTLV-I LTR is
activated by both the HTLV-I and -II *x* proteins, but the
HTLV-II LTR is only activated by the homologous HTLV-II
x protein. However, since the expression vectors used
in these experiments produce *x* from a spliced mRNA, it
was possible that differences in the efficiency of
splicing could affect the amount of *x* protein produced.
If quantitative differences exist between the amounts of
HTLV-I and -II *x* proteins made from these expression
vectors, then it would be difficult to directly compare
the *trans*-activating properties of both *x* proteins on
various promoters. Therefore, we constructed the
91023b *x*-I and 91023b *x*-II expression vectors, which
produce the HTLV-I and HTLV-II *x* proteins,
respectively. These highly efficient expression vectors
contain only the *x* protein coding sequences, with no
intervening sequences, so that an unspliced mRNA is
produced (25). Immunoprecipitation and S_1 nuclease
analysis show that comparable levels of *x* proteins and
mRNA are made from the two vectors. Utilizing the 91023b
vectors in mammalian cells, we found that the HTLV-I LTR
is activated by the *x*I and *x*II proteins, while the
HTLV-II LTR is significantly activated only by *x*II. In

avian cells, however, a much different pattern emerged (25). In quail QT-6 cells and in chicken embryo fibroblasts, the HTLV-I LTR is not detectably activated by xI, even though the HTLV-I LTR is strongly activated by xII. Since it has been hypothesized that the HTLV x proteins may not bind directly to the LTR to activate transcription, these results suggest that the x proteins interact with cellular transcription factors which are highly conserved in mammalian cells. We propose that in avian cells, cellular factors which are required by xI in order to function are either absent or are unable to interact with the xI protein. Even though xI and xII share extensive amino acid homology, these proteins differ dramatically in their ability to *trans*-activate various promoters in different cell types. xII can strongly activate a variety of promoters, while xI only activates the HTLV-I LTR. Future biochemical studies should help to elucidate differences in the mechanism(s) of *trans*-activation by the two HTLV x proteins.

The E1A protein of adenovirus induces transcription from six unique adenovirus promoters (26-28), in addition to some endogenous cellular genes (29,30). E1A is also required for transformation of rodent cells (31-33), and has been shown to repress the activity of some viral and cellular enhancers (34). We have shown by a transient co-transfection assay that the E1A-dependent EIII promoter of adenovirus is *trans*-activated by the HTLV-II x protein at levels comparable to that seen with E1A (35). The HTLV-I x protein does not activate the EIII promoter (unpublished observations).

In order to elucidate similarities in the mechanism of *trans*-activation by E1A and xII, we searched for regions of homology between the two proteins. We found no obvious amino acid sequence homology between E1A and xII. In addition, we found no substantial nucleotide sequence homology between the EIII promoter and the HTLV-II LTR. However, we did find limited nucleotide sequence homology between short regions upstream of the adenovirus EII promoter and regions within the HTLV-I and -II LTRs (Figure 1). Since the adenovirus EII and EIII promoters are adjacent to each other and direct transcription in opposite orientations, it was possible that the upstream regions of the EII promoter were

involved in *trans*-activation of the EIII promoter. We
reasoned that these regions of homology might be
responsible for the similar effects of E1A and *x*II on
the EIII promoter. The area of interest within the HTLV-I
and -II LTRs consists of a sequence of 21 nucleotides
which is repeated three times in the U3 region upstream of
the viral promoter. The region of the EII promoter which
is homologous to the 21 nucleotide repeats of the HTLV
LTRs consists of nine out of twelve nucleotides within an
imperfectly repeated sequence upstream from the EII cap
site. These repeats are located in a region of the EII
promoter which has been shown to act as an enhancer in the
presence of E1A in a transient transfection assay (36).
We tested the role of the EII promoter repeats in
trans-activation of the EIII promoter by deleting these
repeats from an EII/EIII promoter construct. When the
mutant lacking the EII repeats but maintaining the EIII
promoter sequences is co-transfected with either E1A or

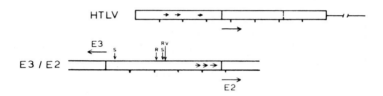

Figure 1A. The LTR of HTLV and the EII/EIII promoters of
adenovirus are shown in schematic form. The separate
regions within the HTLV LTR indicate the U3, R, and U5
regions, proceeding from left to right. Hatchmarks shown
below the sequences are separated by 100 base pairs.
Horizontal arrows located above and below the sequences
indicate the CAP sites of the HTLV, EII, and EIII
promoters. Vertical arrows above the EII/EIII promoter
indicate relevant restriction enzyme sites (S=Sma I, R=Eco
RI, RV=Eco RV). Arrows located within the sequences refer
to the regions of partial homology between the LTR and the
adenovirus promoters. The three arrows within the U3
region of the LTR represent the 21 nucleotide repeats,
while the three arrows upstream of the EII CAP site
represent the adenovirus repeats.

```
E2/E3
  (-121)  A  T  G  G  C  G  C  T  G  A  C  G
  (-108)  C  A  G  G  T  G  C  T  G  G  C  G
  (- 83)  C  T  G  G  A  G  A  T  G  A  C  G

HTLV-II
  (-235)  A  A  G  G  C  T  C  T  G  A  C  G
  (-185)  A  C  G  G  C  C  C  T  G  A  C  G
  (- 92)  A  A  G  G  C  T  C  T  G  A  C  G

HTLV-I
  (-235)  A  A  G  G  C  T  C  T  G  A  C  G
  (-203)  T  A  G  G  C  C  C  T  G  A  C  G
  (-104)  C  A  G  G  C  G  T  T  G  A  C  G
```

Figure 1B. A comparison between the homologous sequences
of the HTLV-I and -II LTRs and the adenovirus EII promoter
is shown. The numbers in parentheses refer to the
location of sequences upstream from the CAP site of the
promoter (+1).

xII, we find that the levels of trans-activation of
the mutant promoter are comparable to that of the parent
EIII promoter (Table 1). Our results demonstrate that the
upstream repeats of the EII promoter are not involved in
activation of the EIII promoter. In addition, in these
experiments, the E1A-dependent EII promoter is not
detectably trans-activated by xII, which supports the
conclusion that the EII upstream repeats are not
responsive to xII, despite some homology to the HTLV
LTRs.

Just as xI does not activate the HTLV-II LTR while
xII activates both the HTLV-I and -II LTRs, E1A and
xII also behave differently when tested on various
promoters. Both E1A and xII activate the EIII promoter,
while the EII promoter is strongly activated only by E1A
and the HTLV-II LTR is strongly activated only by xII.
These results suggest that E1A and the HTLV x proteins

Table 1. *Trans*-activation by E1A and *x*II[a]

Promoter	E1A	*x*II
EIII wild-type[b] (610)[c]	+	+
EIII ERV[d] (271)	+	+
EII[e] (286)	+	−
HTLV-II LTR[f]	−	+

[a]   Results of co-transfections in HeLa cells using 5 ug of a construct with a promoter linked to the CAT gene (21) and 5 ug of an expression vector producing either E1A or *x*II (35). Cells were harvested 48 hours after transfection, and levels of activation were measured by the conversion of ^{14}C-chloramphenicol into acetylated forms. (+) refers to greater than fivefold activation over background; (−) refers to less than twofold activation above background, which was not significant in this assay.

[b]   EIII wild-type is described in (35) and in Figure 1A.

[c]   Numbers in parenthesis refer to the amount of sequences present upstream from the promoter CAP site (see Figure 1A).

[d]   EIII ERV was derived from EIII wild-type by removing adenoviral sequences upstream (away from the EIII promoter) from the EcoRV site (see Figure 1A).

[e]   EII wild-type is described in (35), and consists of upstream sequences bordered by the Eco RI site at the 5' end (see Figure 1A).

[f]   Consists of the entire HTLV-II LTR (20).

may activate transcription by different mechanisms. Different promoters may utilize different cellular co-factors to promote transcription, and multiple pathways may exist within a cell which can activate transcription. Promoters such as EIII and the HTLV-I LTR may undergo activation by several potential pathways within the cell which are stimulated by either E1A or *x*, while a promoter such as EII may be activated via a single pathway which is stimulated by E1A but not by *x*.

Potential Mechanisms of *Trans*-activation and Transformation by the *x* Protein.

The biochemical mechanism of *trans*-activation by the *x* proteins remains unknown. However, based on observations of *x*I and *x*II function *in vitro*, several hypotheses can be formulated to explain how these proteins may activate transcription.

One idea is that the *x* protein may bind directly to the LTR to regulate transcription, although there is currently no experimental evidence to support this mechanism. Similarly, DNA-binding properties of the adenovirus E1A protein have not been demonstrated, although they have been extensively sought (37). Because E1A can activate a variety of unrelated promoters, it has been proposed that E1A activates transcription, not by direct contact with DNA, but by acting through a cellular intermediate. Since *x*II also activates promoters which share no apparent sequence homology, it too may function through a cellular intermediate.

If the *x* proteins interact with cellular factors to regulate transcription, they may do so by either chemically modifying a host factor which then binds the LTR, or by forming a complex with host proteins which together can bind the LTR. From the 82% amino acid homology that exists between *x*I and *x*II, it is not unreasonable to assume that these proteins may interact with the same cellular factors. If this were the case, *x*II would be expected to interact with such factors with greater efficiency than *x*I, since *x*II activates more promoters than *x*I. Alternatively, *x*I and *x*II may interact with different cellular factors. In this instance, *x*II might function with cellular factors which can activate a wide variety of promoters, while *x*I may interact with factors which are more restricted in their actions.

The evidence that *x* may be the transforming gene of the virus is indirect. As previously mentioned, HTLV contains no known oncogene, and integrates without apparent specificity into the human genome. Consequently, the *x* gene emerges as the most likely candidate responsible for transformation. The E1A protein of adenovirus, which is a transcriptional activating protein, can immortalize cells *in vitro*, and in the presence of E1b, can induce neoplastic transformation of cells (39). It has been hypothesized that aberrant transcription of cellular genes mediated by E1A is involved in transformation of cells by adenovirus. Since the *x*II protein, in some respects, is functionally related to E1A (35), *x*II may also be able to induce certain features of the transformed phenotype by aberrant transcriptional activation. *In vitro*, the *x* proteins are able to function in all mammalian cells tested, including HeLa, NIH 3T3, COS, and Epstein-Barr-transformed B cells (25). However, HTLV only transforms cells of T-cell phenotype (9). These results suggest that *x* may promote transformation by activating a T-cell-specific gene involved in proliferation of T-helper/inducer cells. Attempts to transform fresh peripheral blood cells by introducing the *x* gene into the cells by transfection have been hampered by the inability of these cells to survive the transfection procedure. Retroviral vectors which infect peripheral blood cells in a stable fashion may prove more useful in studying the *x* protein's role in T-cell transformation.

Recent studies indicate that a second protein is encoded by a reading frame which overlaps the reading frame utilized for the *x* protein of HTLV-I (38). The second protein was shown to be phosphorylated and located in the nucleus. HTLV-II, BLV, and simian T-cell leukemia virus (STLV) all contain a similar open reading frame in the same region which could encode a putative second protein (19); however, there is as yet no data to show that these other viruses produce a protein from this region. Vectors which only produce *x*I or *x*II, and not the second protein, are capable of transcriptionally activating the HTLV LTRs (25,40) and the adenovirus E3 promoter (25). Consequently, the second protein may not be directly involved in *trans*-activation, although the possibility that it may play an accessory role in this function cannot be ruled out. Another possibility is that the second protein may be involved in transformation of T cells. Future experiments should clarify the role of the *x* protein and the newly discovered second protein in *trans*-activation and transformation by HTLV.

ACKNOWLEDGEMENTS

We thank J. Fujii and C. Nishikubo for technical assistance, and W. Aft for preparation of the manuscript.

REFERENCES

1. Poiesz BJ, Ruscetti FW, Gazdar AF, Bunn PA, Minna JD, Gallo RC (1980). Detection and isolation of type C retrovirus particles from fresh and cultured lymphocytes of a patient with cutaneous T-cell lymphoma. Proc Natl Acad Sci USA 77:7415-7419.
2. Hinuma Y, Nagata K, Hanaoka M, Nakai M, Matsumoto T, Kinoshita K-I, Shirakawa S, Miyoshi I (1981). Adult T-cell leukemia: antigen in an ATL cell line and detection of antibodies to the antigen in human sera. Proc Natl Acad Sci USA 78:6476-6480.
3. Wong-Staal F, Gallo RC (1985). Human T-lymphotropic retroviruses. Nature 317:395-403.
4. Saxon A, Stevens RH, Golde DW (1978). T-lymphocyte variant of hairy-cell leukemia. Ann Intern Med 88:323-326.
5. Kalyanaraman VS, Sarngadharan MG, Robert-Guroff M, Miyoshi I, Blayney D, Golde D, Gallo RC (1982). A new subtype of human T-cell leukemia virus (HTLV-II) associated with a T-cell variant of hairy cell leukemia. Science 218:571-573.
6. Rosenblatt JD, submitted.
7. Miyoshi I, Kubonishi I, Yoshimoto S, Akagi T, Ohtsuki Y, Shiraishi Y, Nagata K, Hinuma Y (1981). Type C virus particles in a cord T-cell line derived by co-cultivating normal human cord leukocytes and human leukaemic T cells. Nature 294:770-771.
8. Popovic M, Lange-Wantzin G, Sarin PS, Mann D, Gallo RC (1983). Transformation of human umbilical cord blood T cells by human T-cell leukemia/lymphoma virus. Proc Natl Acad Sci USA 80:5402-5406.
9. Chen ISY, Quan SG, Golde DW (1983). Human T-cell leukemia virus type II transforms normal human lymphocytes. Proc Natl Acad Sci USA 80:7006-7009.
10. Haseltine WA, Sodroski J, Patarca R, Briggs D, Perkins D, Wong-Staal F (1984). Structure of 3' terminal region of type II human T lymphotropic virus: evidence of new coding region. Science 225:419-421.
11. Slamon DJ, Shimotohno K, Cline MJ, Golde DW, Chen ISY (1984). Identification of the putative transforming protein of the human T-cell leukemia viruses HTLV-I and HTLV-II. Science 226:61-65.

12. Lee TH, Coligan JE, Sodroski JG, Haseltine WA, Salahuddin SZ, Wong-Staal F, Gallo RC, Essex M (1984). Antigens encoded by the 3'-terminal region of human T-cell leukemia virus: evidence for a functional gene. Science 226:57-61.
13. Sagata N, Tsuzuku-Kawamura J, Nagayoshi-Aida M, Shimizu F, Imagawa K, Ikawa Y (1985). Identification and some biochemical properties of the major X_{BL} gene product of bovine leukemia virus. Proc Natl Acad Sci USA 82:7879-7883.
14. Slamon DJ, Press MF, Souza LM, Cline MJ, Golde DW, Gasson JC, Chen ISY (1985). Studies of the putative transforming protein of the type I human T-cell leukemia virus. Science 228:1427-1430.
15. Goh WC, Sodroski J, Rosen C, Essex M, Haseltine WA (1985). Subcellular localization of the product of the long open reading frame of human T-cell leukemia virus type I. Science 227:1227-1228.
16. Wachsman W, Shimotohno K, Clark SC, Golde DW, Chen ISY (1984). Expression of the 3' terminal region of human T-cell leukemia viruses. Science 226:177-179.
17. Wachsman W, Golde DW, Temple PA, Orr EC, Clark SC, Chen ISY (1985). HTLV x gene product: requirement for the env methionine initiation codon. Science 228:1534-1537.
18. Seiki M, Hikikoshi A, Taniguchi T, Yoshida M (1985). Expression of the px gene of HTLV-I: general splicing mechanism in the HTLV family. Science 228:1532-1535.
19. Sagata N, Yasunaga T, Ikawa Y (1985). Two distinct polypeptides may be translated from a single spliced mRNA of the X genes of human T-cell leukemia and bovine leukemia viruses. FEBS Lett 192:37-42.
20. Chen ISY, Slamon DJ, Rosenblatt JD, Shah NP, Quan SG, Wachsman W (1985). The x gene is essential for HTLV replication. Science 229:54-58.
21. Gorman CM, Moffat LF, Howard BH (1982). Recombinant genomes which express chloramphenicol acetyltransferase in mammalian cells. Mol Cell Biol 2:1044-1051.

22. Cann AJ, Rosenblatt JD, Wachsman W, Shah NP, Chen ISY (1985). Identification of the gene responsible for human T-cell leukemia virus transcriptional regulation. Nature (London) 318:571-574.
23. Sodroski J, Rosen C, Goh WC, Haseltine W (1985). A transcriptional activator protein encoded by the x-lor region of the human T-cell leukemia virus. Science 228:1430-1434.
24. Felber BK, Paskalis H, Kleinman-Ewing C, Wong-Staal F, Pavlakis GN (1985). The pX protein of HTLV-I is a transcriptional activator of its long terminal repeats. Science 229:675-679.
25. Shah NP, submitted.
26. Jones N, Shenk T (1979). An adenovirus type 5 early gene function regulates expression of other early viral genes. Proc Natl Acad Sci USA 76:3665-3669.
27. Berk AJ, Lee F, Harrison T, Williams J, Sharp PA (1979). Pre-early adenovirus 5 gene product regulates synthesis of early viral messenger RNAs. Cell 17:935-944.
28. Ricciardi RP, Jones RL, Cepko CL, Sharp PA, Roberts BE (1981). Expression of early adenovirus genes requires a viral encoded acidic polypeptide. Proc Natl Acad Sci USA 78:6121-6125.
29. Nevins JR (1982). Induction of the synthesis of a 70,000 dalton mammalian heat shock protein by the adenovirus E1A gene product. Cell 29:913-919.
30. Stein R, Ziff EB (1984). HeLa cell β-tubulin gene transcription is stimulated by adenovirus 5 in parallel with viral early genes by an E1A-dependent mechanism. Mol Cell Biol 4:2792-2801.
31. Houweling A, van den Elsen P, van der Eb A (1980). Partial transformation of primary rat cells by the leftmost 4.5% fragment of adenovirus 5 DNA. Virology 105:537-550.
32. Ruley HE (1983). Adenovirus early region 1A enables viral and cellular transforming genes to transform primary cells in culture. Nature 304:602-606.
33. Montell C, Courtois G, Eng C, Berk A (1984). Complete transformation by adenovirus 2 requires both E1A proteins. Cell 36:951-961.

34. Hen R, Borrelli E, Chambon P (1985). Repression of the immunoglobulin heavy chain enhancer by the adenovirus-2 E1a products. Science 230:1391-1394.
35. Chen ISY, Cann AJ, Shah NP, Gaynor RB (1985). Functional relation between HTLV-II x and adenovirus E1A proteins in transcriptional activation. Science 230:570-573.
36. Imperiale MJ, Hart RP, Nevins JR (1985). An enhancer-like element in the adenovirus E2 promoter contains sequences essential for uninduced and E1A-induced transcription. Proc Natl Acad Sci USA 82:381-385.
37. Ferguson B, Krippl B, Andrisani O, Jones N, Westphal H, Rosenberg M (1985). E1A 13S and 12S mRNA products made in _Escherichia coli_ both function as nucleus-localized transcription activators but do not directly bind DNA. Mol Cell Biol 5:2653-2661.
38. Kiyokawa T, Seiki M, Iwashita S, Imagawa K, Shimizu F, Yoshida M (1985). p27^{x-III} and p21^{x-III}, proteins encoded by the pX sequence of human T-cell leukemia virus type I. Proc Natl Acad Sci USA 82:8359-8363.
39. Bishop JM (1985). Viral Oncogenes. Cell 42:23-28.
40. Seiki M, Inoue J, Takeda T, Yoshida M (1986). Direct evidence that p40x of human T-cell leukemia virus type I is a _trans_-acting transcriptional activator. EMBO J 5:561-565.

Index